SCIENCE AND PHILOSOPHY IN THE TWENTIETH CENTURY

*Basic Works of
Logical Empiricism*

SERIES EDITOR
SAHOTRA SARKAR
*Dibner Institute at MIT
and McGill University*

A GARLAND SERIES IN
READINGS IN PHILOSOPHY

ROBERT NOZICK, *Advisor*
Harvard University

SERIES CONTENTS

VOLUME

4

LOGICAL EMPIRICISM AND THE SPECIAL SCIENCES

REICHENBACH, FEIGL, AND NAGEL

Edited with introductions by

SAHOTRA SARKAR

Dibner Institute at MIT
and McGill University

Routledge
Taylor & Francis Group

NEW YORK AND LONDON

First published by Garland Publishing, Inc.
This edition published 2013 by Routledge
711 Third Avenue, New York, NY 10017
2 Park Square, Milton Park, Abingdon, Oxfordshire OX14 4RN

Routledge is an imprint of the Taylor & Francis Group, an informa business

Library of Congress Cataloging-in-Publication Data

Logical empiricism and the special sciences / Sahotra Sarkar.
   p. cm. — (Science and philosophy in the twentieth
century ; v. 4)
  Includes bibliographical references.
  ISBN 0-8153-2265-8 (alk. paper)
  1. Social sciences—Philosophy. 2. Logical positivism.
I. Title. II. Series.
H61.S324 1996
300—dc20   95–48265
                    CIP

SET ISBN 9780815322610
POD ISBN 9780415628365
Vol1 9780815322627
Vol2 9780815322634
Vol3 9780815322641
Vol4 9780815322658
Vol5 9780815322665
Vol6 9780815322672

CONTENTS

PHYSICS

PSYCHOLOGY

SERIES INTRODUCTION

The early years of the twentieth century saw remarkable developments in the sciences, particularly physics and biology. The century began with Planck's introduction of what came to be known as the "quantum hypothesis," followed by the work of Einstein, Bohr, and others, which paved the way for the development of quantum mechanics in the 1920s. It remains the most radical departure from the classical worldview that physics has seen. Not only were some physical quantities "quantized," that is, they could only have discrete values, but there were situations in which some of these values were indeterminate. Perhaps even worse, the basic dynamics of physical systems was indeterministic. The mechanical picture of the world, inherited from the seventeenth century, and already under attack during the nineteenth, finally collapsed beyond hope of recovery. Nevertheless, the new physics was unavoidable. Not only did atomic phenomena abide by its rules, but it provided a successful account of chemical bonding and valency. Meanwhile, in 1905, Einstein's special theory of relativity challenged classical notions of space and time. A decade later, general relativity replaced gravitation as a force by the curvature of space-time. Developments in astrophysics confirmed general relativity's unusual claims.

Also around 1900, biologists recovered the laws for the transmission of hereditary factors, or "genes." These laws, though published by Mendel in 1865, had remained largely unknown for a generation. By 1905, a new science called "genetics" had been created. For the first time, the phenomena of heredity were subsumed under exact (mathematical) laws. In the early 1920s, these laws were used by Fisher, Haldane, and Wright to formulate a quantitative, basically testable theory of evolution by natural selection. Around 1900 it also became clear that the transfer of chromosomes mediated the transmission of hereditary characters from parents to offspring. Between 1910 and 1920, genes were shown to be linearly positioned on chromosomes. The rudiments of a physical account of biological inheritance were in place by the mid-1920s. Eventually this work was integrated with other biological subdisciplines, especially biochemistry (itself largely a turn-of-the-century creation), to generate molecular biology, arguably the greatest triumph of science since 1950.

The philosophical response to the advances of early twentieth century science was schizophrenic. Some philosophers, especially in Germany, ignored scientific developments almost altogether and continued to elaborate extensive metaphysical systems having little contact with the physical world. Collectively, these projects came to be called phenomenology. In sharp contrast, another group of philosophers attempted to reform—or, perhaps, even replace—academic philosophy so as to bring it into consonance with modern science. At times, they claimed to have inherited the mantles of Aristotle and Descartes, Newton and Leibniz, Locke and Hume, Kant and Marx. More often, they claimed to be doing something altogether novel.

Most prominent among the latter group of philosophers were those who called themselves "logical positivists" or "logical empiricists." Many of them were associated, in their early years, with a group that met regularly in Vienna (starting in 1922) and called itself the Vienna Circle. The central figure was Moritz Schlick. (A complete list of members of the Vienna Circle will be found in their 1929 manifesto, which is reprinted in Volume 2.) The members of the Vienna Circle had an almost worshipful attitude towards the new physics though, in general, they seemed to have been completely ignorant of the equally fundamental changes taking place in biology. They were impressed by developments in logic, particularly Whitehead and Russell's attempt to carry out Frege's project of constructing mathematics from logic. Kurt Gödel, a member of the Vienna Circle, though hardly a logical empiricist in his philosophical leanings, probed the foundations of logic and showed that any relatively complex system of mathematics must allow statements to be formulated that can neither be proved nor disproved using formalized rules of proof—this is Gödel's famous incompleteness theorem.

Meanwhile, in Berlin, a smaller group around Hans Reichenbach came to a similar philosophical orientation and concentrated on probing the foundations of physics. In Poland, an eminent group of logicians, with Alfred Tarski as the central figure, began equally important investigations of logical notions. There was considerable intellectual exchange between these different groups. These exchanges led to convergence on many points—the philosophical theses that were most commonly advanced will be described below (and in the introductions to Volumes 1–4).

To return to the historical story, most of the logical empiricists had relatively progressive politics. A few, notably Otto Neurath, were avowed Marxists. Others, including Rudolf Carnap and Hans Hahn, were socialists. With the rise of nazism and fascism in Europe in the 1930s, many of the logical empiricists emigrated to Britain and, especially, to the United States. There they eventually came to establish a temporary hegemony over academic philosophy. Reichenbach moved to the University of California at Los Angeles; Herbert Feigl to the University of Minnesota; and Carnap to the University of Chicago. Meanwhile, during his youthful days, W.V.O. Quine was already preaching the logical empiricist gospel at Harvard. Of the major figures, only Neurath remained in Europe. (Hans Hahn had died in 1934 and Schlick had been murdered in 1936—see the introduction to Volume 2.)

Because of its migration to the U.S., logical empiricism became part of the Anglo-American tradition in philosophy, in spite of its European origins. It is at least arguable that as a movement it matured in the U.S. However, in spite of being relatively organized compared to other philosophical movements, the logical empiricists did not present a unified system of universally held theses—a point that seems to elude their modern critics—though they generally exhibited a coherent attitude to the analysis of philosophical problems. This attitude can be traced back to the 1920s. They *generally* accepted an *a priori* faith in logic, though they were sometimes known to disagree on what logic could be. Other than in logic (and in mathematics, which, for most logical empiricists, could be derived from logic), the logical empiricists endorsed a thoroughgoing empiricism—hence their name. All factual (that is, nonlogical) knowledge was ultimately empirical. A sharp distinction between empirical, *a posteriori*, synthetic claims on one hand and *a priori*, analytic claims on the other was a cherished doctrine for most (but not all) logical empiricists. Its rejection by Quine and others in the 1950s was a significant event in the decline of logical empiricism (see Volume 5).

Any claim that was neither logic nor able to be adjudicated by empirical means was rejected by the logical empiricists as "meaningless" or "cognitively insignificant," whatever its noncognitive (for instance, emotional) appeal. Logic escaped this fate by being true by virtue of meaning (of the logical connectives such as "not" and "and" and operators such as "all," "any," and "some") or of conventions. Mathematics was true because it could be reduced to, or constructed from, logic. Besides logic, the logical empiricists generally did not accept any other normative discipline as consisting of meaningful claims. (Ethical claims, according to some of them, were only devices to evoke appropriate emotive responses from others.)

Given these positions, there did not remain much metaphysics to be done (at least insofar as "metaphysics" was interpreted by the academic philosophers). Some logical empiricists, notably Carnap, claimed to have successfully eliminated metaphysics. In practice, metaphysics was replaced by attempts—rarely profound—at the analysis and interpretation of scientific concepts. Those logical empiricists who were particularly enamored of the technical apparatus of mathematical logic, again, most notably Carnap, interpreted this endeavor as describing the syntax and elaborating a semantics for the language of science. (In the case of logic itself, the logical empiricists achieved some important successes in their interpretive efforts in the 1930s—see Volume 3.) Metaphysics cast aside, the logical empiricists turned to epistemology; in particular, to the possibility of quantifying the extent to which different scientific claims were grounded in experience. The project turned out to be far more complex—and convoluted—than initially envisioned. By the time logical empiricism disappeared as an explicit movement within philosophy, little progress had been made towards this end.

An enumeration of positions advanced—or of successes and failures—only barely captures the spirit of logical empiricism. Within their self-proclaimed framework of accepting only logic and empirical knowledge, they venerated a critical attitude. This included continual self-criticism. Much has been written about the untenability of the doctrines espoused by the logical empiricists— what unfortunately goes unrecognized is that the most severe (and the most relevant) criticisms almost always came from within the movement or, at least, from individuals schooled in the movement (notably Quine). There were significant disagreements among the logical empiricists (for instance, between Carnap and Reichenbach on epistemology). There were also significant disagreements within the Vienna Circle: Kurt Gödel probably rejected most of the tenets in the Vienna Circle manifesto; Karl Menger refused to reject metaphysics on logical grounds (see his paper in Volume 2). These cases, however, may only show that not all members of that circle should be regarded as logical empiricists. Nonetheless, and most importantly, the logical empiricists believed philosophy to be a collective enterprise, like the natural sciences, and one in which progress could be made.

The logical empiricists' domination of Anglo-American philosophy was never complete and whatever hegemony they established was brief. Even within their chosen subdisciplines, such as the philosophy of science or logic or mathematics, their positions came under attack in the 1950s. Cherished doctrines such as the analytic-synthetic distinction were abandoned by a new generation of philosophers. The value of their type of conceptual analysis was sometimes derided by the later Wittgenstein's followers and by the so-called "ordinary language" philosophers. Metaphysics returned with a vengeance and, arguably, the influence of the logical empiricists was largely confined to the philosophy of science after the 1950s. But the 1960s saw logical empiricism under attack even among philosophers of science. It is probably reasonable to say that by around 1970, a new generation of philosophers of science had decided that the analyses offered by the logical empiricists were largely superficial and were to be replaced by more sophisticated work. The most popular position of those days was "scientific realism," a return to exactly the kind of metaphysics that the logical empiricists had found devoid of cognitive content.

Significant interest in logical empiricism resurfaced again in the early 1980s. This did not indicate any general return to the positions the logical empiricists advocated. Rather, the source of the interest was largely historical, part of a desire to understand the history of twentieth-century philosophy. It was aided by a new interest among philosophers in the history of the philosophy of science. Carnap and Reichenbach were probably the only prominent logical empiricists who had continued to be read during the 1960s and 1970s; now the works of Schlick and Neurath, among others, were once again read (and, sometimes, translated into English for the first time). Archives began to be mined to expose the intricate details of the relationships between the logical empiricists, and between them and other social and cultural movements of the 1920s and 1930s. This new work took place not only in the U.S., but also in Austria, Germany, and to a lesser extent, elsewhere in Europe. Slowly, as this historical work has progressed, a more positive philosophical assessment of the movement than was usually found in the 1960s and 1970s has also emerged (Sarkar 1992). These developments are far too recent for any assessment to be made of their lasting value. While the historical interest is neither hard to explain nor appreciate, it is less clear why, but perhaps even more interesting that, this positive reassessment is taking place.

There seem to be at least three reasons for the relatively positive reassessment that deserve mention: (1) since more than a generation had passed between the heyday of the movement and the mid- and late-1980s, the new commentators found it easier to have a more balanced view of both the contributions and the failures of logical empiricism than those—especially in the 1960s—who felt that they had to react to its dominance; (2) historical exploration—and exegesis—has revealed that the logical empiricists held a variety of views that are both more complex and more interesting than what their critics attributed to them (see, for example, Suppe 1974); and (3) arguably, the various alternatives to logical empiricism as a philosophy of science that were formulated in the 1960s and 1970s have not delivered on their promises. Going further, and much more controversially, these alternatives (including scientific realism) have proved less fertile and less robust than logical empiricism.

In this new intellectual context, it seems appropriate to make available, to as wide an audience as possible, some of the basic works of logical empiricism, as well as some of the new commentaries that have followed the renewal of interest in the movement. Many of the original pieces are not easily available and there is, at present, neither a detailed history of logical empiricism nor an annotated guide to its most important writings. An important old collection is Ayer (1959), which has a fairly comprehensive bibliography of work up to that point. Many valuable collections devoted to individual figures have been published. Schilpp (1963) collects many important critical pieces on Carnap, with Carnap's responses. The basic works of Feigl (1981), Hahn (1980), Kraft (1981), Menger (1979), Neurath (1973, 1987), Reichenbach (1978), Schlick (1979, 1987), and Waismann (1977) have been published as part of the Vienna Circle Collection. Collections of articles on logical positivism from the 1960s and 1970s include Achinstein and Barker (1969) and Hintikka (1975). Recent works of interest include Coffa (1991), Haller (1982), Menger (1994), and Uebel (1991, 1992). However, a detailed history of logical empiricism remains to be written.

What makes this series different from these works is an attempt to present a global picture of logical empiricism, including the influences that led to its initiation and the criticisms that were responsible for its decline. The emphasis here is on issues rather than on individual figures even though some of the most influential figures—especially Carnap and Reichenbach—feature prominently. However, for most of the topics treated, all the historically and conceptually important exchanges on that topic are collected together. Finally, modern commentaries are also included to bring the series up to date. In general, complete papers (in English whenever translations are available) are included over book sections in an effort to present complete arguments as far as possible. Volume 1 deals with the initial influences on logical empiricism and with the Vienna Circle period. Volume 2 concerns primarily the 1930s, when logical empiricism was at its most confident phase, when its adherents truly believed that they were reforming philosophy for all future times. Volume 3 includes pieces that reflect logical empiricism in its mature phase, after self-criticism and technical developments induced more sophisticated doctrines than those produced in the 1930s. Volume 4 shows how logical empiricism analyzed the special sciences. Volume 5 consists of the most important criticisms of logical empiricism and its responses. It marks the decline of logical empiricism. All of these volumes, except Volume 4, include a concluding section with modern commentaries. Volume 6 consists entirely of these commentaries. Each volume is introduced with an editorial note that puts the contents in perspective. Thanks are due to Richard Creath and Alan Richardson for advice on selecting the pieces for this series, and to Gregg Jaeger for help in assembling them and for commenting on the introductions. Work on these volumes was done while the editor was a Fellow at the Dibner Institute for the History of Science at MIT. Thanks are due to it for its support.

FURTHER READING

Achinstein, P. and Barker, S., eds. 1969. *The Legacy of Logical Positivism*. Baltimore: Johns Hopkins University Press.

Ayer, A.J., ed. 1959. *Logical Positivism*. New York: Free Press.

Coffa, A. 1991. *The Semantic Tradition from Kant to Carnap: To the Vienna Station*. Cambridge, UK: Cambridge University Press.

Feigl, H. 1981. *Inquiries and Provocations: Selected Writings, 1929 – 1974*. Dordrecht: Kluwer.

Hahn, H. 1980. *Empiricism, Logic, and Mathematics: Philosophical Papers*. Dordrecht: Kluwer.

Haller, R., ed. 1982. *Schlick und Neurath—Ein Symposion*. *Grazer philosophische Studien* 16 –17.

Hintikka, J., ed. 1975. *Rudolf Carnap, Logical Empiricist*. Dordrecht: Reidel.

Kraft, V. 1981. *Foundations for a Scientific Analysis of Value*. Dordrecht: Kluwer.

Menger, K. 1979. *Selected Papers in Logic, Foundations, Didactics, Economics*. Dordrecht: Kluwer.

Menger, K. 1994. *Reminiscences of the Vienna Circle and the Mathematical Colloquium*. Dordrecht: Kluwer.

Neurath, O. 1973. *Empiricism and Sociology*. Dordrecht: Kluwer.

Neurath, O. 1987. *Unified Science*. Dordrecht: Kluwer.

Reichenbach, H. 1978. *Selected Writings, 1909–1953*. Vols. 1, 2. Dordrecht: Kluwer.

Sarkar, S., ed. 1991. *Rudolf Carnap—A Centenary Reappraisal. Synthese* 93.

Schilpp, P. A., ed. 1963. *The Philosophy of Rudolf Carnap*. La Salle, IL: Open Court.

Schlick, M. 1979. *Philosophical Papers*. Vols. 1, 2. Dordrecht: Kluwer.

Schlick, M. 1987. *The Problems of Philosophy in Their Interconnection*. Dordrecht: Kluwer.

Suppe, F., ed. 1974. *The Structure of Scientific Theories*. Urbana: University of Illinois Press.

Uebel, T. E., ed. 1991. *Rediscovering the Forgotten Vienna Circle*. Dordrecht: Kluwer.

Uebel, T. E. 1992. *Overcoming Logical Positivism from Within: The Emergence of Neurath's Naturalism in the Vienna Circle's Protocol Sentence Debate*. Amsterdam: Rodopi.

Waismann, F. 1977. *Philosophical Papers*. Dordrecht: Kluwer.

INTRODUCTION

What did logical empiricism contribute to the philosophy of science? There are at least two reasons why an answer to this question is necessary for an evaluation of logical empiricism as a philosophical movement:

(1) Much more stridently than any other philosophical movement, logical empiricism claimed to have drawn its inspiration from the natural sciences. As noted in the series introduction, the early developments in physics, notably quantum mechanics and the special and general theories of relativity, were critical to the movement's development. Cherished doctrines, including empiricism itself (over realism), as well as the various types of conventionalism, were allegedly induced by these developments. If logical empiricism cannot even provide an adequate interpretation of modern science (at the very least, physics), then there is good reason to doubt its value.

(2) At least since the 1950s, the main explicit influence of logical empiricism has been in what has come to be called "philosophy of science," a hybrid of epistemological and metaphysical questions raised by the empirical sciences. (Logical empiricism's contributions to the philosophy of logic and mathematics have largely permeated those fields without continued controversy— see Volume 3.) For a generation, logical empiricism—that is, the issues that were raised and the analyses that were offered by its adherents—dominated discussions of science within academic philosophy. Since the 1960s, however, after its decline as a school had set in, logical empiricism became just one of several possible philosophies of science (see also Volume 5). Determining how it fares in comparison with these alternatives must form part of judging the value of logical empiricism.

The current philosophical consensus seems to be that logical empiricism did not contribute much of significance to the philosophy of science. Whether it contributed anything of value at all or even impeded the development of the field remains controversial. This consensus persists even after the more positive reassessment of logical empiricism that has emerged from the late 1980s. The

positive reassessment has been about general formal and epistemological issues, rather than about the logical empiricists' conceptual analyses of the empirical sciences. It is even possible that by
focusing philosophical attention on narrow formal issues, logical
empiricism might even have impeded the development of philosophy of science. It is not the purpose of this introduction—or of this
volume—to adjudicate these disputes. Rather, this volume collects
together some of the most prominent and most representative
writings of the logical empiricists on specific issues raised by the
"special sciences." However, it should be noted that on more than
one occasion, critics of logical empiricism have assumed that the
logical empiricists presented a monolithic doctrine (which was
easily criticized) and, worse, have attributed positions to the
logical empiricists that they generally did not espouse. A classic
example of this is the widespread view that logical empiricism
endorsed reductionism in both psychology and biology. A perusal
of the selections in this volume will do much to keep the record
straight by attributing to the logical empiricists only what they
actually said.

Selecting pieces for this volume has been particularly difficult,
given the sheer diversity and magnitude of logical empiricist work
on the special sciences. In accordance with the general policy of
this series, complete papers have been preferred over sections
from books. An attempt has been made to present the diversity of
points of view within the logical empiricist camp while still paying
most attention to the work that was particularly influential.
Further, an attempt has also been made to present the logical
empiricist analyses of different sciences and to avoid exclusive—or
almost exclusive—attention to physics, even though it is this
science that attracted most attention from the logical empiricists
in their heyday. The rationale for this is to encourage the exploration of the possibility that logical empiricism provides reasonable
interpretations of some, though not all, of the sciences.

There are four sections: physics, psychology, the social sciences, and biology. In each section, as in the other volumes in the
series, the selections are reproduced in the chronological order in
which they were written. However, since many of the selections
extend or reproduce earlier analyses, this order often does not
reflect conceptual history. The most systematic—and, in many
ways, brilliant—account of the logical empiricist philosophy of
science is Nagel's (1961) *The Structure of Science*. It can be very
profitably used with this volume for those interested in more detail.

The first section concerns physics. The pieces included here
are from the late (post-1940) period of logical empiricism. Earlier

pieces in the philosophy of physics were reprinted in Volume 1. Reichenbach, until his death in 1953, was the most important logical empiricist contributor to the philosophy of physics and his influence continues to be felt today. The two papers by Reichenbach that are included here, one of relativity theory and the other on quantum mechanics, provide an introduction to his most influential work. (Book-length treatments are Reichenbach 1944, 1958, and 1971.) Many of the theses that have dominated discussions of space and time in the philosophy of physics are found in the first of these two pieces: the determination of space-time topology by causal considerations, the conventionality of space-time geometry, and the conventionality of simultaneity in Minkowski space-time. Reichenbach's insights have been developed and extended by Adolf Grünbaum. The piece reprinted here provides an introduction to that work. More detail can be found in Grünbaum (1963, 1968). Most of these positions (though not the connection between space-time topology and causality) have been controversial, and have proved to be fertile topics of subsequent philosophical work— see Earman *et al.* (1977) and Torretti (1983). Meanwhile, Hempel's piece, with which this section begins, elaborates some standard logical empiricist themes, the *a priori* and analytic nature of mathematical knowledge (see Volumes 2 and 3) as well as the distinction between pure geometry and physical geometry (that is, the geometry of physical space or space-time). Frank's piece provides an overview of how logical empiricists approached physical theory. A more detailed account can be found in Carnap and Gardner (1966).

The second section concerns psychology. The first three pieces all defend physicalism in psychology as it was understood by the logical empiricists in the early 1930s. None of them shows any serious contact with experimental psychology. Carnap's piece is an attempt to show how psychology can be done in a physicalist language. A subtheme throughout the piece is the unity of science, which is to be achieved through the adoption of a unified language for all the sciences. Hempel's piece contains a more careful defense of behaviorism (which Carnap had also endorsed). Schlick attempts to show how physicalism in psychology disposes of classical mind-body problems. Feigl's first piece is an account of the development of the logical empiricists' views on psychology. It also endorses a view more sympathetic to neurophysiological reductionism than the earlier physicalist pieces. Feigl's second piece briefly discusses how logical empiricism could approach functionalism in psychology. Its interest is mainly historical. A fuller account of Feigl's (1958) views, his well-known paper "The 'Mental' and the 'Physical,'" proved to be too long to reproduce—it can be profitably consulted along with the pieces in this section.

The third section takes up the social sciences. In the Vienna Circle, the importance of these was most consistently emphasized by Neurath. The first piece in this section is one of Neurath's most concerted attempts to show exactly how sociology was to be done in a physicalist language (thereby maintaining the unity of science). A desire to have a predictive sociology leads Neurath to endorse what he calls "social behaviorism." Hempel, in his first piece in this section, tries to show how general laws play the same role in history as in the natural sciences. Grünbaum argues that an acceptance of deterministic laws in the social sciences does not preclude social activism. Watkins, though not strictly from within the logical empiricist camp, argues for methodological individualism (that is, individual determinants provide the starting point for the explanation of social phenomena). This piece is included because the same point of view is implicit in almost all writings on the social sciences from the later period of logical empiricism. Finally, the second piece by Hempel consists of an attempt to show how functional analysis (including explanation) in the social sciences can be incorporated into the logical empiricist framework. The basic idea is that functional attributions pose no genuine problem: They are just an alternative (and fruitful) description of self-regulating systems, which are, ultimately, only ordinary physical systems. Nagel (e.g., 1961), on whose work Hempel draws, used the same strategy to justify functional explanations in biology.

Finally, biology is the focus of the last section. As was mentioned before in the series introduction, the logical empiricists largely ignored biology in spite of its remarkable achievements from 1900 to 1930. The first systematic attempt to apply logical empiricist tenets to biology was Woodger's (1937) *The Axiomatic Method in Biology*, a book marred by its devotion to operationalism. Besides Woodger, who is also important in the history of logical empiricism as the translator of Tarski's papers, almost no other logical empiricist paid significant attention to biology. Two of Woodger's papers are reprinted here. The first consists of Woodger's most sophisticated attempt to axiomatize basic Mendelian genetics. The second is an attempt to formalize the concept of a spatial hierarchy (as found in biology). Though this piece was supposed to be about the relation of biology to physics, it is not. Furthermore, it contains no claim about the reducibility of biology to physics even though Woodger (1952) had developed—apparently independently of Nagel—the standard logical empiricist model of theory reduction. Nagel's paper, which was reproduced with minor changes as part of *The Structure of Science* (Nagel 1961), was far more influential than any of Woodger's work. It is an informal application of his analysis of reduction (see Volume 3) to biology. It is characteristically restrained about the prospects of reductionism and continues to have much to offer to philosophical discussions of reduction (see Sarkar [forthcoming]).

FURTHER READING

Carnap, R., and Gardner, M. 1966. *Philosophical Foundations of Physics: An Introduction to the Philosophy of Science*. New York: Basic Books.

Earman, J., Glymour C., and Stachel, J., eds. 1977. *Foundations of Space-Time Theories*. Minneapolis: University of Minnesota Press.

Feigl, H. 1958. "The 'Mental' and the 'Physical.'" *Minnesota Studies in the Philosophy of Science* 2: 390–497.

Grünbaum, A. 1963. *Philosophical Problems of Space and Time*. New York: Knopf.

Grünbaum, A. 1968. *Geometry and Chronometry in Philosophical Perspective*. Minneapolis: University of Minnesota Press.

Nagel, E. 1961. *The Structure of Science*. New York: Harcourt, Brace.

Reichenbach, H. 1944. *Philosophical Foundations of Quantum Mechanics*. Berkeley: University of California Press.

Reichenbach, H. 1958. *The Philosophy of Space and Time*. New York: Dover.

Reichenbach, H. 1971. *The Direction of Time*. Berkeley: University of California Press.

Sarkar, S. Forthcoming. *Genetics and Reductionism: A Primer*. Cambridge, UK: Cambridge University Press.

Torretti, R. 1983. *Relativity and Geometry*. Oxford: Pergamon Press.

Woodger, J. H. 1937. *The Axiomatic Method in Biology*. Cambridge, UK: Cambridge University Press.

Woodger, J. H. 1952. *Biology and Language*. Cambridge, UK: Cambridge University Press.

GEOMETRY AND EMPIRICAL SCIENCE

C. G. HEMPEL, Queens College

1. Introduction. The most distinctive characteristic which differentiates mathematics from the various branches of empirical science, and which accounts for its fame as the queen of the sciences, is no doubt the peculiar certainty and necessity of its results. No proposition in even the most advanced parts of empirical science can ever attain this status; a hypothesis concerning "matters of empirical fact" can at best acquire what is loosely called a high probability or a high degree of confirmation on the basis of the relevant evidence available; but however well it may have been confirmed by careful tests, the possibility can never be precluded that it will have to be discarded later in the light of new and disconfirming evidence. Thus, all the theories and hypotheses of empirical science share this provisional character of being established and accepted "until further notice," whereas a mathematical theorem, once proved, is established once and for all; it holds with that particular certainty which no subsequent empirical discoveries, however unexpected and extraordinary, can ever affect to the slightest extent. It is the purpose of this paper to examine the nature of that proverbial "mathematical certainty" with special reference to geometry, in an attempt to shed some light on the question as to the validity of geometrical theories, and their significance for our knowledge of the structure of physical space.

The nature of mathematical truth can be understood through an analysis of the method by means of which it is established. On this point I can be very brief: it is the method of mathematical demonstration, which consists in the logical deduction of the proposition to be proved from other propositions, previously established. Clearly, this procedure would involve an infinite regress unless some propositions were accepted without proof; such propositions are indeed found in every mathematical discipline which is rigorously developed; they are the *axioms* or *postulates* (we shall use these terms interchangeably) of the theory. Geometry provides the historically first example of the axiomatic presentation of a mathematical discipline. The classical set of postulates, however, on which Euclid based his system, has proved insufficient for the deduction of the well-known theorems of so-called euclidean geometry; it has therefore been revised and supplemented in modern times, and at present various adequate systems of postulates for euclidean geometry are available; the one most closely related to Euclid's system is probably that of Hilbert.

2. The inadequacy of Euclid's postulates. The inadequacy of Euclid's own set of postulates illustrates a point which is crucial for the axiomatic method in modern mathematics: Once the postulates for a theory have been laid down, every further proposition of the theory must be proved exclusively by logical deduction from the postulates; any appeal, explicit or implicit, to a feeling of self-evidence, or to the characteristics of geometrical figures, or to our experiences concerning the behavior of rigid bodies in physical space, or the like, is

1

strictly prohibited; such devices may have a heuristic value in guiding our efforts to find a strict proof for a theorem, but the proof itself must contain absolutely no reference to such aids. This is particularly important in geometry, where our so-called intuition of geometrical relationships, supported by reference to figures or to previous physical experiences, may induce us tacitly to make use of assumptions which are neither formulated in our postulates nor provable by means of them. Consider, for example, the theorem that in a triangle the three medians bisecting the sides intersect in one point which divides each of them in the ratio of 1:2. To prove this theorem, one shows first that in any triangle ABC (see figure) the line segment MN which connects the centers of AB and AC is parallel to BC and therefore half as long as the latter side. Then the lines BN and CM are drawn, and an examination of the triangles MON and BOC leads to the proof of the theorem. In this procedure, it is usually taken for granted that BN and CM intersect in a point O which lies between B and N as well as between C

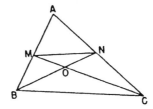

and M. This assumption is based on geometrical intuition, and indeed, it cannot be deduced from Euclid's postulates; to make it strictly demonstrable and independent of any reference to intuition, a special group of postulates has been added to those of Euclid; they are the postulates of order. One of these—to give an example—asserts that if A, B, C are points on a straight line l, and if B lies between A and C, then B also lies between C and A.—Not even as "trivial" an assumption as this may be taken for granted; the system of postulates has to be made so complete that all the required propositions can be deduced from it by purely logical means.

Another illustration of the point under consideration is provided by the proposition that triangles which agree in two sides and the enclosed angle, are congruent. In Euclid's Elements, this proposition is presented as a theorem; the alleged proof, however, makes use of the ideas of motion and superimposition of figures and thus involves tacit assumptions which are based on our geometric intuition and on experiences with rigid bodies, but which are definitely not warranted by—i.e. deducible from—Euclid's postulates. In Hilbert's system, therefore, this proposition (more precisely: part of it) is explicitly included among the postulates.

3. Mathematical certainty. It is this purely deductive character of mathematical proof which forms the basis of mathematical certainty: What the rigorous proof of a theorem—say the proposition about the sum of the angles in a triangle—establishes is not the truth of the proposition in question but rather a conditional insight to the effect that that proposition is certainly true *provided that* the postulates are true; in other words, the proof of a mathematical proposition establishes the fact that the latter is logically implied by the postulates of the theory in question. Thus, each mathematical theorem can be cast into the form

$$(P_1 \cdot P_2 \cdot P_3 \cdot \cdots \cdot P_N) \rightarrow T$$

where the expression on the left is the conjunction (joint assertion) of all the postulates, the symbol on the right represents the theorem in its customary formulation, and the arrow expresses the relation of logical implication or entailment. Precisely this character of mathematical theorems is the reason for their peculiar certainty and necessity, as I shall now attempt to show.

It is typical of any purely logical deduction that the conclusion to which it leads simply re-asserts (a proper or improper) part of what has already been stated in the premises. Thus, to illustrate this point by a very elementary example, from the premise, "This figure is a right triangle," we can deduce the conclusion, "This figure is a triangle"; but this conclusion clearly reiterates part of the information already contained in the premise. Again, from the premises, "All primes different from 2 are odd" and "n is a prime different from 2," we can infer logically that n is odd; but this consequence merely repeats part (indeed a relatively small part) of the information contained in the premises. The same situation prevails in all other cases of logical deduction; and we may, therefore, say that logical deduction—which is the one and only method of mathematical proof—is a technique of conceptual analysis: it discloses what assertions are concealed in a given set of premises, and it makes us realize to what we committed ourselves in accepting those premises; but none of the results obtained by this technique ever goes by one iota beyond the information already contained in the initial assumptions.

Since all mathematical proofs rest exclusively on logical deductions from certain postulates, it follows that a mathematical theorem, such as the Pythagorean theorem in geometry, asserts nothing that is *objectively* or *theoretically new* as compared with the postulates from which it is derived, although its content may well be *psychologically new* in the sense that we were not aware of its being implicitly contained in the postulates.

The nature of the peculiar certainty of mathematics is now clear: A mathematical theorem is certain *relatively* to the set of postulates from which it is derived; *i.e.* it is necessarily true *if* those postulates are true; and this is so because the theorem, if rigorously proved, simply re-asserts part of what has been stipulated in the postulates. A truth of this conditional type obviously implies no assertions about matters of empirical fact and can, therefore, never get into conflict with any empirical findings, even of the most unexpected kind; consequently, unlike the hypotheses and theories of empirical science, it can never suffer the fate of being disconfirmed by new evidence: A mathematical truth is irrefutably certain just because it is devoid of factual, or empirical content. Any theorem of geometry, therefore, when cast into the conditional form described earlier, is analytic in the technical sense of logic, and thus true *a priori*; *i.e.* its truth can be established by means of the formal machinery of logic alone, without any reference to empirical data.

4. **Postulates and truth.** Now it might be felt that our analysis of geometrical truth so far tells only half of the relevant story. For while a geometrical proof no doubt enables us to assert a proposition conditionally—namely on condition that the postulates are accepted—, is it not correct to add that geometry also unconditionally asserts the truth of its postulates and thus, by virtue of the deductive relationship between postulates and theorems, enables us unconditionally to assert the truth of its theorems? Is it not an unconditional assertion of geometry that two points determine one and only one straight line that connects them, or that in any triangle, the sum of the angles equals two right angles? That this is definitely not the case, is evidenced by two important aspects of the axiomatic treatment of geometry which will now be briefly considered.

The first of these features is the well-known fact that in the more recent development of mathematics, several systems of geometry have been constructed which are incompatible with euclidean geometry, and in which, for example, the two propositions just mentioned do not necessarily hold. Let us briefly recollect some of the basic facts concerning these *non-euclidean geometries*. The postulates on which euclidean geometry rests include the famous postulate of the parallels, which, in the case of plane geometry, asserts in effect that through every point P not on a given line l there exists exactly one parallel to l, *i.e.*, one straight line which does not meet l. As this postulate is considerably less simple than the others, and as it was also felt to be intuitively less plausible than the latter, many efforts were made in the history of geometry to prove that this proposition need not be accepted as an axiom, but that it can be deduced as a theorem from the remaining body of postulates. All attempts in this direction failed, however; and finally it was conclusively demonstrated that a proof of the parallel principle on the basis of the other postulates of euclidean geometry (even in its modern, completed form) is impossible. This was shown by proving that a perfectly self-consistent geometrical theory is obtained if the postulate of the parallels is replaced by the assumption that through any point P not on a given straight line l there exist at least two parallels to l. This postulate obviously contradicts the euclidean postulate of the parallels, and if the latter were actually a consequence of the other postulates of euclidean geometry, then the new set of postulates would clearly involve a contradiction, which can be shown not to be the case. This first non-euclidean type of geometry, which is called hyperbolic geometry, was discovered in the early 20's of the last century almost simultaneously, but independently by the Russian N. I. Lobatschefskij, and by the Hungarian J. Bolyai. Later, Riemann developed an alternative geometry, known as elliptical geometry, in which the axiom of the parallels is replaced by the postulate that no line has any parallels. (The acceptance of this postulate, however, in contradistinction to that of hyperbolic geometry, requires the modification of some further axioms of euclidean geometry, if a consistent new theory is to result.) As is to be expected, many of the theorems of these non-euclidean geometries are at variance with those of euclidean theory; thus, *e.g.*, in the hyperbolic geometry of two dimensions, there exist, for each straight line l, through any point P not on l, infinitely many straight lines which do not meet l; also, the sum of the angles in any triangle is less than two right angles. In elliptic geometry, this angle sum is always greater than two right angles; no two straight lines are parallel; and while two different points usually determine exactly one straight line connecting them (as they always do in euclidean geometry), there are certain pairs of points which are connected by infinitely many different straight lines. An illustration of this latter type of geometry is provided by the geometrical structure of that curved two-dimensional space which is represented by the surface of a sphere, when the concept of straight line is interpreted by that of great circle on the sphere. In this space, there are no parallel lines since any two great circles intersect; the endpoints of any diameter of the sphere are points connected by infinitely many different "straight lines," and the sum of the angles in a triangle is always in excess of two right angles. Also, in this space, the ratio between the circumference and the diameter of a circle (not necessarily a great circle) is always less than 2π.

Elliptic and hyperbolic geometry are not the only types of non-euclidean geometry; various other types have been developed; we shall later have occasion to refer to a much more general form of non-euclidean geometry which was likewise devised by Riemann.

The fact that these different types of geometry have been developed in modern mathematics shows clearly that mathematics cannot be said to assert the truth of any particular set of geometrical postulates; all that pure mathematics is interested in, and all that it can establish, is the deductive consequences of given sets of postulates and thus the necessary truth of the ensuing theorems relatively to the postulates under consideration.

A second observation which likewise shows that mathematics does not assert the truth of any particular set of postulates refers to *the status of the concepts in geometry*. There exists, in every axiomatized theory, a close parallelism between the treatment of the propositions and that of the concepts of the system. As we have seen, the propositions fall into two classes: the postulates, for which no proof is given, and the theorems, each of which has to be derived from the postulates. Analogously, the concepts fall into two classes: the primitive or basic concepts, for which no definition is given, and the others, each of which has to be precisely defined in terms of the primitives. (The admission of some undefined concepts is clearly necessary if an infinite regress in definition is to be avoided.) The analogy goes farther: Just as there exists an infinity of theoretically suitable axiom systems for one and the same theory—say, euclidean geometry—, so there also exists an infinity of theoretically possible choices for the primitive terms of that theory; very often—but not always—different axiomatizations of the same theory involve not only different postulates, but also different sets of primitives. Hilbert's axiomatization of plane geometry contains six primitives: point, straight line, incidence (of a point on a line), betweenness (as a relation of three points on a straight line), congruence for line segments, and congruence for angles. (Solid geometry, in Hilbert's axiomatization, requires two further primitives, that of plane and that of incidence of a point on a plane.) All other concepts of geometry, such as those of angle, triangle, circle, *etc.*, are defined in terms of these basic concepts.

But if the primitives are not defined within geometrical theory, what meaning are we to assign to them? The answer is that it is entirely unnecessary to connect any particular meaning with them. True, the words "point," "straight line," *etc.*, carry definite connotations with them which relate to the familiar geometrical figures, but the validity of the propositions is completely independent of these connotations. Indeed, suppose that in axiomatized euclidean geometry, we replace the over-suggestive terms "point," "straight line," "incidence," "betweenness," *etc.*, by the neutral terms "object of kind 1," " object of kind 2," "relation No. 1," "relation No. 2," *etc.*, and suppose that we present this modified wording of geometry to a competent mathematician or logician who, however, knows nothing of the customary connotations of the primitive terms. For this logician, all proofs would clearly remain valid, for as we saw before, a rigorous proof in geometry rests on deduction from the axioms alone without any reference to the customary interpretation of the various geometrical concepts used. We see therefore that indeed no specific meaning has to be attached to the primitive terms of an axiomatized theory; and in a precise logical presentation of axiomatized geometry the primitive concepts are accordingly treated as so-called logical variables.

As a consequence, geometry cannot be said to assert the truth of its postulates, since the latter are formulated in terms of concepts without any specific meaning; indeed, for this very reason, the postulates themselves do not make any specific assertion which could possibly be called true or false! In the terminology of modern logic, the postulates are not sentences, but sentential functions with the primitive concepts as variable arguments.—This point also shows that the postulates of geometry cannot be considered as "self-evident truths," because where no assertion is made, no self-evidence can be claimed.

5. Pure and physical geometry. Geometry thus construed is a purely formal discipline; we shall refer to it also as *pure geometry*. A pure geometry, then,—no matter whether it is of the euclidean or of a non-euclidean variety—deals with no specific subject-matter; in particular, it asserts nothing about physical space. All its theorems are analytic and thus true with certainty precisely because they are devoid of factual content. Thus, to characterize the import of pure geometry, we might use the standard form of a movie-disclaimer: No portrayal of the characteristics of geometrical figures or of the spatial properties or relationships of actual physical bodies is intended, and any similarities between the primitive concepts and their customary geometrical connotations are purely coincidental.

But just as in the case of some motion pictures, so in the case at least of euclidean geometry, the disclaimer does not sound quite convincing: Historically speaking, at least, euclidean geometry has its origin in the generalization and systematization of certain empirical discoveries which were made in connection with the measurement of areas and volumes, the practice of surveying, and the development of astronomy. Thus understood, geometry has factual import; it is an empirical science which might be called, in very general terms, the theory of the structure of physical space, or briefly, *physical geometry*. What is the relation between pure and physical geometry?

When the physicist uses the concepts of point, straight line, incidence, *etc.*, in statements about physical objects, he obviously connects with each of them a more or less definite physical meaning. Thus, the term "point" serves to designate physical points, *i.e.*, objects of the kind illustrated by pin-points, cross hairs, *etc.* Similarly, the term "straight line" refers to straight lines in the sense of physics, such as illustrated by taut strings or by the path of light rays in a homogeneous medium. Analogously, each of the other geometrical concepts has a concrete physical meaning in the statements of physical geometry. In view of this situation, we can say that physical geometry is obtained by what is called, in contemporary logic, a semantical interpretation of pure geometry, Generally speaking, a semantical interpretation of a pure mathematical theory, whose primitives are not assigned any specific meaning, consists in giving each primitive (and thus, indirectly, each defined term) a specific meaning or designatum. In the case of physical geometry, this meaning is physical in the sense just illustrated; it is possible, however, to assign a purely arithmetical meaning to each concept of geometry; the possibility of such an arithmetical interpretation of geometry is of great importance in the study of the consistency and other logical characteristics of geometry, but it falls outside the scope of the present discussion.

By virtue of the physical interpretation of the originally uninterpreted primitives of a geometrical theory, physical meaning is indirectly assigned also to every defined concept of the theory; and if every geometrical term is now taken in its physical interpretation, then every postulate and every theorem of the theory under consideration turns into a statement of physics, with respect to which the question as to truth or falsity may meaningfully be raised—a circumstance which clearly contradistinguishes the propositions of physical geometry from those of the corresponding uninterpreted pure theory.—Consider, for example, the following postulate of pure euclidean geometry: For any two objects x, y of kind 1, there exists exactly one object l of kind 2 such that both x and y stand in relation No. 1 to l. As long as the three primitives occurring in this postulate are uninterpreted, it is obviously meaningless to ask whether the postulate is true. But by virtue of the above physical interpretation, the postulate turns into the following statement: For any two physical points x, y there exists exactly one physical straight line l such that both x and y lie on l. But this is a physical hypothesis, and we may now meaningfully ask whether it is true or false. Similarly, the theorem about the sum of the angles in a triangle turns into the assertion that the sum of the angles (in the physical sense) of a figure bounded by the paths of three light rays equals two right angles.

Thus, the physical interpretation transforms a given pure geometrical theory—euclidean or non-euclidean—into a system of physical hypotheses which, if true, might be said to constitute a theory of the structure of physical space. But the question whether a given geometrical theory in physical interpretation is factually correct represents a problem not of pure mathematics but of empirical science; it has to be settled on the basis of suitable experiments or systematic observations. The only assertion the mathematician can make in this context is this: If all the postulates of a given geometry, in their physical interpretation, are true, then all the theorems of that geometry, in their physical interpretation, are necessarily true, too, since they are logically deducible from the postulates. It might seem, therefore, that in order to decide whether physical space is euclidean or non-euclidean in structure, all we have to do is to test the respective postulates in their physical interpretation. However, this is not directly feasible; here, as in the case of any other physical theory, the basic hypotheses are largely incapable of a direct experimental test; in geometry, this is particularly obvious for such postulates as the parallel axiom or Cantor's axiom of continuity in Hilbert's system of euclidean geometry, which makes an assertion about certain infinite sets of points on a straight line. Thus, the empirical test of a physical geometry no less than that of any other scientific theory has to proceed indirectly; namely, by deducing from the basic hypotheses of the theory certain consequences, or predictions, which are amenable to an experimental test. If a test bears out a prediction, then it constitutes confirming evidence (though, of course, no conclusive proof) for the theory; otherwise, it disconfirms the theory. If an adequate amount of confirming evidence for a theory has been established, and if no disconfirming evidence has been found, then the theory may be accepted by the scientist "until further notice."

It is in the context of this indirect procedure that pure mathematics and logic acquire their inestimable importance for empirical science: While formal logic and pure mathematics do not in themselves establish any assertions about matters of empirical fact, they provide an efficient and entirely indispensable machinery for deducing, from abstract theoretical assumptions, such as the laws of Newtonian mechanics or the postulates of euclidean geometry in physical interpretation, consequences concrete and specific enough to be accessible to direct experimental test. Thus, *e.g.*, pure euclidean geometry shows that from its postulates there may be deduced the theorem about the sum of the angles in a triangle, and that this deduction is possible no matter how the basic concepts of geometry are interpreted; hence also in the case of the physical interpretation of euclidean geometry. This theorem, in its physical interpretation, is accessible to experimental test; and since the postulates of elliptic and of hyperbolic geometry imply values different from two right angles for the angle sum of a triangle, this particular proposition seems to afford a good opportunity for a crucial experiment. And no less a mathematician than Gauss did indeed perform this test; by means of optical methods—and thus using the interpretation of physical straight lines as paths of light rays—he ascertained the angle sum of a large triangle determined by three mountain tops. Within the limits of experimental error, he found it equal to two right angles.

6. On Poincaré's conventionalism concerning geometry. But suppose that Gauss had found a noticeable deviation from this value; would that have meant a refutation of euclidean geometry in its physical interpretation, or, in other words, of the hypothesis that physical space is euclidean in structure? Not necessarily; for the deviation might have been accounted for by a hypothesis to the effects that the paths of the light rays involved in the sighting process were bent by some disturbing force and thus were not actually straight lines. The same kind of reference to deforming forces could also be used if, say, the euclidean theorems of congruence for plane figures were tested in their physical interpretation by means of experiments involving rigid bodies, and if any violations of the theorems were found. This point is by no means trivial; Henri Poincaré, the great French mathematician and theoretical physicist, based on considerations of this type his famous *conventionalism concerning geometry*. It was his opinion that no empirical test, whatever its outcome, can conclusively invalidate the euclidean conception of physical space; in other words, the validity of euclidean geometry in physical science can always be preserved—if necessary, by suitable changes in the theories of physics, such as the introduction of new hypotheses concerning deforming or deflecting forces. Thus, the question as to whether physical space has a euclidean or a non-euclidean structure would become a matter of convention, and the decision to preserve euclidean geometry at all costs would recommend itself, according to Poincaré, by the greater simplicity of euclidean as compared with non-euclidean geometrical theory.

It appears, however, that Poincaré's account is an oversimplification. It rightly calls attention to the fact that the test of a physical geometry G always presupposes a certain body P of non-geometrical physical hypotheses (including the physical theory of the instruments of measurement and observation used in the test), and that the so-called test of G actually bears on the combined theoretical system $G \cdot P$ rather than on G alone. Now, if predictions derived from $G \cdot P$ are contradicted by experimental findings, then a change in the theoretical structure becomes necessary. In classical physics, G always was euclidean geometry in its physical interpretation, GE; and when experimental evidence required a modification of the theory, it was P rather than GE which was changed. But Poincaré's assertion that this procedure would always be distinguished by its greater simplicity is not entirely correct; for what has to be taken into consideration is the simplicity of the total system $G \cdot P$, and not just that of its geometrical part. And here it is clearly conceivable that a simpler total theory in accordance with all the relevant empirical evidence is obtainable by going over to a non-euclidean form of geometry rather than by preserving the euclidean structure of physical space and making adjustments only in part P.

And indeed, just this situation has arisen in physics in connection with the development of the general theory of relativity: If the primitive terms of geometry are given physical interpretations along the lines indicated before, then certain findings in astronomy represent good evidence in favor of a total physical theory with a non-euclidean geometry as part G. According to this theory, the physical universe at large is a three-dimensional curved space of a very complex geometrical structure; it is finite in volume and yet unbounded in all directions. However, in comparatively small areas, such as those involved in Gauss' experiment, euclidean geometry can serve as a good approximative account of the geometrical structure of space. The kind of structure ascribed to physical space in this theory may be illustrated by an analogue in two dimensions; namely, the surface of a sphere. The geometrical structure of the latter, as was pointed out before, can be described by means of elliptic geometry, if the primitive term "straight line" is interpreted as meaning "great circle," and if the other primitives are given analogous interpretations. In this sense, the surface of a sphere is a two-dimensional curved space of non-euclidean structure, whereas the plane is a two-dimensional space of euclidean structure. While the plane is unbounded in all directions, and infinite in size, the spherical surface is finite in size and yet unbounded in all directions: a two-dimensional physicist, travelling along "straight lines" of that space would never encounter any boundaries of his space; instead, he would finally return to his point of departure, provided that his life span and his technical facilities were sufficient for such a trip in consideration of the size of his "universe." It is interesting to note that the physicists of that world, even if they lacked any intuition of a three-dimensional space, could empirically ascertain the fact that their two-dimensional space was curved. This might be done by means of the method of traveling along straight lines; another, simpler test would consist in determining the angle sum in a triangle; again another in determining, by means of measuring tapes, the ratio of the circumference of a circle (not necessarily a great circle) to its diameter; this ratio would turn out to be less than π.

The geometrical structure which relativity physics ascribes to physical space is a three-dimensional analogue to that of the surface of a sphere, or, to be more exact, to that of the closed and finite surface of a potato, whose curvature varies from point to point. In our physical universe, the curvature of space at a given point is determined by the distribution of masses in its neighborhood; near large masses such as the sun, space is strongly curved, while in regions of low mass-density, the structure of the universe is approximately euclidean. The hypothesis stating the connection between the mass distribution and the curvature of space at a point has been approximately confirmed by astronomical observations concerning the paths of light rays in the gravitational field of the sun.

The geometrical theory which is used to describe the structure of the physical universe is of a type that may be characterized as a generalization of elliptic geometry. It was originally constructed by Riemann as a purely mathematical theory, without any concrete possibility of practical application at hand. When

Einstein, in developing his general theory of relativity, looked for an appropriate mathematical theory to deal with the structure of physical space, he found in Riemann's abstract system the conceptual tool he needed. This fact throws an interesting sidelight on the importance for scientific progress of that type of investigation which the "practical-minded" man in the street tends to dismiss as useless, abstract mathematical speculation.

Of course, a geometrical theory in physical interpretation can never be validated with mathematical certainty, no matter how extensive the experimental tests to which it is subjected; like any other theory of empirical science, it can acquire only a more or less high degree of confirmation. Indeed, the considerations presented in this article show that the demand for mathematical certainty in empirical matters is misguided and unreasonable; for, as we saw, mathematical certainty of knowledge can be attained only at the price of analyticity and thus of complete lack of factual content. Let me summarize this insight in Einstein's words:

"As far as the laws of mathematics refer to reality, they are not certain; and as far as they are certain, they do not refer to reality."

THE PHILOSOPHICAL SIGNIFICANCE OF THE THEORY OF RELATIVITY

I

THE philosophical significance of the theory of relativity has been the subject of contradictory opinions. Whereas many writers have emphasized the philosophical implications of the theory and have even tried to interpret it as a sort of philosophical system, others have denied the existence of such implications and have voiced the opinion that Einstein's theory is merely a physical matter, of interest only to the mathematical physicist. These critics believe that philosophical views are constructed by other means than the methods of the scientist and are independent of the results of physics.

Now it is true that what has been called the philosophy of relativity represents, to a great extent, the fruit of misunderstandings of the theory rather than of its physical content. Philosophers who regard it as an ultimate wisdom that everything is relative are mistaken when they believe that Einstein's theory supplies evidence for such a sweeping generalization; and their error is even deeper when they transfer such a relativity to the field of ethics, when they claim that Einstein's theory implies a relativism of men's duties and rights. The theory of relativity is restricted to the cognitive field. That moral conceptions vary with the social class and the structure of civilization is a fact which is not derivable from Einstein's theory; the parallelism between the relativity of ethics and that of space and time is nothing more than a superficial analogy, which blurs the essential logical differences between the fields of volition and cognition. It appears understandable that those who were trained in the precision of mathematico-physical methods wish to divorce physics from such blossoms of philosophizing.

Yet it would be another mistake to believe that Einstein's theory is not a philosophical theory. This discovery of a physicist has radical consequences for the theory of knowledge. It compels us to revise certain traditional conceptions that have played an important part in the history of philosophy, and it offers solutions for certain questions which are as old as the history of philosophy and which could not be answered earlier. Plato's attempt to solve the problems of geometry by a theory of ideas, Kant's attempt to account for the nature of space and time by a *"reine Anschauung"* and by a transcendental philosophy, these represent answers to the very questions to which Einstein's theory has given a different answer at a later time. If Plato's and Kant's doctrines are philosophical theories, then Einstein's theory of relativity is a philosophical and not a merely physical matter. And the questions referred to are not of a secondary nature but of primary import for philosophy; that much is evident from the central position they occupy in the systems of Plato and Kant. These systems are untenable if Einstein's answer is put in the place of the answers given to the same questions by their authors; their foundations are shaken when space and time are not the revelations of an insight into a world of ideas, or of a vision grown from pure reason, which a philosophical apriorism claimed to have established. The analysis of knowledge has always been the basic issue of philosophy; and if knowledge in so fundamental a domain as that of space and time is subject to revision, the implications of such criticism will involve the whole of philosophy.

To advocate the philosophical significance of Einstein's theory, however, does not mean to make Einstein a philosopher; or, at least, it does not mean that Einstein is a philosopher of primary intent. Einstein's primary objectives were all in the realm of physics. But he saw that certain physical problems could not be solved unless the solutions were preceded by a logical analysis of the fundamentals of space and time, and he saw that this analysis, in turn, presupposed a philosophic readjustment of certain familiar conceptions of knowledge. The physicist who wanted to understand the Michelson experiment had to commit himself to a philosophy for which the meaning of a statement is reducible to its verifiability, that is, he had to adopt the verifiability theory of meaning if he wanted to escape a maze of ambiguous questions and gratuitous complications. It is this positivist, or let me rather say, empiricist commitment which determines the philosophical position of Einstein. It was not necessary for him to elaborate on it to any great extent; he merely had to join a trend of development characterized, within the generation of physicists before him, by such names as Kirchhoff, Hertz, Mach, and to carry through to its ultimate consequences a philosophical evolution documented at earlier stages in such principles as Occam's razor and Leibnitz' identity of indiscernibles.

Einstein has referred to this conception of meaning in various remarks, though he has never felt it necessary to enter into a discussion of its grounds or into an analysis of its philosophical position. The exposition and substantiation of a philosophical theory is nowhere to be found in his writings. In fact, Einstein's philosophy is not so much a philosophical system as a philosophical attitude; apart from occasional remarks, he left it to others to say what philosophy his equations entail and thus remained a philosopher by implication, so to speak. That is both his strength and his weakness; his strength, because it made his physics so conclusive; his weakness, because it left his theory open to misunderstandings and erroneous interpretations.

It seems to be a general law that the making of a new physics precedes a new philosophy of physics. Philosophic analysis is more easily achieved when it is applied to concrete purposes, when it is done within the pursuit of research aimed at an interpretation of observational data. The philosophic results of the procedure are often recognized at a later stage; they are the fruit of reflection about the methods employed in the solution of the concrete problem. But those who make the new physics usually do not have the leisure, or do not regard it as their objective, to expound and elaborate the philosophy implicit in their constructions. Occasionally, in popular presentations, a physicist attempts to explain the logical background of his

15

theories; thus many a physicist has been misled into believing that philosophy of physics is the same as a popularization of physics. Einstein himself does not belong to this group of writers who do not realize that what they achieve is as much a popularization of philosophy as it is one of physics, and that the philosophy of physics is as technical and intricate as is physics itself. Nevertheless, Einstein is not a philosopher in the technical sense either. It appears to be practically impossible that the man who is looking for new physical laws should also concentrate on the analysis of his method; he will perform this second task only when such analysis is indispensable for the finding of physical results. The division of labor between the physicist and the philosopher seems to be an inescapable consequence of the organization of the human mind.

It is not only a limitation of human capacities which calls for a division of labor between the physicist and the philosopher. The discovery of general relations that lend themselves to empirical verification requires a mentality different from that of the philosopher, whose methods are analytic and critical rather than predictive. The physicist who is looking for new discoveries must not be too critical; in the initial stages he is dependent on guessing, and he will find his way only if he is carried along by a certain faith which serves as a directive for his guesses. When I, on a certain occasion, asked Professor Einstein how he found his theory of relativity, he answered that he found it because he was so strongly convinced of the harmony of the universe. No doubt his theory supplies a most successful demonstration of the usefulness of such a conviction. But a creed is not a philosophy; it carries this name only in the popular interpretation of the term. The philosopher of science is not much interested in the thought processes which lead to scientific discoveries; he looks for a logical analysis of the completed theory, including the relationships establishing its validity. That is, he is not interested in the context of discovery, but in the context of justification. But the critical attitude may make a man incapable of discovery; and, as long as he is successful, the creative physicist may very well prefer his creed to the logic of the analytic philosopher.

16

The philosopher has no objections to a physicist's beliefs, so long as they are not advanced in the form of a philosophy. He knows that a personal faith is justified as an instrument of finding a physical theory, that it is but a primitive form of guessing, which is eventually replaced by the elaborate theory, and that it is ultimately subject to the same empirical tests as the theory. The philosophy of physics, on the other hand, is not a product of creed but of analysis. It incorporates the physicist's beliefs into the psychology of discovery; it endeavors to clarify the meanings of physical theories, independently of the interpretation by their authors, and is concerned with logical relationships alone.

Seen from this viewpoint it appears amazing to what extent the logical analysis of relativity coincides with the original interpretation by its author, as far as it can be constructed from the scanty remarks in Einstein's publications. In contradistinction to some developments in quantum theory, the logical schema of the theory of relativity corresponds surprisingly with the program which controlled its discovery. His philosophic clarity distinguishes Einstein from many a physicist whose work became the source of a philosophy different from the interpretation given by the author. In the following pages I shall attempt to outline the philosophical results of Einstein's theory, hoping to find a friendly comment by the man who was the first to see all these relations, even though he did not always formulate them explicitly. And the gratitude of the philosopher goes to this great physicist whose work includes more implicit philosophy than is contained in many a philosophical system.

II

The logical basis of the theory of relativity is the discovery that many statements, which were regarded as capable of demonstrable truth or falsity, are mere definitions.

This formulation sounds like the statement of an insignificant technical discovery and does not reveal the far-reaching implications which make up the philosophical significance of the theory. Nonetheless it is a complete formulation of the *logical* part of the theory.

Consider, for instance, the problem of geometry. That the unit of measurement is a matter of definition is a familiar fact; everybody knows that it does not make any difference whether we measure distances in feet or meters or light-years. However, that the comparison of distances is also a matter of definition is known only to the expert of relativity. This result can also be formulated as the definitional character of congruence. That a certain distance is congruent to another distance situated at a different place can never be proved to be true; it can only be maintained in the sense of a definition. More precisely speaking, it can be maintained as true only after a definition of congruence is given; it therefore depends on an original comparison of distances which is a matter of definition. A comparison of distances by means of the transport of solid bodies is but one definition of congruence. Another definition would result if we regarded a rod, once it had been transported to another location, as twice as long, thrice transported as three times as long, and so on. A further illustration refers to time: that the simultaneity of events occurring at distant places is a matter of definition was not known before Einstein based his special theory of relativity on this logical discovery.

The definitions employed for the construction of space and time are of a particular kind: they are co-ordinative definitions. That is, they are given by the co-ordination of a physical object, or process, to some fundamental concept. For instance, the concept "equal length" is defined by reference to a physical object, a solid rod, whose transport lays down equal distances. The concept "simultaneous" is defined by the use of light-rays which move over equal distances. The definitions of the theory of relativity are all of this type; they are co-ordinative definitions.

In the expositions of the theory of relativity the use of different definitions is often illustrated by a reference to different observers. This kind of presentation has led to the erroneous conception that the relativity of space-time measurements is connected with the subjectivity of the observer, that the privacy of the world of sense perception is the origin of the relativity maintained by Einstein. Such Protagorean interpretation of Einstein's relativity is utterly mistaken. The definitional character of simultaneity, for instance, has nothing to do with the perspective variations resulting for observers located in different frames of reference. That we co-ordinate different definitions of simultaneity to different observers merely serves as a simplification of the presentation of logical relationships. We could as well interchange the co-ordination and let the observer located in the "moving" system employ the time definition of the observer located in the system "at rest," and vice versa; or we could even let both employ the same time definition, for instance that of the system "at rest." Such variations would lead to different transformations; for instance, the last mentioned definition would lead, not to the Lorentz transformation, but to the classical transformation from a system at rest to a moving system. It is convenient to identify one definitional system with one observer; to speak of different observers is merely a mode of speech expressing the plurality of definitional systems. In a logical exposition of the theory of relativity the observer can be completely eliminated.

Definitions are arbitrary; and it is a consequence of the definitional character of fundamental concepts that with the change of the definitions various descriptional systems arise. But these systems are equivalent to each other, and it is possible to go from each system to another one by a suitable transformation. Thus the definitional character of fundamental concepts leads to a plurality of equivalent descriptions. A familiar illustration is given by the various descriptions of motion resulting when the system regarded as being at rest is varied. Another illustration is presented by the various geometries resulting, for the same physical space, through changes in the definition of congruence. All these descriptions represent different languages saying the same thing; equivalent descriptions, therefore, express the same physical content. The theory of equivalent descriptions is also applicable to other fields of physics; but the domain of space and time has become the model case of this theory.

The word "relativity" should be interpreted as meaning "relative to a certain definitional system." That relativity implies plurality follows because the variation of definitions leads to the plurality of equivalent descriptions. But we see that the plurality implied is not a plurality of different views, or of systems of contradictory content; it is merely a plurality of equivalent languages and thus of forms of expression which do not contradict each other but have the same content. Relativity does not mean an abandonment of truth; it only means that truth can be formulated in various ways.

I should like to make this point quite clear. The two statements "the room is 21 feet long" and "the room is 7 yards long" are quivalent descriptions; they state the same fact. That the simple truth they express can be formulated in these two ways does not eliminate the concept of truth; it merely illustrates the fact that the number characterizing a length is relative to the unit of measurement. All relativities of Einstein's theory are of this type. For instance, the Lorentz transformation connects different descriptions of space-time relations which are equivalent in the same sense as the statements about a length of 21 feet and a length of 7 yards.

Some confusion has arisen from considerations referring to the property of simplicity. One descriptional system can be simpler than another; but that fact does not make it "truer" than the other. The decimal system is simpler than the yard-foot-inch system; but an architect's plan drawn in feet and inches is as true a description of a house as a plan drawn in the decimal system. A simplicity of this kind, for which I have used the name of *descriptive simplicity*, is not a criterion of truth. Only within the frame of inductive considerations can simplicity be a criterion of truth; for instance, the simplest curve between observational data plotted in a diagram is regarded as "truer," i.e., more probable, than other connecting curves. This *inductive simplicity*, however, refers to non-equivalent descriptions and does not play a part in the theory of relativity, in which only equivalent descriptions are compared. The simplicity of descriptions used in Einstein's theory is therefore always a descriptive simplicity. For instance, the fact that non-Euclidean geometry often supplies a simpler description of physical space than does Euclidean geometry does not make the non-Euclidean description "truer."

Another confusion must be ascribed to the theory of conventionalism, which goes back to Poincaré. According to this theory, geometry is a matter of convention, and no empirical meaning can be assigned to a statement about the geometry of physical space. Now it is true that physical space can be described by both a Euclidean and a non-Euclidean geometry; but it is an erroneous interpretation of this relativity of geometry to call a statement about the geometrical structure of physical space meaningless. The choice of a geometry is arbitrary only so long as no definition of congruence is specified. Once this definition is set up, it becomes an empirical question *which* geometry holds for a physical space. For instance, it is an empirical fact that, when we use solid bodies for the definition of congruence, our physical space is practically Euclidean within terrestrial dimensions. If, in a different part of the universe, the same definition of congruence were to lead to a non-Euclidean geometry, that part of universal space would have a geometrical structure different from that of our world. It is true that a Euclidean geometry could also be introduced for that part of the universe; but then the definition of congruence would no longer be given by solid bodies.[1] The combination of a statement about a geometry with a statement of the co-ordinative definition of congruence employed is subject to empirical test and thus expresses a property of the physical world. The conventionalist overlooks the fact that only the incomplete statement of a geometry, in which a reference to the definition of congruence is omitted, is arbitrary; if the statement is made complete by the addition of a reference to the definition of congruence, it becomes empirically verifiable and thus has physical content.

Instead of speaking of conventionalism, therefore, we should speak of the relativity of geometry. Geometry is relative in precisely the same sense as other relative concepts. We might call it a convention to say that Chicago is to the left of New York; but we should not forget that this conventional statement can be made objectively true as soon as the point of refer-

[1] Poincaré believed that the definition of a solid body could not be given without reference to a geometry. That this conception is mistaken, is shown in the present author's *Philosophie der Raum-Zeit-Lehre* (Berlin, 1928) §5.

ence is included in the statement. It is not a convention but a physical fact that Chicago is to the left of New York, seen, for instance, from Washington, D.C. The relativity of simple concepts, such as left and right, is well known. That the fundamental concepts of space and time are of the same type is the essence of the theory of relativity.

The relativity of geometry is a consequence of the fact that different geometries can be represented on one another by a one-to-one correspondence. For certain geometrical systems, however, the representation will not be continuous throughout, and there will result singularities in individual points or lines. For instance, a sphere cannot be projected on a plane without a singularity in at least one point; in the usual projections, the North Pole of the sphere corresponds to the infinity of the plane. This peculiarity involves certain limitations for the relativity of geometry. Assume that in one geometrical description, say, by a spherical space, we have a normal causality for all physical occurrences; then a transformation to certain other geometries, including the Euclidean geometry, leads to violations of the principle of causality, to *causal anomalies*. A light signal going from a point A by way of the North Pole to a point B in a finite time will be so represented within a Euclidean interpretation of this space, that it moves from A in one direction towards infinity and returns from the other side towards B, thus passing through an infinite distance in a finite time. Still more complicated causal anomalies result for other transformations.[*] If the principle of normal causality, i.e., a continuous spreading from cause to effect in a finite time, or *action by contact*, is set up as a necessary prerequisite of the description of nature, certain worlds cannot be interpreted by certain geometries. It may well happen that the geometry thus excluded is the Euclidean one; if Einstein's hypothesis of a closed universe is correct, a

[*] Cf. the author's *Philosophie der Raum-Zeit-Lehre* (Berlin, 1928), §12. It has turned out that within the plurality of descriptions applicable to quantum mechanics the problem of causal anomalies plays an even more important part, since we have there a case where no description exists which avoids causal anomalies. (Cf. also the author's *Philosophic Foundations of Quantum Mechanics*, Berkeley, 1944), §§5-7, §26.

Euclidean description of the universe would be excluded for all adherents of a normal causality.

It is this fact which I regard as the strongest refutation of the Kantian conception of space. The relativity of geometry has been used by Neo-Kantians as a back door through which the apriorism of Euclidean geometry was introduced into Einstein's theory: if it is always possible to select a Euclidean geometry for the description of the universe, then the Kantian insists that it be this description which should be used, because Euclidean geometry, for a Kantian, is the only one that can be visualized. We see that this rule may lead to violations of the principle of causality; and since causality, for a Kantian, is as much an *a priori* principle as Euclidean geometry, his rule may compel the Kantian to jump from the frying pan into the fire. There is no defense of Kantianism, if the statement of the geometry of the physical world is worded in a complete form, including all its physical implications; because in this form the statement is empirically verifiable and depends for its truth on the nature of the physical world.[*]

It should be clear from this analysis that the plurality of equivalent description does not rule out the possibility of true empirical statements. The empirical content of statements about space and time is only stated in a more complicated way.

III

Though we now possess, in Einstein's theory, a complete statement of the relativity of space and time, we should not forget that this is the result of a long historical development. I mentioned above Occam's razor and Leibnitz' identity of indiscernibles in connection with the verifiability theory of meaning. It is a matter of fact that Leibnitz applied his principle successfully to the problem of motion and that he arrived at a relativity of motion on logical grounds. The famous correspondence between Leibnitz and Clarke,—the latter a contemporary defender of Newton's absolutism,—presents us with the same type of discussion which is familiar from the modern discussions

[*] This refutation of Kantianism was presented in the author's *Relativitätstheorie und Erkenntnis Apriori* (Berlin, 1920).

of relativity and reads as though Leibnitz had taken his arguments from expositions of Einstein's theory. Leibnitz even went so far as to recognize the relationship between causal order and time order.[*] This conception of relativity was carried on at a later time by Ernst Mach, who contributed to the discussion the important idea that a relativity of rotational motion requires an extension of relativism to the concept of inertial force. Einstein has always acknowledged Mach as a forerunner of his theory.

Another line of development, which likewise found its completion through Einstein's theory, is presented by the history of geometry. The discovery of non-Euclidean geometries by Gauss, Bolyai, and Lobachewski was associated with the idea that physical geometry might be non-Euclidean; and it is known that Gauss tried to test the Euclidean character of terrestrial geometry by triangular measurements from mountain tops. But the man to whom we owe the philosophical clarification of the problem of geometry is Helmholtz. He saw that physical geometry is dependent on the definition of congruence by means of the solid body and thus arrived at a clear statement of the nature of physical geometry, superior in logical insight to Poincaré's conventionalism developed several decades later. It was Helmholtz, too, who clarified the problem of a visual presentation of non-Euclidean geometry by the discovery that visualization is a fruit of experiences with solid bodies and light-rays. We find in Helmholtz' writings the famous statement that imagining something visually means depicting the series of sense perceptions which one would have if one lived in such a world. That Helmholtz did not succeed in dissuading contemporary philosophers from a Kantian apriorism of space and time is not his fault. His philosophical views were known only among a small group of experts. When, with Einstein's theory, the public interest turned toward these problems, philosophers began to give in and to depart from Kant's apriorism. Let us hope that this development will continue and eventually include even those philosophers who in our day still defend an apriorist philosophy against the attacks of the mathematical physicist.

[*] For an analysis of Leibnitz' views see the author's "Die Bewegungslehre bei Newton, Leibnitz und Huyghens," *Kantstudien* [vol. 29, 1924], 416.

Although there exists a historical evolution of the concepts of space and motion, this line of development finds no analogue in the concept of time. The first to speak of a relativity of the measure of time, i.e., of what is called the uniform flow of time, was Mach. However, Einstein's idea of a relativity of simultaneity has no forerunners. It appears that this discovery could not be made before the perfection of experimental methods of physics. Einstein's relativity of simultaneity is closely associated with the assumption that light is the fastest signal, an idea which could not be conceived before the negative outcome of such experiments as that by Michelson.

It was the combination of the relativity of time and of motion which made Einstein's theory so successful and led to results far beyond the reach of earlier theories. The discovery of the special theory of relativity, which none of Einstein's forerunners had thought of, thus became the key to a general theory of space and time, which included all the ideas of Leibnitz, Gauss, Riemann, Helmholtz, and Mach, and which added to them certain fundamental discoveries which could not have been anticipated at an earlier stage. In particular, I refer to Einstein's conception according to which the geometry of physical space is a function of the distribution of masses, an idea entirely new in the history of geometry.

This short account shows that the evolution of philosophical ideas is guided by the evolution of physical theories. The philosophy of space and time is not the work of the ivory tower philosopher. It was constructed by men who attempted to combine observational data with mathematical analysis. The great synthesis of the various lines of development, which we owe to Einstein, bears witness to the fact that philosophy of science has taken over a function which philosophical systems could not perform.

IV

The question of what is space and time has fascinated the authors of philosophical systems over and again. Plato answered it by inventing a world of "higher" reality, the world of ideas, which includes space and time among its ideal objects and reveals their relations to the mathematician who is able to perform the necessary act of vision. For Spinoza space was an attribute of God. Kant, on the other hand, denied the reality of space and time and regarded these two conceptual systems as forms of visualization, i.e., as constructions of the human mind, by means of which the human observer combines his perceptions so as to collect them into an orderly system.

The answer we can give to the question on the basis of Einstein's theory is very different from the answers of these philosophers. The theory of relativity shows that space and time are neither ideal objects nor forms of order necessary for the human mind. They constitute a relational system expressing certain general features of physical objects and thus are descriptive of the physical world. Let us make this fact quite clear.

It is true that, like all concepts, space and time are inventions of the human mind. But not all inventions of the human mind are fit to describe the physical world. By the latter phrase we mean that the concepts refer to certain physical objects and differentiate them from others. For instance, the concept "centaur" is empty, whereas the concept "bear" refers to certain physical objects and distinguishes them from others. The concept "thing," on the other hand, though not empty, is so general that it does not differentiate between objects. Our examples concern one-place predicates, but the same distinction applies to two-place predicates. The relation "telepathy" is empty, whereas the relation "father" is not. When we say that non-empty one-place predicates like "bear" describe real objects, we must also say that non-empty many-place predicates like "father" describe real relations.

It is in this sense that the theory of relativity maintains the reality of space and time. These conceptual systems describe relations holding between physical objects, namely, solid bodies, light-rays, and watches. In addition, these relations formulate physical laws of great generality, determining some fundamental features of the physical world. Space and time have as much reality as, say, the relation "father" or the Newtonian forces of attraction.

The following consideration may serve as a further explanation why geometry is descriptive of physical reality. As long as only one geometry, the Euclidean geometry, was known, the fact that this geometry could be used for a description of the physical world represented a problem for the philosopher; and Kant's philosophy must be understood as an attempt to explain why a structural system derived from the human mind can account for observational relations. With the discovery of a plurality of geometries the situation changed completely. The human mind was shown to be capable of inventing all kinds of geometrical systems, and the question, which of the systems is suitable for the description of physical reality, was turned into an empirical question, i.e., its answer was ultimately left to empirical data. Concerning the empirical nature of this answer we refer the reader to our considerations in Section II; it is the combined statement of geometry and co-ordinative definitions which is empirical. But, if the statement about the geometry of the physical world is empirical, geometry describes a property of the physical world in the same sense, say, as temperature or weight describe properties of physical objects. When we speak of the reality of physical space we mean this very fact.

As mentioned above, the objects whose general relationship is expressed in the spatio-temporal order are solid bodies, light-rays, and natural watches, i.e., closed periodic systems, like revolving atoms or revolving planets. The important part which light-rays play in the theory of relativity derives from the fact that light is the fastest signal, i.e., represents the fastest form of a causal chain. The concept of causal chain can be shown to be the basic concept in terms of which the structure of space and time is built up. The spatio-temporal order thus must be regarded as the expression of the causal order of the physical world. The close connection between space and time on the one hand and causality on the other hand is perhaps the most prominent feature of Einstein's theory, although this feature has not always been recognized in its significance. Time order, the order of *earlier* and *later,* is reducible to causal order; the cause is always earlier than the effect, a relation which cannot be reversed. That Einstein's theory admits of a reversal of time order for certain events, a result known from the relativity of simultaneity, is merely a consequence of this fundamental fact.

Since the speed of causal transmission is limited, there exist events of such a kind that neither of them can be the cause or the effect of the other. For events of this kind a time order is not defined, and either of them can be called earlier or later than the other.

Ultimately even spatial order is reducible to causal order; a space point B is called closer to A than a space point C, if a direct light-signal, i.e., a fastest causal chain, from A to C passes by B. For a construction of geometry in terms of light-rays and mass-points, i.e., a light-geometry, I refer to another publication.[*]

The connection between time order and causal order leads to the question of the direction of time. I should like to add some remarks about this problem which has often been discussed, but which has not always been stated clearly enough. The relation between cause and effect is an asymmetrical relation; if P is the cause of Q, then Q is not the cause of P. This fundamental fact is essential for temporal order, because it makes time a serial relation. By a serial relation we understand a relation that orders its elements in a linear arrangement; such a relation is always asymmetrical and transitive, like the relation "smaller than." The time of Einstein's theory has these properties; that is necessary, because otherwise it could not be used for the construction of a serial order.

But what we call the direction of time must be distinguished from the asymmetrical character of the concepts "earlier" and "later." A relation can be asymmetrical and transitive without distinguishing one direction from the opposite one. For instance, the points of a straight line are ordered by a serial relation which we may express by the words "before" and "after." If A is before B, then B is not before A, and if A is before B and B is before C, then A is before C. But which direction of the line we should call "before" and which one "after" is not indicated by the nature of the line; this definition can only be set up by an arbitrary choice, for instance, by pointing into one direction and calling it the direction of "before." In other words, the relations "before" and "after" are structurally indistinguish-

[*] H. Reichenbach, *Philosophie der Raum-Zeit-Lehre* (Berlin, 1928), §27.

able and therefore interchangeable; whether we say that point *A* is before point *B* or after point *B* is a matter of arbitrary definition. It is different with the relation "smaller than" among real numbers. This relation is also a serial relation and thus asymmetrical and transitive; but in addition, it is structurally different from its converse, the relation "larger than," a fact expressible through the difference of positive and negative numbers. The square of a positive number is a positive number, and the square of a negative number is also a positive number. This peculiarity enables us to define the relation "smaller than:" a number which cannot be the square of another number is smaller than a number which is the square of another number. The series of real numbers possesses therefore a direction: the direction "smaller than" is not interchangeable with the direction "larger than;" these relations are therefore not only asymmetrical but also *unidirectional*.

The problem of the time relation is whether it is unidirectional. The relation "earlier than" which we use in everyday life is structurally different from the relation "later than." For instance, we may make up our mind to go to the theatre tomorrow; but it would be nonsensical to make up our mind to go to the theatre yesterday. The physicist formulates this distinction as the *irreversibility of time:* time flows in one direction, and the flow of time cannot be reversed. We see that, in the language of the theory of relations, the question of the irreversibility of time is expressed, not by the question of whether time is an asymmetrical relation, but by the question of whether it is a unidirectional relation.

For the theory of relativity, time is certainly an asymmetrical relation, since otherwise the time relation would not establish a serial order; but it is not unidirectional. In other words, the irreversibility of time does not find an expression in the theory of relativity. We must not conclude that that is the ultimate word which the physicist has to say about time. All we can say is that, as far as the theory of relativity is concerned, we need not make a qualitative distinction between the two directions of time, between the "earlier" and "later." A physical theory may very well abstract from certain properties of the physical world; that does not mean that these properties do not exist. The irreversibility of time has so far been dealt with only in thermodynamics, where it is conceived as being merely of a statistical nature, not applicable to elementary processes. This answer is none too satisfactory; particularly in view of the fact that it has led to certain paradoxes. Quantum physics so far, however, has no better answer. I would like to say that I regard this problem as at present unsolved and do not agree with those who believe that there is no genuine problem of the direction of time.

It is an amazing fact that the mathematico-physical treatment of the concept of time formulated in Einstein's theory has led to a clarification which philosophical analysis could not achieve. For the philosopher such concepts as time order and simultaneity were primitive notions inaccessible to further analysis. But the claim that a concept is exempt from analysis often merely springs from an inability to understand its meaning. With his reduction of the time concept to that of causality and his generalization of time order toward a relativity of simultaneity, Einstein has not only changed our conceptions of time; he has also clarified the meaning of the classical time concept which preceded his discoveries. In other words, we know better today what absolute time means than anyone of the adherents of the classical time conceptions. Absolute simultaneity would hold in a world in which there exists no upper limit for the speed of signals, i.e., for causal transmission. A world of this type is as well imaginable as Einstein's world. It is an empirical question to which type our world belongs. Experiment has decided in favor of Einstein's conception. As in the case of geometry, the human mind is capable of constructing various forms of a temporal schema; the question which of these schemes fits the physical world, i.e., is true, can only be answered by reference to observational data. What the human mind contributes to the problem of time is not one definite time order, but a plurality of possible time orders, and the selection of one time order as the real one is left to empirical observation. Time is the order of causal chains; that is the outstanding result of Einstein's discoveries. The only philosopher who anticipated this result was Leibnitz; though, of course, in his day it was impossible to conceive of a relativity of simultaneity. And Leibnitz was a mathematician as well as a philosopher. It appears that the solution of the problem of time and space is reserved to philosophers who, like Leibnitz, are mathematicians, or to mathematicians who, like Einstein, are philosophers.

V

From the time of Kant, the history of philosophy shows a growing rift between philosophical systems and the philosophy of science. The system of Kant was constructed with the intention of proving that knowledge is the resultant of two components, a mental and an observational one; the mental component was assumed to be given by the laws of pure reason and conceived as a synthetic element different from the merely analytic operations of logic. The concept of a *synthetic a priori* formulates the Kantian position: there is a *synthetic a priori* part of knowledge, i.e., there are non-empty statements which are absolutely necessary. Among these principles of knowledge Kant includes the laws of Euclidean geometry, of absolute time, of causality and of the conservation of mass. His followers in the 19th century took over this conception, adding many variations.

The development of science, on the other hand, has led away from Kantian metaphysics. The principles which Kant regarded as *synthetic a priori* were recognized as being of a questionable truth; principles contradictory to them were developed and employed for the construction of knowledge. These new principles were not advanced with a claim to absolute truth but in the form of attempts to find a description of nature fitting the observational material. Among the plurality of possible systems, the one corresponding to physical reality could be singled out only by observation and experiment. In other words, the synthetic principles of knowledge which Kant had regarded as *a priori* were recognized as *a posteriori*, as verifiable through experience only and as valid in the restricted sense of empirical hypotheses.

It is this process of a dissolution of the *synthetic a priori* into which we must incorporate the theory of relativity, when we desire to judge it from the viewpoint of the history of philosophy. A line which began with the invention of non-Euclidean geometries 20 years after Kant's death runs uninterruptedly right up and into Einstein's theory of space and time. The laws of geometry, for 2000 years regarded as laws of reason, were recognized as empirical laws, which fit the world of our environment to a high degree of precision; but they must be abandoned for astronomic dimensions. The apparent self-evidence of these laws, which made them seem to be inescapable presuppositions of all knowledge, turned out to be the product of habit; through their suitability to all experiences of everyday life these laws had acquired a degree of reliability which erroneously was taken for absolute certainty. Helmholtz was the first to advocate the idea that human beings, living in a non-Euclidean world, would develop an ability of visualization which would make them regard the laws of non-Euclidean geometry as necessary and self-evident, in the same fashion as the laws of Euclidean geometry appear self-evident to us. Transferring this idea to Einstein's conception of time, we would say that human beings, in whose daily experiences the effects of the speed of light would be noticeably different from those of an infinite velocity, would become accustomed to the relativity of simultaneity and regard the rules of the Lorentz-transformation as necessary and self-evident, just as we regard the classical rules of motion and simultaneity self-evident. For instance, if a telephone connection with the planet Mars were established, and we would have to wait a quarter of an hour for the answer to our questions, the relativity of simultaneity would become as trivial a matter as the time difference between the standard times of different time zones is today. What philosophers had regarded as laws of reason turned out to be a conditioning through the physical laws of our environment; we have ground to assume that in a different environment a corresponding conditioning would lead to another adaptation of the mind.

The process of the dissolution of the *synthetic a priori* is one of the significant features of the philosophy of our time. We should not commit the mistake of considering it a breakdown of human abilities, if conceptions which we regarded as absolutely true are shown to be of limited validity and have to be abandoned in certain fields of knowledge. On the contrary, the fact that we are able to overcome these conceptions and to replace them by better ones reveals unexpected abilities of the human mind, a versatility vastly superior to the dogmatism of a pure reason which dictates its laws to the scientist.

Kant believed himself to possess a proof for his assertion that his *synthetic a priori* principles were necessary truths: According to him these principles were necessary conditions of knowledge. He overlooked the fact that such a proof can demonstrate the truth of the principles only if it is taken for granted that knowledge within the frame of these principles will always be possible. What has happened, then, in Einstein's theory is a proof that knowledge within the framework of Kantian principles is not possible. For a Kantian, such a result could only signify a breakdown of science. It is a fortunate fact that the scientist was not a Kantian and, instead of abandoning his attempts of constructing knowledge, looked for ways of changing the so-called *a priori* principles. Through his ability of dealing with space-time relations essentially different from the traditional frame of knowledge, Einstein has shown the way to a philosophy superior to the philosophy of the *synthetic a priori*.

It is the philosophy of empiricism, therefore, into which Einstein's relativity belongs. It is true, Einstein's empiricism is not the one of Bacon and Mill, who believed that all laws of nature can be found by simple inductive generalizations. Einstein's empiricism is that of modern theoretical physics, the empiricism of mathematical construction, which is so devised that it connects observational data by deductive operations and enables us to predict new observational data. Mathematical physics will always remain empiricist as long as it leaves the ultimate criterion of truth to sense perception. The enormous amount of deductive method in such a physics can be accounted for in terms of analytic operations alone. In addition to deductive operations there is, of course, an inductive element included in the physics of mathematical hypotheses; but even the principle of induction, by far the most difficult obstacle to a radical empiricism, can be shown today to be justifiable without a belief in a *synthetic a priori.* The method of modern science can be completely accounted for in terms of an empiricism which recognizes only sense perception and the analytic principles of logic as sources of knowledge. In spite of the enormous mathematical apparatus, Einstein's theory of space and time is the triumph of such a radical empiricism in a field which had always been regarded as a reservation for the discoveries of pure reason.

The process of the dissolution of the *synthetic a priori* is going on. To the abandonment of absolute space and time quantum physics has added that of causality; furthermore, it has abandoned the classical concept of material substance and has shown that the constituents of matter, the atomic particles, do not possess the unambiguous nature of the solid bodies of the macroscopic world. If we understand by metaphysics the belief in principles that are non-analytic, yet derive their validity from reason alone, modern science is anti-metaphysical. It has refused to recognize the authority of the philosopher who claims to know the truth from intuition, from insight into a world of ideas or into the nature of reason or the principles of being, or from whatever super-empirical source. There is no separate entrance to truth for philosophers. The path of the philosopher is indicated by that of the scientist: all the philosopher can do is to analyze the results of science, to construe their meanings and stake out their validity. Theory of knowledge is analysis of science.

I said above that Einstein is a philosopher by implication. That means that making the philosophic implications of Einstein's theory explicit is the task of the philosopher. Let us not forget that it is implications of an enormous reach which are derivable from the theory of relativity, and let us realize that it must be an eminently philosophical physics that lends itself to such implications. It does not happen very often that physical systems of such philosophical significance are presented to us; Einstein's predecessor was Newton. It is the privilege of our generation that we have among us a physicist whose work occupies the same rank as that of the man who determined the philosophy of space and time for two centuries. If physicists present us with implicational philosophies of such excellence, it is a pleasure to be a philosopher. The lasting fame of the philosophy of modern physics will justly go to the man who made the physics rather than to those who have been at work deriving the implications of his work and who are pointing out its position in the history of philosophy. There are many who have contributed to the philosophy of Einstein's theory, but there is only one Einstein.

HANS REICHENBACH

DEPARTMENT OF PHILOSOPHY
UNIVERSITY OF CALIFORNIA AT LOS ANGELES

PHILIPP FRANK

Harvard University

INTRODUCTION TO THE PHILOSOPHY OF

PHYSICAL SCIENCE, ON THE BASIS

OF LOGICAL EMPIRICISM [1]

CHAPTER I

Preliminary Remarks about Physics and Philosophy

The advances in physical science have been hailed for their applications in technology. But, on the other hand, we know too that lively and even passionate debates have centered around the Copernican theory, although there was obviously no technical application of the motion of our earth. Copernicus initiated a radical change in the picture of the universe that man had developed through the ages. In a similar way, men like Galileo or Newton have not been primarily interested in technological progress. They improved the procedure by which man has gradually evolved his picture of the world and has formed his conception about the position of man in the universe. In the twentieth century physics continued its philosophical message to mankind with unabated force. To appreciate this fact it is sufficient to remind you of Einstein's Theory of Relativity, and Bohr's mechanics of the atom and the nucleus. When you open nowadays any book or magazine that deals with our contemporary culture, you will read about the impact of physical science upon our present world picture. We are told that the attitude of physical science toward materialism, toward free will, even toward religion has in our time fundamentally altered the nineteenth century attitude.

In his traditional curriculum the science student himself learns very little about the implication of his science in our general world

[1] Summary of a course, "Introduction to the Philosophy of Physical Science", held at Harvard University between 1940 and 1950.

picture. This means that he becomes insufficiently informed about the place of science in our general civilization. To understand the place of physical science within our general pattern of knowledge is the principle goal of any course in the philosophy of physical science.

Physical science is based upon observation and experiment. The record of the most sophisticated experiment in nuclear physics does not contain words or grammatical forms which are not used in the description of our breadfast table. We use such expressions as "a red spot touches a green spot", "the edge of a hard body coincides with a mark of a scale", "a warm body cools off", and so on. These expressions are understood without ambiguity by anybody who is trained in the English language. We call these expressions "terms of the observational language" or, because of its wide use, the "language of everyday life".

But the general picture of the universe and man's position in it is formulated by using words of quite a different type. One uses words like "force", "energy", "entropy", "electromagnetic field", and even "physical reality", "materialism", and so forth. In order to reach its goal, the philosophy of science has to investigate the relationship between these two languages: the observational language and the symbolic language of our world picture.

The relations between the observational language of experimental physics and the symbolic language of our general world picture are examined by two lines of inquiry, or, we may even say, by two fields of science; the logical-empirical and the psychological-sociological.

Firstly, we set up a symbolic language in order to formulate simple propositions from which the observed facts can be derived logically. Newton introduced the symbolic terms "mass" and "force" in order to formulate his "laws of motion" from which it became possible to derive all our experience about motion of observable bodies. This way of relating the observed phenomena to general statements is investigated by "logico-empirical analysis". But, as a matter of fact, the symbols used in our general statements are not only determined by the requirement that the observed facts can be derived by a simple logical chain from these principles.

The second field is the psychological-sociological. Copernicus' system was originally not recognized by the authorities although even the Roman Church did not doubt that the observed facts can be adequately derived from it. But the "motion of the earth" was a symbol which was not in harmony with the symbols used by the Church to express its religion. The political and religious doctrines of a period have an influence on the choice of symbols. In addition, then, there

29

37

are also requirements of individual psychology. Among other things, any choice of symbols which are in too great a discrepancy with the every-day life language, the so called "common sense language", are often discouraged. This research belongs rather in psychology and sociology. So we need besides the first mentioned analysis also this latter type.

If we apply both types of analysis to physical science we obtain what one may call a "philosophy of physical science". From our considerations it seems obvious that the philosophy of science helps us to understand the place of science in human activity. But the philosophy of science is also necessary for a satisfactory understanding of science itself.

The words and rules of the observational language have changed little through the centuries. But the words and grammar of the symbolic language of the world picture have undergone a distinct evolution. We notice periods of rapid changes. Before Newton, words like "mass" or "force" had not been used in the sense that he gave them. They now became words of the symbolic language which had little to do with what the observational language calls "force" — if this word is used at all. But the expressions "length of a table" or "length of a time interval" were used by Newton in the same sense as in the every day language. In the twentieth century, however, Einstein pointed out that the word "length" can no longer be used in the common sense meaning if we want to give a simple and adequate description of facts newly discovered by physical science.

This introduction of a new language accounts for the fact that new theories like Einstein's Relativity have been often termed obscure and contradicting common sense. And even today there are quite a few physicists who do not believe that these new theories are as honest and "down to earth" physical theories as, for example Newton's mechanics has been. This failure to grasp the meaning of radical changes in physics has its root in the failure of our traditional college training to teach the students with sufficient precision and strength the clear distinction between observational language and symbolic language. In other words, the students of science get little or no training in the philosophy of science. Obviously the word "philosophy" does not mean in this context a realm of knowledge beyond science, since physics and logico-empirical analysis of physics be really separated from each other. This analysis is in fact an essential part of every systematic presentation of theoretical physics. Socio-psychological analysis, on the other hand, is certainly not a part of physics in the traditional sense of this word, but it is part of the sciences of psychology and sociology.

If we mean by "science" not only physical science but include all fields, we have to say that the philosophy of science is a part of science proper.

The traditional way of teaching science and philosophy all over the world has, through many centuries, been to draw a line of separation between scientific and philosophical insight, or, more specifically, between the scientific and the philosophical criterion of truth.

There are a great many ways of formulating this distinction. It seems instructive to start from a statement of the great mediaeval philosopher, Thomas Aquinas. In his *Summa Theologica*, (I. Part. Qu. 32), he distinguishes two ways of proving a statement. As an example of the first way he quotes the proof by which it is shown that the movement of the celestial bodies is always of uniform velocity along circular orbits. The starting point of this proof is the principle that the celestial bodies are the most perfect beings. From this statement it follows that they must traverse the most perfect orbit and this orbit is obviously the circumference. From their eternally unchangeable nature it follows that their velocity cannot change. Therefore, this means that the first way of proving a physical theorem is to derive it logically from self-evident principles.

It was known already at the time of St. Thomas and even in the antiquity, that there are many observed phenomena in astronomy which seemed to be in disagreement with the uniform velocity of the heavens. We know that the velocity of the apparent motion of the sun around the earth is oscillating in the course of each year between a maximum and a minimum value. The Greek astronomers had already accounted for this fact by the hypotheses of the eccentrics and the epicycles. According to the first hypothesis, the circular orbit of the sun has a centrum which is not coincident with the earth. The second hypothesis claims that not the sun but a "fictitious sun" performs a circular uniform motion around the earth while the sun itself traverses a small circle (epicycle) around this fictitious sun. From the superposition of these two circular motions of uniform velocity, there results a circular motion of the sun with a velocity which is oscillating between a maximum and a minimum.

Now, according to the formulation of St. Thomas, the hypothesis of the eccentrics or the epicycles cannot be derived from any self-evident principles. We believe them to be true only because we can derive from these hypotheses statements which can be confirmed by astronomical observations. The second way of proving a statement, therefore, is to show that its logical consequences are in agreement with our experience.

This second way of proving is the method which is today gener-

31

ally used in science, while the first one is what one would call today a "philosophical proof". St. Thomas emphasized that the scientific proof is not so reliable because it does not exclude the possibility that the same observed phenomena could be later derived from a different hypothesis. The philosophical proof, however, guarantees us the unique truth of a principle. As a matter of fact, the two hypotheses of the eccentric and the epicycles show us that in ancient Greece it was commonly known already that these same phenomena can be derived from several hypotheses.

This distinction between the philosophical and the scientific truth played a paramount role on the famous conflict between Copernicus and the Roman Church. The Church claimed that the Copernican theory (motion of the earth) is, perhaps, scientifically true, but, certainly, philosophically false. It has to be understood that I am here using the word "scientific" in the modern sense. The Church used instead of it the term "mathematically true", while, what we call today "philosophical" was called "physical" since at that time "physics" was a part of "philosophy".

The distinction between philosophical and scientific truth is, of course, only a real distinction if there are principles which can be demonstrated in a way which is different from the scientific way. For if all statements could only be proved by the agreement of their consequences with experience, this would also hold for the basic principles of the philosophical derivation (like the statement that the celestial bodies are the most perfect beings). Then the scientific way would be the only way of proof. People who hold this opinion follow the doctrine of "empiricism" or "positivism" or, as it is sometimes called, "scientism".

The belief in the existence of the Thomist distinction is perhaps the best way of formulating what one may call the "belief in metaphysics" as opposite to empiricism. This belief presupposes the belief in the existence of principles which can be proved by a method which is essentially different from the scientific method.

However, even if one does not believe in such principles, the Thomistic distinction does have a certain meaning in modern science, although in a weakened form. Every student of physics knows that even the most elementary textbooks distinguish between physical theorems which can be "proved", and those which are just immediate generalizations of a series of observations. "Proved" means "derived from a general law", for example, the Law of the Conservation of Energy. This law may, in a broader aspect, also be justified only by the agreement of its consequences with experience. But it has greater

32

stability, it gives more confidence than the curve drawn as the result of a special series. The distinction has now become only a gradual or relative distinction, but we may say even today that a statement derived from the principle of relativity is "more philosophically proved" than a statement which is only based on a curve plotting the observed volumes of seawater against pressures.

CHAPTER II

The Laws of Motion, An Example of Logico-Empirical Analysis

In order to gain a certain perspective for the understanding of the logical, historical, and psychological argument in this course, it is advisable to start from the procedure of actual science. The best way seems to me to be the discussion of a special but characteristic example of scientific approach.

Our observations and experiments teach us about the complex situations under which observable motions of bodies occur. If we observe a small piece of paper dropped from our hand to the floor, we soon understand how hard it would be to predict how long it takes for it to hit the floor. The prediction seems to us at first glance a hopeless task. The slightest breeze, the slightest change of temperature seems to be relevant to the outcome of the experiment.

The phenomenon itself can be described, of course, in the observational language, which is simultaneously vague and clear. There will be no difference in opinion whether a piece of paper is ascending or descending if we exclude the domain in which it is nearly at rest. There will never be a difference in opinion as to what direction the wind blows if we are not asking for exact information. This means that for approximate information, observational language is unambiguous. Everybody who is trained in the common use of the English language understands it in the same way. The vagueness of this language consists in the fact that there are not words enough to express small differences in position, in velocity, in color, or in temperature.

In this language we are not able to formulate precise statements about uniformities in behavior or to make reliable predictions.

In order to appreciate this difficulty we must understand that even the expression "length of a time interval" has only a very vague meaning in the observational language. If we say that the fall of the piece of paper takes a certain time, we have to describe the observable

33

facts by which we estimate this length. We may do it by our "instinctive sense of time", which means by a psychological estimation. We may also use the number of our pulse-beats. Both ways yield only very vague regularities. If, instead of these psychological and physiological methods, we make use of "physical methods" of measuring time intervals, we can measure them by the number of oscillations of a pendulum, by the rotation produced by the unwinding of a spring, by the running of sand in an "egg clock", by the space interval traversed by a rolling billiard ball if no force is acting, etc.

If we tried to precisely formulate the duration of the fall of the paper in terms of the pulse beats, or the pendulum, or the rotation of the earth, we would never get a clear cut result. As a matter of fact, we would never know which of all these phenomena should be used to measure the other one. You could just as well use the fall of the paper as a measure.

If we now examine how our actual physical science tackles this problem, we notice that the laws of motion do not speak of the position of the piece of paper, as we would in the every day English. Nor do they speak of a comparison of our paper with pulse beats or running sand. The laws contain on the one hand the "distance" of the paper from a certain coordinate system, on the other hand the "time" which has passed during the fall. In the observational language the words "distance" and "time" do not occur, except in the sense of results of special measuring operations. If we may measure "time" by the oscillation of a pendulum, or by the quantity of sand running through a leak, we always get the result that equal distances are traversed in equal time intervals if no force is acting on a body. Because differents methods of measurement can be used to formulate the law of inertia, we introduce a symbolic expression "t" or "time" which is a kind of average of all the described methods of measuring time. Because there are many ways of measuring time, we can use this symbol to describe a great many phenomena of the physical world in a simple way. We say now that the number of pulse-beats and the number of the oscillations of a pendulum or the quantity of sand running from a leak are all proportionate to "time t".

The laws of motion used in the science of mechanics make use of symbols which are similar to those used to introduce time. In this way we introduce symbols like "force", "acceleration".

Before we discuss the relation of these laws with the observations which are formulated in the observational language, we may point out by an example what these laws look like.

If we consider a very small piece of paper, we can derive its

34

motion from the laws which Newton set up for the motion of a "mass point". To simplify our presentation we restrict ourselves to motions along a vertical straight line. Then "Newtons Law of Motion" is expressed by the formula m a = f, or, in the words: "mass times acceleration is equal to the force acting upon the mass." In the case of our paper, we can derive the laws by regarding the force f as the sum of two forces acting upon our mass: $f = f_1 + f_2$. In this formula f_1 is the force of gravity and f_2 the force of the air resistance. About the force f_1 the usual hypothesis is that the force of gravity is proportional to the mass $f_1 = mg$ where g is dependent on the distance from the center of the earth. Within this room g can be regarded as a constant. The resistance of air f_2 is assumed to be (for a small speed) proportional to the velocity v. Its direction points upward, opposite to the direction of gravity. Hence the law of motion from which our observations about the falling paper should be derived is

$$ma = f_1 + f_2 = mg - Rv \qquad (I)$$

R is a constant, the coefficient of resistance of the air for a mass of a certain size and shape. To these relations we have to add the definitions of a and v (acceleration and velocity). If we denote the distance traversed in the small time interval dt by dx and the increment of velocity during dt by dv we have to add to the relations (I) the relations

$$v = \frac{dx}{dt} \qquad\qquad a = \frac{dv}{dt} \qquad (II)$$

If we make use of the "operational definitions" of the symbols x, t, f, R, g ... we can translate the relations (I) into a statement about observable facts formulated in the observational language. It will be a statement which is "testable" by experiments. But because of the complex nature of this statement, it will in practice hardly be possible to test it by actual experiments. The principles themselves are simple before they are translated into the observational language, but after the translation they become involved and to remote from the actual possibilities of the experimentor.

For this reason it is important not to test the validity of the principles by immediate translation into the observational language. The fact that the principles are formulated in a symbolic language makes it possible to use a different way of testing which has become the characteristic method of testing the principles in the most developed parts of science, in astronomy, mechanics, and mathematical physics. This method makes use of the simple fact that from principles, like the equations (I), one can derive conclusions without

35

knowing what the symbols mean in the observational world. We have only to start from the fact that the symbols are mathematical quantities.

Then we can derive logically from (I) and (II) the following statements:

If mg is greater than RV the accelerations a $=$ dv/dt is positive. This means that with increasing t the velocity v will increase.

If we start with v $=$ 0 we start with a case in which v will increase. But then it will, eventually, reach the value Rv $=$ mg and a $=$dv/dt becomes zero too, and the velocity v no longer changes, but stays constant whatever may be the increase of t. This final constant value of the velocity is given by

$$v = \frac{mg}{R} \tag{III}$$

This result is reached from (I) and (II) without knowing anything about the physical meaning of the symbols. We use purely mathematical conclusions. But when we have reached the conclusion (III), we know that after t has reached a certain value, the velocity v remains constant and we can learn that after a certain number of oscillations of a pendulum our piece of paper will start falling in a very simple way. The pendulum will perform an equal number of oscillations for each inch traversed. From this number of oscillations we can compute the velocity v and check whether the result of this computation checks with the value of v computed from the formula (III). If there is agreement we say that our principles are confirmed by experiment or by observations. It is obvious that the result which is actually confirmed by observation is not derived from the principles (I) and (II) alone, but also makes use of a specific translation into the observational language. Only by using such a translation can the mathematical result (III) of our principles be tested by experiments.

In ordinary speech one calls the rules of translation from the symbolic into the observational language "semantical rules". They define the "meaning" of the symbols in the realm of our experience. Since in physics these rules consist in the description of the physical operations by which we measure a quantity (for instance: distance of time) we call these rules "operational definitions". If we know the operational definition of an expression we know its "operational meaning". We must say, therefore, that our experiments do not confirm the principles themselves (for instance not the Newtonian law ma $=$ f or the law of resistance f $=$ −Rv itself) but these laws, which

36

44

are statements of the symbolic language, plus the operational definitions of the symbols. We test, briefly, the principles including their operational meaning.

In these considerations we took for granted that we know the mathematical rule by which we derived $v = mg/R$ from $Rv = mg$. In order to enumerate all parts of a scientific theory, we have to mention, besides the principles in the symbolic language and their operational definitions, also the "logical and mathematical rules" by means of which we can draw conclusions without knowing the operational meaning of the symbols.

Hence, every scientific theory consists of three parts: first, the "principles" or the "symbolic structure", second, the "logical rules" of transformation, and third, the "semantical rules". The principles are the work of creative imagination based on more or less elaborate series of experiments which are formulated without using the symbolic language which has to be created by the scientist. The second part of the work is deductive, logical or mathematical. It is only concerned with conclusion within the symbolic structure. If conclusions are reached which can be translated into statements within the reach of the experimentor, his work starts and he has to check these conclusions by elaborate experimental technique.

To sum up; the symbolic structure is introduced as a link between the observations which face us directly in our daily life, and the systematic experimentation of the scientist.

Between the original observations which are not guided by theories, and the experiments made under the direction of theories we have the large field of deductive operating with symbols. This is the domain of the mathematician which fills large volumes already in fields like analytical mechanics or mathematical economics or statistical genetics.

CHAPTER III

The Historical Background of Our Actual Scientific Thought

A. Organismic Science

The symbolic structure of science has gone through fundamental changes in the course of human history. To understand this evolution the most important thing is to have a clear comprehension of the turning points where the most radical changes occurred.

We are only going to consider the turning points the impact

37

45

of which is still noticeable in our actual science. Roughly speaking, we shall direct our attention to the years 1600 and 1900. Before 1600, in the Hellenistic period and in the middle ages, the "organismic conception of physics" prevailed, while the era between 1600 and 1900 can be regarded as the period of "mechanistic physics". After 1900, in the twentieth century, the modern conception of physics developed which we may call the "logico-empirical conception of physics".

It is important to know something about organismic physics. For only by understanding the birth of mechanistic physics are we able to correctly appreciate its decline, and the rise of a new conception. The most characteristic feature of "organismic physics" is the attempt to understand the behavior of inanimate bodies (stones, fire, etc.) in analogy to the behavior of organisms (animals, human beings) which was regarded as immediately understandable. Every animal has its proper way of moving: the fish swims, the bird flies, the worm creeps, etc. In the same way the bodies which are not organisms, in the language of everyday life, follow laws of motion which are determined by the "nature" of the body concerned.

The most conspicuous division between the bodies was the one into celestial bodies (sun, moon, stars) and terrestrial bodies (stones, smoke, etc.). The celestial bodies were regarded as spiritual, divine beings. Their nature was perfection and eternal life without change. Hence they were assumed to move with constant speed in circular orbits around the center of the universe which coincided with the center of our earth. The only purpose of their motion was dignity and eternity.

On the other hand the terrestrial bodies were moving for a certain purpose. They regularly were on the search for their "natural place". When abandoned to themselves they approached their natural place on the shortest possible way. The location of this "natural place" depends upon the "nature" of the body, what we would now call their chemical nature. Heavy bodies, consisting of earth and water, move in their „natural place" near the center of the earth. Light bodies, consisting of air and fire, have their natural place above the surface of the earth near the celestial spheres. Therefore, a stone abandoned to itself falls downwards while smoke or fire ascends upward. If a body does not move towards its natural place in the shortest way, we know that "violence" has been brought to bear upon it. A launched projectile, for instance, moves in a horizontal direction before it falls to the ground.

Generally speaking, the motion of the terrestrial bodies was

38

46

regarded as much less regular than the motion of the heavenly spheres. The simple laws of the orbits of sun, moon and stars were in sharp contrast to the complex laws according to which a piece of paper falls to the ground.

We can now better formulate the distinction between scientific and philosophical criterion of „truth" of which we spoke in Ch. I. If a phenomenon could be derived from the principles of organismic physics, from the idea of purposiveness, from the distinction between terrestrial and celestial bodies, one could say that this phenomenon is "philosophically proved" or "this phenomenon is explained" or "can be understood by the human mind". If the phenomenon could only be derived from principles which were non conclusions from this organismic physics, this phenomenon was only "scientifically proved". It was also said that "it was not explained" but "only described" by science.

One of the oldest examples, mentioned already in Chapter I is the inequality in the angular velocity of the sun in its annual motion around the earth. This inequality could not be derived from the perfection and eternity of the divine celestial bodies but only from hypotheses which were set up "ad hoc" for the purpose of accounting for this phenomenon. Therefore, these inequalities could only be derived from such hypotheses as the epicycles or eccentrics which in turn could not be derived from the organismic principles. Hypotheses like the "epicycles" provided, according to the ancient distinction, "only a description" but not "real explanation" of such phenomenon as the inequalities in the motion of the sun.

B. *Mechanistic Science*

The rise of mechanistic science is connected with the name of Galileo who published his most important writings around 1600. He gave simple and adequate quantitative descriptions of some phenomena of motion which had been familiar to everyday experience: free falling body, launched stones, motion on the inclined plane, oscillation of a pendulum, etc. This achievement was the basis of a new scheme of deriving the phenomena of nature from general principles. Instead of taking the "organism" as the basis of "understanding" one took the "mechanism" as exemplified by launched stones, oscillating pendulum, etc. Every phenomenon was regarded as "understood" or "philosophically proved" if it could be derived from the statements which had served to derive the laws of those simple mechanisms. The behavior of organisms which seemed in the old period to be immediately understandable became now hard to explain. For "ex-

39

47

plain" meant now to "derive from the laws of these simple mechanisms". These laws were brought by Newton (about 1700) into their definite form. From the beginning of the 18th century "mechanistic science" meant the theory that all phenomena of nature can be derived from the Newtonian laws.

In the 18th and still more in the 19th century it had become a general opinion that to "understand a phenomenon" means to derive it from the Newtonian Laws of Motion. The mediaeval distinction between "philosophical and scientific truth" appeared now in a new version. Now, a statement of physics or any other science was only "scientifically true" if its consequences agreed with the observation. But to be "philosophically true" it had, in addition to it, to be derivable from the Newtonian laws. This seemed to be particularly hard in the case of the biological and "psychological sciences" where the derivation from mechanical laws seemed to have little prospect.

Towards the end of the 19th century this concept of understanding led to a crisis in the belief in science as a guidance of life. A mechanistic understanding seemed to be not around the corner, and another kind of understanding was believed not to exist. However, the conclusion was drawn by many people that "no scientific understanding of nature at all is possible." This feeling led to the slogan "ignorabimus" (we shall never know) and to the belief that "science is bankrupt" and other sources of knowledge have to be called for. A great many people inferred that a return to the organismic conception of science has to occur and along with it a return to the mediaeval ideas in religion and government. This longing for the Middle Ages was connected with disappointment in the results of mechanistic science in the field of social and economic problems. Industrialization, free enterprise and liberalism were commonly regarded as the fruits of the mechanistic conception of science. These developments, however, turned out not to have destroyed poverty and misery. The failure of the mechanistic aspect within science seemed to be accompanied with a failure in the wider field of human affairs,

But besides this feeling about "bankruptcy" of science and liberalism new trends had developed which denied that the failure of mechanistic science to provide an "understanding" of nature meant a permanent "ignorabimus". There may be other methods of scientific understanding. A return to mediaeval ideas may, after all, not be necessary or even advisable.

C. *Twentieth Century Science (Logico-empirical concept of science)*
     The criticism directed against the mechanistic conception of

40

science towards the end of the 19th century pointed out that Newton's Laws of Motion by themselves are not more self-evident than the laws of any other field of physics, say, optics or electricity. For this reason "understanding" is not synonymous to "derivation from Newton's Laws of Mechanics". In this period the "Electromagnetic Theory of Matter" was advanced, according to which a moving electric charge shows all observable properties of ordinary matter, above all the property of inertia. Therefore, one could assume that all mass in the world was only an apparent mass. Actually, only electric charges without mass exist. This theory was regarded as an argument against the "materialistic" philosophy of the 18th and 19th century. It was argued that by the introduction of electrical energy instead of mass a spiritual element was introduced into physics. This argument was used as a "refutation of materialism by physical science" and, moreover, was used to bolster up the political groups fighting for the return to mediaeval conceptions in economics, government, and religion. Besides the electromagnetic theory of matter, other physical theories developed which were not based on Newton's Laws of Motion. Thermodynamics was based on the law of conservation of energy and the increase of entropy. This means the special observable phenomena were derived from very general statements about observable phenomena. In the statistical theory of matter, the basic principles were no longer hypothesis about single observable phenomena, but only about the average character of observable events.

Under the influence of these physical theories on the one side and the logical criticism directed against Newton's mechanics as the unique "explanation" on the other side, a new conception of physical science arose around the turn of the century.

It was no longer required that the basic principles be statements which use a specific kind of symbols, like the organismic or mechanistic ones. The principles could be formulated in any symbolic language. Two requirements only had to be met. Firstly, that the assumptions should be logically consistent and not of too great a number or of too great a complexity. (Logical requirement). Secondly, that by means of operational definitions from these basic principles (or at least from some conclusions drawn from them) statements about observable facts could be derived which agreed with the observation and experiments of the physicist. (Empirical requirement). As only these two requirements had to be met by a physical theory we can also give to this 20th century conception of physics the name "Logico-empirical conception".

This conception has been the background of all the 20th century

41

theories like the relativity theory and the quantum theory (atomic and nuclear physics). The basic conceptions are now, in the relativity theory, the general components of a gravitational or other field of forces between which Einstein's field equations state certain relations. In the quantum theory, Schroedinger's wave function (the average number of particles in a certain region of space and time) is subjected to some differential equations which are the basic principle of the theory. In both cases there is a rather complex system of mathematical conclusions leading from the equations to the relations which can be checked by actual observation. In addition to it, the operational definitions of these gravitational potentials and wave functions is also very different from the operational definitions which had been used in Newtonian mechanics.

The role of this "logico-empirical conception" which has been so characteristic for 20th century physics can in our general history of thought perhaps be described, briefly, as follows: this new idea of a scientific theory held prevented the feeling of bankruptcy to spread. It provided a new basis of science which was compatible with the increasing evidence of the insufficiency of Newtonian mechanics as a universal basis; it provided, on the other hand, a firm rampart against all attempts to push human thought back to the level of organismic and mediaeval science.

CHAPTER IV

The Philosophical Interpretations of Physical Science

A. *Straight Metaphysics*

The conflict between the metaphysical and the positivistic interpretations of science can be formulated as follows: according to the second one the symbolic principles of science have no other meaning besides that they allow to derive statements about observable facts. The metaphysical interpretation claims that those principles can be true or false by themselves. They can be proved to be true by methods which are different from testing their results by observations.

These two interpretations have been held by various schools of thought. We discuss them in their most radical form, which we call "straight metaphysics" and "straight positivism".

As an example of radical metaphysics we discuss the philosophy of St. Thomas Aquinas and, in particular, a textbook of cosmology (by McWilliams) in which the physical sciences are interpreted according to the Thomistic doctrine. We discuss the conception of physical law (genuine law in contrast to mere uniformity), the conflict between atomism and dynamism with the compromise solution given by the Thomists (Hylozoism). Eventually we discuss Thomas' doctrine of prime matter and substantial form.

The school of thought called Neothomism has claimed that science and philosophy can develop independent of each other. None of them can disprove the results of the other one. (Separated levels of knowledge. First flight: science; second flight: philosophy; third flight: theology).

B. *Straight Positivism*

We can start with the British philosopher, David Hume, who as early as 1788 claimed that no statement has any scientific meaning except it states observable facts or presents a mathematical conclusion.

In 1829 the French philosopher and scientist, Auguste Comte, published his system of "positive philosophy". He advanced his law of the three stages in the evolution of the human mind in any field of science: the theological, the metaphysical and the positive or scientific stage. Metaphysics, according to Comte, is only a provisional stage between theology and science.

In the last quarter of the 19th century Ernst Mach and Henri Poincaré were the most successful advocates of a positivistic approach to science. According to Mach, "science is an economic summary of the results of our observations", which according to Poincaré, "the most general principles of science are pure conventions which are neither true nor false but are useful to bring order into the results of our experiments and observations."

These two conceptions of science are not contradictory but emphasize different aspects of the positivistic conception of science. The doctrines of Mach and Poincaré were brought into a more coherent logical system by the school of "logical empiricism" or "logical positivism" which emerged around 1925. This most recent concept of science is based not only on the positivistic tradition of men like Mach and Poincaré but also on the new trends in formal logic represented by men like Bertrand Russell, D. Hilbert, etc.

According to the scheme of logical empiricism as advanced by L. Wittgenstein, R. Carnap, H. Hahn, O. Neurath, etc., we have, (as already suggested by D. Hume), sharply to distinguish between "tautological" statements which do not say anything about the real world and "factual" statements which can be checked by experiments. Every system of principles which does not allow the derivation of factual

43

statements is excluded from science and called "metaphysical". Among these metaphysical principles would belong the statements about the reality or non-reality of the external world, the statement that "everything is mind" or "everything is matter", briefly the traditional doctrines of realism, idealism, materialism, etc.

According to this conception of science, the general principles of science have to be reformulated in order to remove their conventional or tautological character. Instead of saying "the sum of energy in the world remains constant" we have to say: "There are few types of energy in the world. Each of them can be measured according to describable methods. Then the sum of these few and measurable energies remains constant during every experiment in an isolated system."

C. Between Metaphysics and Positivism

Most scientists and philosophers do neither advocate straight metaphysics nor straight positivism. They try to make a certain compromise. The most important suggestion of a compromise is the philosophy of Immanuel Kant, a German philosopher of Scotch parentage. He lived at the end of the 18th century and recommended a "Copernican turn" in philosophy. The general statements of science as the law of causality, the laws of pure geometry and arithmetic, etc., are, according to Kant, of eternal validity because they are not based upon the observation of the physical world. We find them by examining our own mind. Our rough sense observation cannot become "experience" (in the ordinary sense of this word) if our mind does not organize this raw material. The mind throws a frame over our pure sensations. The mind produces a "frame" through which we see the world. We can, e.g. not experience the world without the "frame of time and space". Therefore, pure geometry contains statements which say something about the frame through which we experience the world. Therefore, we needed only to explore our own mind to discover and prove the theorems of geometry. Since the mind can do it on the lowest step of scientific evolution the laws of geometry are of eternal validity and cannot be modified by the advance of science.

Among physicists, A. Eddington adhered to the Kantian view and illustrated it by the parabola about the fisherman and his net.

Other authors released the Kantian conception of its rigidity and assumed that the frame produced by human mind experiences itself an evolution along with the advance of science. In this way the conception becomes more similar to the positivistic conception, and in

44

particular to the conception of H. Poincaré, according to whom the general laws are creations of the human mind. But they are not of eternal validity, according to the positivistic approach of H. Poincaré the human mind moulds its products according to the needs of actual science.

Obviously, by attributing less or more rigidity to the "creations of the human mind" that appear as the principles of actual science, we can bring about a continuous transition from straight metaphysics to straight positivism.

45

50. THE LOGICAL FOUNDATIONS OF QUANTUM MECHANICS* +

[1952d]

TABLE OF CONTENTS

1. INDETERMINISM

The indeterminism associated with the quantum theory has been considered, as well by philosophers as by physicists, as the strongest deviation from classical physics that can be imagined. I do not believe that this judgment is defensible. It rather seems to me that this deviation is not as great as it seems at first sight, because there exist developments in classical physics which allow one to suppose that determinism is not the last word of the man of science regarding that which concerns natural phenomena. Thus, some other results of quantum physics contradict the classical conception of the physical world to a greater degree than does the abandonment of strict causality. I wish to speak of the change in the concept of matter, of the substance which makes up physical bodies, of a profound change which has resulted from the discovery of M. de Broglie regarding a certain equivalence of waves and particles and which has completely overturned philosophical theories (which had) come forth from classical science, as well as from our everyday experiences. Permit me, by way of introduction, to tell you about these developments from a comparative point of view, that is to say, by comparing them with classical ideas from which they have parted, and which they have transformed into a new picture of the physical world. It is only after this study of the philosophical background that we will be able to understand the significance of the discoveries of physicists during the thirty years elapsed since the first publications of M. de Broglie, and that we will be in shape to examine some other consequences of the new physics, consequences comparable if not superior to those which I have just cited.

* Lectures given at the Institute Henri Poincaré, June 4, 6, 7, 1952.
+Translated from 'Les fondements logiques de la mécanique des quanta', *Annales de l'Institut Henri Poincaré* 13, part 2: 109–158 (1952/53). Copyright © 1952 by Gauthier-Villars, Paris.

Classical determinism originated in astronomy. Astronomers have the advantage of occupying themselves with large objects, far apart from each other, objects which thus are neither disturbed by one another, nor by human observation. These celestial objects move according to very precise laws which allow prediction of future positions if present positions are known, or rather, which allow very precise forecasts if one possesses a knowledge, sufficiently exact, of their present state. This extreme regularity had been considered as the ideal of nature, so to speak, and if one could not always find astronomical precision among terrestrial phenomena, one concluded that it is only the limitations of human capacities that limits our predictions to the uncertainty of statistical forecasts. The supposed ideal of nature had thus become the ideal of science, and the search for strict causal laws had taken the form of a moral obligation whose validity was beyond doubt.

It is hardly necessary to cite here the famous passage in which Laplace assigns to a superior intelligence the capacity to gather together "in the same formula the movements of the largest bodies of the universe and those of the lightest atom: nothing would be uncertain for it, and the future like the past would be present to its eyes".[1] To a modern physicist, these words seem like a profession of faith of a long gone age, of an age when one attributed to the stars the power to reveal the laws of the lightest atoms, of an age of confidence in the harmony of nature as well as in the capacity of man to translate her into mathematical formulas. The physicist of today no longer shares this confidence; he can scarcely imagine a time when it was believed that the large-scale world offered the picture of the small-scale world.

However, the question of determinism is not a question of faith, a question of the psychology of the physicist or of his epoch. It is a physical or epistemological question; it is a question which demands a logical analysis and an answer based on reason and experience. The opinion that all natural phenomena are governed by strict laws is the expression of a physical theory which, as such, ought to be subject to the critical rules generally accepted for the discussion of theories. Let us begin this analysis by attempting to give a precise formulation to the theory of determinism, a formulation which allows (us) to judge it as true or false, or, at least, as probable or improbable.

The causality hypothesis can be put in two different forms: a conditional form and a categorical form. In the conditional form, the statement of causality begins with the word 'if': if the situation A is completely described at time t_1 by the values $u_1, \ldots, u_n$ of certain parameters, it will be followed at time t_2 by the situation B. In the categorical form, we consider it established that the values $u_1, \ldots, u_n$ exist; and, omitting the phrase which begins with 'if', we conclude that the situation B will be produced. It is the second form, the categorical form, which leads to determinism, for the determinist maintains that it is admissible to separate the consequent from the antecedent and to affirm the conclusion.

One would be unable to find fault with this procedure if the situation were as simple as I have just described. Unfortunately it is much more complicated.

Even if we are justified in considering the conditional form as true, we are well aware that the condition expressed in the premise is not fulfilled. The description of the situation A by the parameters $u_1, \ldots, u_n$ is not complete. Despite this, we use it; and we are obliged to use it because we do not possess a better description. Fortunately, the error committed in using an incorrect description is not too great; the prediction of the situation B will be approximately valid. This means that it will be true within some narrow numerical limits in most cases. The laws of probability come to our rescue; if we have to forego an exact prediction, these laws offer us as a replacement a statistical prediction.

The solution seems simple, but it shows that the problem of causality is inseparable from the problem of probabilities. It permits us to separate the conclusion from the conditional form at the cost of sacrificing the pretension of arriving at a true statement: the conclusion is only probable. The solution thus substitutes the concept of probability for the concept of truth; and it requires that we reformulate the principle of causality in such a way that it takes into account its link with the principles of probabilities. Here is the answer that a logical analysis gives to the question of determinism: if a determinism can be sustained, it must first be formulated as a theorem concerning probabilities.

If each prediction is restricted to a degree of probability, then the causal analysis can only increase this degree of probability. The principle of causality is based on the notion that this increase of probability is always possible; and the idea of determinism refers to a limiting process regarding the predictive probabilities. If the description D^1 predicts the future state of B with a probability p^1, one could increase this probability by employing a more detailed description D^2. This new description of the initial situation differs from the first in the following respects:

1. The new description includes an account of the physical conditions in the spatial surroundings of the situation A.
2. The new description uses more precise measurements of parameters, and adds some new parameters to describe A, which were hitherto neglected.
3. The new description uses causal laws which have been perfected.

Item 1 serves to diminish some unexpected interventions coming from the exterior of the volume v in which the situation A at time t_1 and the situation B at time t_2 fit together. Items 2 and 3 serve to render more precise the account of the physical relations in the interior of v.

By repeating this process, one arrives at a sequence of descriptions D^i and of probabilities p^i, which allow us to formulate the hypothesis of determinism: this hypothesis claims that the sequence of the p^i converges toward the value 1 and that at the same time the sequence D^i converges toward an ultimate description D. The idea of *determinism* can then be symbolized by the following schema:

$$\begin{cases} D^1, D^2, D^3, \ldots D^i, \ldots \rightarrow D, \\ p^1, p^2, p^3, \ldots p^i, \ldots \rightarrow 1. \end{cases} \tag{1}$$

The question now arises as to whether this schema is corroborated by the testimony of experience.

For the classical physicist, the convergence of the probabilities p^i was accepted without doubt. Let us then postpone the discussion of this point and examine the question of the existence of an ultimate description. As regards points 2 and 3 (above), classical physics gave the answer that the definitive natural laws had been established in Newton's mechanics, and that the definitive parameters had been given by the mechanical model of the atom. Yet there remained the difficulties in that which concerns point 1. If space is infinite, the ultimate description D should include an infinite number of parameters; a logician would have difficulty in accepting such an idea, doubting that it is really meaningful.

The theory of relativity seemed to offer a remedy: the speed of causal transmission being limited to that of light, all the parameters that can influence the situation B at time t_2 are included, at time t_1, in a sphere of finite volume V. Yet another difficulty arises: by the same token, one will conclude that it is impossible to know the values of all these parameters before the instant t_2. Hence one can not give at time t_1 an ultimate description which allows prediction of the situation B.

The difficulty would be eliminated if the universe were spatially finite. In this case, at least, the logician would not have objections against the existence of the ultimate description D. Let us then accept this supposition in order to arrive at a form of determinism which is not susceptible to logical objections.

Yet, even in this case, determinism implies a doubtful enough hypothesis. In fact it is quite possible that the sequence of probabilities p^i converges to a limit, while the sequence of descriptions does not. This conception, which may be called *classical indeterminism*, can be symbolized by the following schema:

$$D^1, D^2, D^3, \ldots, D^i, \ldots$$
$$p^1, p^2, p^3, \ldots, p^i, \ldots \to 1. \tag{2}$$

Is there a possibility of distinguishing between the two conceptions (1) and (2) by means of an empirical criterion?

I suggest employment of the following criterion: that we examine the change in the probabilities when the prediction refers to a time instant $t_3 > t_2$. Having been given the same description D^i of the situation A at time t_1, we are well aware that the probability will diminish if t_2 is replaced by t_3. By repeating this consideration we arrive at a probability lattice of the following form:

$$\left\{ \begin{array}{l} D_1^1 \, D_1^2 \, D_1^3 \ldots D_1^i \ldots \\[4pt] p_2^1 \, p_2^2 \, p_2^3 \ldots p_2^i \ldots \to 1 \\[4pt] p_3^1 \, p_3^2 \, p_3^3 \ldots p_3^i \ldots \to 1 \\[4pt] \cdots \cdots \cdots \cdots \cdots \cdots \cdots \cdots \\[4pt] p_k^1 \, p_k^2 \, p_k^3 \ldots p_k^i \ldots \to 1 \\[4pt] \cdots \cdots \cdots \cdots \cdots \cdots \cdots \cdots \\[4pt] \downarrow \ \downarrow \ \downarrow \quad \ \downarrow \\[4pt] p \ \ p \ \ p \quad \ \ p \end{array} \right. \tag{3}$$

The subscripts refer to time; the superscripts refer to the degree of detail of the description. In each line, the probabilities converge toward the value 1. In each column, the probabilities diminish and converge toward a mean probability of the appearance of the situation B, independently of the situation A.

Although this lattice is convergent, it does not present a uniform convergence. Given an interval of magnitude ϵ, one can find, for the time t_k, a description D_1^i such that $p_k^i \geqslant 1 - \epsilon$; but for the same description D_1^i, there is a time t_m, $m > k$, such that $p_m^i < 1 - \epsilon$. Hence there is no description D_1^i which predicts the future state for all times with a probability $p^i \geqslant 1 - \epsilon$. Consequently, a definite value cannot be assigned to the lower right-hand corner: in following the last column, one would assign it the value 1; in following the last line, one would assign it the value p.

It seems to me that this probability lattice indicates that the hypothesis of determinism is not confirmed by experience, even in classical physics. It is true that the notion of an ultimate description D does not logically contradict the structure of the lattice. But if this ultimate description D existed, it would be essentially different from all the descriptions D^i, in that it would establish a column of non-decreasing probabilities, while the values of the probabilities in each column starting with a D^i decrease. The non-uniformly convergent lattice can thus be considered as an *inductive proof* against the hypothesis of determinism, or at least, as the expression of the absence of any inductive proof in favor of this hypothesis.

To be sure, a *deductive proof* against this hypothesis cannot be given; in other words, one does not have a logical contradiction if one adds the supposition of the existence of an ultimate description to the observational facts included in this lattice. But this hypothesis will be introduced at the price of sacrificing continuity, because there is no continuous transition from the observable columns D^i to the non-observable ultimate column D. And continuity is the essence of inductive inference. The situation would be different if the lattice exhibited a uniform convergence: if so, one could consider this as an inductive proof of the existence of an ultimate description.

Employing the language of the verifiability theory of meaning, one can translate this result as follows. Determinism, if it exists, is not manifested in relations among observable quantities. These relations, to be sure, do not exclude determinism; but neither do they confirm it. Determinism, in classical physics, represents a vacuous addition to the system of observable relations; if it is omitted, if one renounces speaking of an ultimate description, nothing changes in the ensemble of verifiable statements.

Finally, we come to consider the situation in quantum mechanics. It is well known that, because of Heisenberg's uncertainty relation, it is impossible to increase the probability of a prediction beyond a certain value $p < 1$. We thus arrive at the following schema:

$$D^1, D^2, D^3, \ldots, D^i, \ldots, \rightarrow D$$

$$p^1, p^2, p^3, \ldots, p^i, \ldots, \rightarrow p \tag{4}$$

If the descriptions D^i approach a certain limiting description, given by the function ψ characterizing the state A, the probabilities p^i increase and converge to a value $p < 1$. It is this schema which formulates quantum indeterminism.

Comparing the three schemata (1), (2), (4), one sees clearly that one is concerned here with a continuous extension of concepts. The determinist believes in the existence of an ultimate description with the help of which he would arrive at perfect knowledge, such that "nothing would be uncertain for him, and the future would be seen just like the past". The classical indeterminist abandons the notion of this ultimate description; he rests content to improve his predictions bit by bit, without ever pretending to arrive by this process at a perfect prediction. The indeterminist of quantum theory restricts the power of predictions to a limit below the probability 1; and he can admit the existence of an ultimate description because this limiting description offers no predictive certainty.

This logical picture of the situation is presented here in order to show that one is concerned with a question of physics, and not a question of philosophical views or of world-conception. I do not believe that there are supposedly philosophical questions exempt from scientific treatment. If philosophy deals with problems of the structure of the world, it should abandon the idea of deriving this structure from an intuition allegedly of eternal truths. The philosophical truths of yesterday have become the errors of today. The philosopher who wants to contribute toward making the universe intelligible can aid the physicist only in searching for the correct form of a question, but not in searching for the answer. That is to say, his contribution will consist in a logical analysis of the problems, which separates the physical content of a theory from the additions in the guise of definitions, and which clarifies meanings of terms instead of prescribing the ways of thought. The philosopher can define determinism, and also indeterminism; but he cannot choose one of these structural forms as the existing one. It is the physicist who ascertains the structure which corresponds to the facts of observations.

According to quantum physics, it is schema (4) which describes the causal relations governing the physical world; that is to say, quantum physics has decided in favor of indeterminism. This result derives from the principle that the ψ-function includes all that can be furnished by observations. If this principle, which may be called the 'synoptic principle of quantum theory', is accepted, then all that remains is mathematics; this means that one can derive the uncertainty relations from it mathematically. These relations uniquely express the fact that the Fourier analysis of a wave packet furnishes a greater number of frequencies, or harmonic oscillations, the more the extension of the packet is reduced. Once the synoptic principle is accepted, there is no more need to discuss the existence of hidden variables, precisely because such variables would constitute physical quantities not subordinated to this principle. Hence the study of the problem of determinism leads to the question: by what right does quantum physics maintain the synoptic principle?

The answer can be given that this principle has been confirmed by many observations. A case has never been found where it was possible to take measurements outside the limits fixed by Heisenberg's relation; and it has always been possible to express the sum total of all observations by a ψ-function. Experiences of this kind surely offer reasons to accept the principle; they serve to indicate that it is advantageous to employ the principle, but they cannot establish the principle as reasonable beyond doubt. In other words, they can not confer upon this principle the status of a natural law. One could always look forward to encountering some day an experimental situation which admits of observations that cannot be included in a ψ-function. This is, for example, Einstein's belief; he considers quantum mechanics as a statistical theory comparable to the statistical interpretation of classical thermodynamics, a theory which allows us to predict mean values of certain quantities which can be verified by measurements, but which does not exclude the construction of a detailed theory which determines the individual values of all physical quantities with the help of strict causal laws.

That is, today the synoptic principle has two sides, a positive side and a negative side; it allows advance calculation of results of certain experimental arrangements, but, on the other hand, it excludes the existence of observations which violate Heisenberg's law. It is this second aspect which is attacked by the determinists who are not in doubt about the legitimacy of the first aspect.

To find a solution, let us proceed by a method which may be called the method of the reversed situation. Suppose that the positive part of the principle is correct, while the negative part has to be abandoned. What would be the derivable results for physical objects, for particles which would have a well defined existence like the molecules of classical physics?

One must now recall the logical state of the classical statistics of Boltzmann and Gibbs. The existence of probabilistic laws does not contradict the notion of a determinism governing the trajectory of the individual molecule. Boltzmann established his famous H-theorem while supposing that Newtonian mechanics determines the motion of each molecule and their collisions. In fact, Boltzmann was able to derive the probability metric existing in phase space from the canonical equations of motion with the help of Liouville's theorem, which furnishes equal probabilities for equal volumes in this space. And, more recently, von Neumann and Birkhoff have succeeded in deducing the ergodic theorem from these same equations, a theorem which Boltzmann had added to his calculations as an indispensible axiom, but which he could not prove. Now the classical statistics of gases and determinism are quite compatible; they are the children of the same father, so to speak, the father being Newton, and the children live in pre-established harmony thanks to the canonical equations. Classical physics, even if it were incapable of adducing many reasons in favor of indeterminism, could certainly not invoke decisive proofs against its existence.

Let us now examine the logical state of quantum statistics. The statistics of Boltzmann has been replaced by those of Bose and of Fermi, according to which elementary particles are indistinguishable, that is to say they do not possess any individuality. As a consequence, this theory yields different numbers for possible combinations of particles, a result which in turn furnishes new values for thermodynamical quantities. Now since these values have been confirmed by observations, one concludes inversely that the experiments inform us that the particles are indistinguishable, or at least they behave as if they were. One must study the significance of this result.

No one would have said, at the time of classical physics, that he was able to really distinguish one molecule from another, or that he would ever be able to do it. An individual molecule cannot be observed, much less be marked as the zoologists mark fish or birds. The claim of the classical physicist according to which each molecule possesses individuality should be understood indirectly: molecules have individuality because their statistical behavior indicates that two arrangements of molecules should be counted as different, if one of the arrangements differs from the other only by a change of position of several molecules. Global inference replaces direct verification. Now, applying the same inference, we shall say that in quantum physics, particles do not have individuality.

The determinists, by contrast, believe in the individuality of particles; they refuse to close the door to the possibility that some day we might succeed in localizing and observing in continuous fashion an individual particle. Let us examine what the consequence of this would be. We would have to interpret the Bose or the Fermi statistics, which would remain valid, as produced by an ensemble of individualized particles. This is possible, but it leads to very strange consequences.

Let us suppose that we are playing heads or tails. In throwing two coins simultaneously, we distinguish four possible combinations, which can be symbolized. Let A stand for heads, and B for tails, in the following way, letting the two coins be distinguished by subscripts:

$$A_1 A_2, \quad A_1 B_2, \quad B_1 A_2, \quad B_1 B_2 \tag{5}$$

The heads-tails combination, whatever be the order, is produced in half of the cases, while each of the two other combinations is produced only in one case out of four.

Let us now suppose that the game of heads or tails is governed by Bose statistics. We would then have the case heads-tails once every three times; and the same fraction would be observed for the case heads-heads and for the case tails-tails. Would we have to conclude that the coins are indistinguishable?

Not at all. They can be easily distinguished by direct observation. We would arrive at a different conclusion: we would conclude that the throws are not independent of each other. We would say that if the first coin shows heads, there is a tendency for the other coin to show the same side, such that the probability of the combination $A_1 B_2$ is reduced to $1/6$. The same reasoning applies to the combination $B_1 A_2$, and the probability of the disjunction $A_1 B_2$ or $B_1 A_2$ takes the observed value $1/3$. This interpretation can be extended, in a consistent manner, to a disjunction of r possible cases.[2]

The interpretation will be more complicated if one considers three outcomes which furnish Bose statistics. In this case, there exists not only a dependence linking one outcome to each of the others; there exists also a dependence between one outcome and the other two taken in combination. This means that we must consider relative probabilities which have a two-term reference class, such as the expression $P(A_1 . B_2, C_3)$, which denotes the probability of getting outcome C for the third item if the first shows outcome A and the second shows outcome B. We have here

$$P(A_1 . B_2, C_3) \neq P(A_1, C_3)$$
$$P(A_1 . B_2, C_3) \neq P(B_2, C_3)$$

$$(6)$$

Let us examine the consequences of this probability consideration for the problem of the quantum statistics of gases. The fact that the statistics of gases furnishes the distribution calculated by Bose or by Fermi is well confirmed by experiment. If we are to put this in accord with the idea of distinguishable particles, we will have to introduce inter-particle forces which attract or repel them in such a way that their motions are no longer mutually independent. Each particle would depend not only on each other particle, but also on the totality of positions of the other particles. These forces would have a rather strange nature because they would be transmitted instantaneously across space and could be observed only by their effects on the statistics.

We clearly see that the problem of determinism is posed, in quantum theory, in a form essentially different from that of classical physics. The classical statistics of gases is compatible with determinism; and although determinism can be considered, in classical physics, as a hardly necessary addition, this physics offers no contrary indication. Quantum statistics, however, can scarcely be reconciled with determinism. This does not mean that determinism is absolutely excluded. If there were methods permitting one to observe the individual motions of the particles without disturbing them, and if these observations substantiated the existence of strict laws with the help of which one could exactly predict the particle trajectories, then determinism would be established. But what kind of physics would then have been bestowed on us! It would be a physics of mystical forces, far from the accepted idea of force in classical physics; and the resulting physics would hardly resemble that of Newton or of Laplace. In fact, this physics would be further removed from the principles of common sense than the indeterministic physics of quanta accepted today.

Here is the reason why the synoptic principle, according to which the content of observations can always be included in a ψ-function, is solidly established by quantum physics, in its negative part as well as in its positive part. It is not only the absence of other forms of observation which has persuaded the physicist to accept this principle. The system of quantum physics, in its entirety, constitutes an intrinsic proof in favor of the synoptic principle and hence against the validity of determinism. Any merger of the positive part of quantum physics and determinism would lead us to consequences so absurd that they cannot be accepted as plausible; conversely, the deduction of these consequences presents an inductive argument against determinism.

This situation can be compared to the one which exists in the theory of relativity regarding the principle of limiting speed of signals, which is equal to the speed of light. This principle of Einstein is not based solely on the fact that signals having a speed greater than that of light have not been found. Rather, it is founded on the testimony of a complete theory, crowned with success, whose consequences would be absurd if Einstein's principle were false. The negative part of this principle, the exclusion of signals speedier than light, is thus based on positive reasons; this is why the principle has been accepted.

Quantum statistics and determinism are natural enemies; they do not come from the same father. In the above, I have presented several ideas which can serve to clarify this divergence. Nevertheless, there is much to add. The difficulties that I have just described for the Bose and Fermi statistics form only one part of a more general problem, a problem which has taken an unexpected turn in quantum physics: the problem of unobserved objects. The following section is devoted to several ideas concerning this general problem.

2. UNOBSERVED OBJECTS AND THREE-VALUED LOGIC

The period when Newton and Huyghens were discussing the nature of light marks the beginning of a historical development which is clearly divided into three phases. During about a hundred years following Newton's work, the corpuscular theory was generally accepted. The second phase began when, with the discoveries of Young and Fresnel, the wave theory received unexpected support; and since Maxwell and Hertz have demonstrated the existence of electric waves, the wave theory became "a certainty, humanly speaking", if I may be permitted to use the words of Heinrich Hertz himself. The development that followed has shown that the profound intuition of this great physicist was never so lucidly manifested than in these words "humanly speaking". These words anticipated the third phase, the phase of the duality of waves-corpuscles, which is linked to the discoveries of M. de Broglie and which represents what we can call today 'the definitive form, humanly speaking'. Let us try to make precise the meaning of this solution, which is not restricted to the interpretation of light, but applies as well to that of matter.

On several occasions it has been the case that physical questions have invited the physicist to become a philosopher. To be sure, his philosophy is not of the kind of systems which have in stock ready-made answers to all questions that can be posed. The physicist is satisfied if he can find an answer to the one question which occupies him at the moment; but he insists that the answer be given in a precise language, a language as precise as the equations of physics, and which does not disappear in a fog of words and pictures. Fortunately, such a language exists today, and it has been furnished with tools similar to mathematical equations, namely the formulas of mathematical logic.

The problem of waves and particles is a problem of the logic of knowledge. It refers to the relation between observation and inferred object, to the method of extending the knowledge acquired by observation of macroscopic objects to non-observable objects, a method of crucial importance for physics, by the employment of which it has made its great discoveries and which finally has drawn it into some unprecedented difficulties. Let us study this method, at first in a completely general form, and then in its application to the problem of waves and particles.

The exigencies of practical life require us to add to our observations a theory of unobserved objects. This theory is quite simple: we suppose that unobserved objects are similar to those that we do observe. No one doubts that a house remains the same whether or not we observe it visually. The act of observation does not change things — this principle seems to be a truism. But a moment's thought shows that it is wholly impossible to verify it. A verification would require us to compare the unobserved object with the observed object, and hence to observe the unobserved object, which amounts to a contradiction.

It follows that the principle does not have the logical status of a true proposition. Rather, it is here a matter of convention; one introduces the definition that the unobserved object is governed by the same laws as those which have been verified for observed objects. This definition constitutes a rule which allows us to extend the observation language to unobserved objects; let us then call it an *extension rule of language*. Once this rule is established, one can find out whether the unobserved object is the same as the observed object or not; for example, we can conclude that the house stays at its place when we do not look at it, while the young girl does not remain in the box when the magician saws the box in two parts. Similar statements would have no meaning if a rule of extension of the observation language had not been added.

The last example exhibits a case where the rule allows us to infer that a change occurred in the unobserved object. Using a more scientific example, I could speak of the change in temperature which occurs when a thermometer is placed in a reservoir of water. We have here a case where the act of observation changes the object; despite this change, we have no difficulty in calculating the value of this change with the help of the laws of thermodynamics. Thus it seems that the extension rule of language allows us always to speak of unobserved objects in the form of meaningful propositions.[3]

Nevertheless, we must examine this thesis more closely. A rule of language is arbitrary, to be sure! But one can ask whether it is always possible to apply it. In other words, one must study the question of knowing whether the physical system constructed with the help of this rule is coherent. In classical physics, the answer is evidently affirmative; but this fact should be regarded as a result of experiment. It cannot be shown by purely logical considerations that it is always possible to construct a physical language satisfying the rule of extension. Here is the point at which quantum physics differs from the physics of macroscopic objects, a possibility which was overlooked in logical investigations prior to wave mechanics.

In this new mechanics, one must distinguish two different questions. The first concerns the disturbance of the object by the act of observation. The second concerns the state of the unobserved object. We would be mistaken if we wanted to conclude that the disturbance by the observation necessarily entails the indeterminacy of the unobserved object. The example of the thermometer placed in the water-reservoir shows that the disturbance by the observation does not exclude the determination of the state of the object before the observation: this state can be calculated with the help of physical theory. If the same method does not apply to quantum physics, one must adduce special reasons for this impossibility; it is here that the problem of the extension rule of language comes in.

Let us study the first question first. There is a very simple way to demonstrate the fact of the disturbance by the observation. This fact is deducible in a well-known manner from the theorem of the addition of probability amplitudes. If we have three non-commutative quantities u, v, w, this theorem gives the value

$$P(u_i, w_m) = \left| \sum_k \alpha_{ik} \beta_{km} \right|^2 \tag{7}$$

for the probability of measuring the value w_m after the observation of the value u_i. The terms α_{ik} and β_{km} represent the transformation matrices. On the other hand, the probability calculus yields a theorem of addition of probabilities in the following form:

$$P(u_i, w_m) = \sum_k P(u_i, v_k)P(v_k, w_m) = \sum_k |\alpha_{ik}|^2 |\beta_{km}|^2 \tag{8}$$

Since the two expressions (7) and (8) are not identical, it is necessary that we interpret the expression $P(u_i, w_m)$ in two different ways: in the case of relation (8), this expression denotes the probability of observing the value w_m after observation of the quantity u_i if a measurement of the quantity v has been made in between, a measurement the result of which has not been registered, while (7) yields this probability calculated for the case where no measurement has been made between the two observations. Now any measurement changes the physical conditions; this fact is completely expressed in the relations linking observable quantities.

Allow me to add a logical remark. Although it may be possible to show that observation disturbs objects, it can never be shown that observation does not disturb them. This means that in a different physics, which would furnish us with identical results for the two cases (7) and (8), we would not be able to conclude that the unobserved quantity is equal to the observed quantity; we could conclude only that it is permissible for us to define these two quantities as equal. There is then a certain asymmetry between the positive case and the negative case. In other words, we can demonstrate only, either the admissibility, or the inadmissibility of an extension of language such that the unobserved object is equal to the observed object. We shall encounter the same asymmetry again soon in a more general question.

Let us tackle the second question. Given that the act of observation disturbs the object, is it possible to calculate the value of the quantity in question before the observation? The answer depends on the admissibility of the rule of extension of language, on the convention that natural laws are the same for both observed objects and for those which are not observed. For, if this rule is admissible, we are permitted to employ physical theory for a relevant inference of the state of the unobserved object, as in the example of the thermometer inserted in the water-reservoir.

It is sometimes said that the disturbance by the observation excludes the possibility of meaningful statements concerning the unobserved object. This opinion is not correct. One is concerned here with a problem of the extension of language which exists as well for classical physics, because in this physics there are also observations which change the object. A similar extension offers no difficulty if the rule of extension is admissible. What distinguishes quantum physics from that of the classical statistics of molecules is the fact that this rule is inadmissible, that it can be shown that the laws of unobserved objects deviate from those that govern observed objects. Here we meet again the logical situation that we have just studied; on the basis of only observable facts, one can demonstrate a certain difference between observed objects and unobserved objects.

For simplicity of notation, let us call observable phenomena simply *phenomena*; those that cannot be observed we shall call *interphenomena*, because they are interposed between the phenomena with the help of an inference procedure. We shall use the term 'observable' in a broad sense, in such a way that it includes as phenomena the coincidence of two particles, indicated by a measuring instrument. Such quantum phenomena are inferred from macroscopic observations by the use of classical theory only; this is why they can be treated on an equal footing with macroscopic phenomena.

While phenomena always have the character of narrowly localized objects, interphenomena offer two divergent possibilities of interpretation: it is here that the alternative appears, waves or particles. These concepts have no significance if we speak only of phenomena; they belong to a language which comprises interphenomena. In speaking of particles, we attribute to interphenomena a narrow localization similar to that of the phenomena; in speaking of waves, we consider the interphenomena as extended over a large space.

The two interpretations represent two different rules concerning the extension of language, and the question arises: which of the two satisfies the principle that physical laws are the same for both the phenomena and the interphenomena?

The answer is quite definitive: neither the one nor the other satisfies this principle. The attempt to assign to interphenomena a definite existence, that is to say to attribute to them precise values of position and speeds which exist simultaneously, necessarily brings us to strange consequences regarding the principle of causality. I am not speaking here of the substitution of probability in place of certitude of prediction. This change seems minor compared to certain deviations of a different kind. Classical causality included the idea of gradual propagation, of a continuous transmission of the causal effect from one point in space to another, a transmission which occurs as time goes on. By contrast, the interphenomena of quantum mechanics exhibit causal relations which occur abruptly, which take no time to propagate, and which thus represent a true action at a distance. In this regard, the interphenomena differ essentially from the quantum phenomena, because observable quantities are always governed by causal laws which, except for their probabilistic character, resemble those of classical physics and satisfy the requirement of continuous propagation.

I do not need to give all the details, because these results have often been discussed. I am thinking of diffraction experiments, of apparatus in which a beam of electrons passes through two slits and is projected onto a screen. A pattern of interference fringes is observed, and it is well known that the pattern produced by two slits is not made up of the superposition of the patterns produced individually by each slit if the other is closed. One has a macroscopic observation which can be translated into the language of interphenomena only under pain of violation of the principle of causality: a particle passing through one of the slits knows if the other slit is open or closed. Or rather, omitting the anthropomorphism of the word 'knows', we would say that the motion of a particle passing through one slit depends on the physical condition of the other slit, a causal dependence which contradicts the principle of gradual action. I have suggested that, in such a case, one speak of a *causal anomaly*, a term which indicates the deviation from the causal behavior of observable phenomena.

The wave interpretation, it is true, solves this difficulty in that it exhibits the pattern on the screen as a train of waves passing simultaneously through both slits and subject to interference. However, this interpretation brings us to causal anomalies of another kind. The wave arrives at the screen all extended; but its effect consists of a little flash localized at a single point of the screen. As soon as the flash occurs, the extended wave disappears. This means that the process at a point of the screen exercises an influence on the wave at each point of the wave front, an instantaneous influence which contradicts the principle of gradual action and which represents a form of causal anomaly. The transition from the particle interpretation to the wave interpretation does not eliminate causal anomalies; the anomalies merely change places, and the interphenomena called 'waves' are not more reasonable than those that we call 'particles'.

The word 'reasonable', it is true, also represents an anthropomorphism. Let us then say that neither of the two interpretations can satisfy the extension rule of language, according to which unobserved objects are governed by the same physical laws as observable objects. These latter do not present any causal anomalies; it is only in the world of interphenomena that such anomalies subsist.

This result is equally applicable to a third interpretation which has been studied in a critical way by M. Louis de Broglie, more than 20 years ago, and which has been advanced recently with many promises and little logical analysis in an article of D. Bohm:[4] the interpretation of the pilot wave. This interpretation represents a combination of the two others; it speaks of a field of waves which guide the particles. Insofar as it employs the idea of waves, this interpretation avoids the anomalies arising for particles which pass through a diffraction grating; interference is a property only of waves and the particle is guided along the trajectories of the field resulting from the superposition of waves.

Nevertheless, the anomalies of the wave interpretation are repeated for this interpretation; moreover, it entails another sort of anomaly which has been very clearly pointed out by M. Louis de Broglie ever since his first studies of this subject. In this theory, there is a dependence of the statistical distribution characterizing the assembly of molecules − on the path of the particle, a dependence resulting from the fact that the ψ-function, which represents the wave, expresses at the same time the probability of observing a particle at a certain place. The consequences are quite strange, in particular, if one considers radiation of weak intensity such that the particles are sent one at a time. The path traversed by the first particle would here indicate the statistical distribution of the particles which are to follow it. The dependence can be interpreted in two ways: either the first particle causally determines the behavior of the particles to follow, or the first particle is behaviorally determined by future events. This difficulty is not avoided by attributing to the wave a physical reality, as M. Bohm believes. This real wave can account for the individual path of the particle; but the fact that the wave reflects the distribution of the particles to follow expresses a pre-established harmony which would have enchanted Leibniz, but which, in today's physics, can only be called a causal anomaly.

Allow me to add here several words concerning the question of the interpretation of a physical theory. I do not believe that there are any forbidden interpretations. In favor of those who would want to interpret the quantum equations in a certain way, I would like to say that I do not allow myself the locution 'forbidden meaning' in physics. I would propose that this be replaced by another maxim which says: "If you like to choose an interpretation, go ahead and do it; but draw the consequences". Here is the point which distinguishes logical analysis from the language of pictures. Let those who prefer to conceive the microcosm as made up of waves, or of particles, or of the two together, do so! But let them not forget to describe in a precise language the properties of this physical reality which they have created. They will be surprised by the strange aspect that the world of their construction presents in a close-up view.

The problem of physical reality is posed, for today's logician, in the form of an analysis of language. The world admits of a plurality of descriptions; each is true, but each requires, in order to be verifiable, a statement of the conditions on which it is based. Hence there is a class of equivalent descriptions, among which one can make distinctions only according to descriptive simplicity, that is, a simplicity that concerns only the form of the description and which has no significance whatever regarding the question of truth. Yet one can consider the question of whether the class includes a *normal system*, that is to say, a description which satisfies the rule of extension according to which unobserved objects are governed by the same laws as observable objects. For classical physics, such a system exists; it is everyday language. By contrast, quantum physics does not admit of a normal system of description; each admissible description includes some causal anomalies. I have proposed the name *principle of anomaly* for this result.

Here is the decisive difference between the two physics, and here is the way in which modern logic treats some surprising properties of the microcosm revealed in quantum physics. A property of the world of small dimensions is expressed in the form of a property of the class of admissible descriptions. To illustrate this method, one could cite those geometric methods which characterize a space in terms of the invariants of a class of transformations; and one could recall the methods of the theory of general relativity which describe a gravitational field by some invariant relations regarding the transformations of systems of reference.

A description which attributes a well-defined existence to interphenomena can be called *exhaustive*; it gives an answer to each question that can be asked regarding their state at each moment. However, since these answers often seem disagreeable to us, dissolving into causal anomalies, it has been proposed that statements concerning the interphenomena be entirely omitted. One thus arrives at a *restrictive* interpretation, an interpretation which does not speak at all of interphenomena, and which is reduced to observable phenomena.

The interpretation advanced by Bohr and Heisenberg is of this kind. It seems to me, nevertheless, that the prohibition against speaking of the inter-phenomena cannot be justified by logical reasons; rather it is based on physical reasons. It is because of the absence of a normal system that it would be prudent not to speak of interphenomena; and that is all that can be said in favor of this prohibition. And the absence of a normal system expresses a physical fact. Therefore the restrictive interpretation arises from a physical fact.

In the language of the logician, the theory of Bohr and Heisenberg takes the form of a theory of meaning. A certain group of propositions is considered as meaningless and is thus omitted from the domain of admissible statements. The question arises as to whether this radical remedy does not go beyond the limits of a reasonable restriction. In fact, it seems doubtful that physics can completely renounce a description of the interphenomena. For example, if a beam of electrons passes through two slits and produces a pattern of inter-ference fringes on a screen, one would like to say that something or other traverses the slits; no one doubts that this "something or other" cannot penetrate the solid matter of the diaphragm. Is it necessary to sacrifice this commonplace statement? If we take the conception of Bohr and Heisenberg seriously, then we would have to.

Mathematical logic offers a method of avoiding this unhappy consequence. It has presented us with systems of multi-valued logic, that is to say, systems in which the duality of truth-values, 'true' and 'false', is replaced by a multi-plicity of values. In particular, it is the three-valued logic which presents itself for the interpretation of quantum physics. This logic includes an intermediate category between 'true' and 'false' which can be considered as denoting 'indeterminate', a category which can serve to include the statements con-cerning the interphenomena.

A very interesting interpretation of this kind has been constructed by Mme. P. Fevrier, who was the first to apply the results of multi-valued logic to quantum physics. The interpretation which I have proposed differs on certain points from that of Mme. Fevrier; and I permit myself to present here the general ideas on which this interpretation is based.[5]

The use of truth-tables, a method developed originally for two-valued logic, is very convenient for the construction of a logic. These tables indicate the relations between the truth-values of elementary propositions and those of propositions compounded from the latter with the help of logical operations, such as those expressed by the terms 'not', 'or', 'and', 'implies', etc. Their form is given in tables I and II. The letter "T" denotes 'true'; the letter "F" denotes 'false'; the letter "I" denotes 'indeterminate'. The theorems, or *tautologies*, of a logic are those formulas which possess a T in each place of their column. A comparative list of some tautologies of the two logics is given in table III.

TABLE IA
Two valued logic

a	negation $-a$
T	F
F	T

TABLE IB
Two-valued logic

a	b	disjunction $a \lor b$	conjunction $a \cdot b$	implication $a \supset b$	equivalence $a \equiv b$
T	T	T	T	T	T
T	F	T	F	F	F
F	T	T	F	T	F
F	F	F	F	T	T

I do not wish to enter into a detailed discussion of these tables. Allow me, nevertheless, to show you the method by which the three-valued logic treats the problem of radiation passing through two slits.

If no observation is made at the slits, one would say, using two-valued logic, that an individual particle observed on the screen has passed either through one slit, or through the other. This statement, which belongs to an exhaustive description, leads us into causal anomalies, as we have just discussed. The restricted description, in the form of a three valued logic, replaces this disjunction by another, constructed with the help of three values, and called *diametrical disjunction*. It has the following properties: if the particle has been observed in the vicinity of one of the slits, the disjunction is true;

TABLE IIA
Three-valued logic

	negation		
	cyclical	diametrical	complete
A	$\sim A$	$-A$	$\bar{A}$
T	I	F	I
I	F	I	T
F	T	T	T

TABLE IIB
Three-valued logic

				implication			equivalence	
			con-		alter-	quasi-		
		disjunction	junction	normal	native	implication	normal	alternative
A	B	$A \lor B$	$A \cdot B$	$A \supset B$	$A \rightarrow B$	$A \Rightarrow B$	$A \equiv B$	$A \triangleq B$
T	T	T	T	T	T	T	T	T
T	I	T	I	I	F	I	I	F
T	F	T	F	F	F	F	F	F
I	T	T	I	T	T	I	I	F
I	I	I	I	T	T	I	T	T
I	F	I	F	I	T	I	I	F
F	T	T	F	T	T	I	F	F
F	I	I	F	T	T	I	I	F
F	F	F	F	T	T	I	T	T

and if no observation has been made at the slits and the statement of passage is indeterminate, the disjunction is also true. By employing this meaning for the word 'or' we can thus say that the particle passes through one slit or the other, without arriving at causal anomalies. The diametrical disjunction, which can also be called the equivalence of contraries, is written in the form:

$$B_1 \equiv -B_2 \tag{9}$$

This statement, which is true, characterizes the passage of radiation across the two slits. But we cannot derive from it the ordinary disjunction

$$B_1 \lor B_2 \tag{10}$$

because this last disjunction may be indeterminate. It would be different in two-valued logic: here one could deduce relation (10) from relation (9).

The language of three-valued logic thus permits us to formulate, as admissible statements, all that we know as regards phenomena and interphenomena, without bringing forth causal anomalies. It is only the relation between the

TABLE III
Tautologies

	Two-valued logic	Three-valued logic
T1. Law of identity	$a \equiv a$	$A \equiv A$
T2. Law of double negation	$\bar{\bar{a}} \equiv a$	$A \equiv - - A$ $\bar{A} \equiv \bar{\bar{\bar{A}}}$
T3. Law of triple negation	$- - -$	$A \equiv \sim \sim \sim A$
T4. Relation between negations	$- - -$	$\bar{A} \equiv \sim A \vee \sim \sim A$
T5. *Tertium non datur*	$a \vee \bar{a}$	$- - -$
T6. *Quantum non datur*	$- - -$	$A \vee \sim A \vee \sim \sim A$
T7. Law of contradiction	$\overline{a \cdot \bar{a}}$	$\overline{A \cdot \bar{A}}$ $\overline{A \cdot \sim A}$ $\overline{A \cdot - A}$
T8. Laws of de Morgan	$\overline{a \cdot b} \equiv \bar{a} \vee \bar{b}$ $\overline{a \vee b} \equiv \bar{a} \cdot \bar{b}$	$-(A \cdot B) \equiv -A \vee -B$ $-(A \vee B) \equiv -A \cdot -B$
T9. First distributive law	$a \cdot (b \vee c) \equiv a \cdot b \vee a \cdot c$	$A \cdot (B \vee C) \equiv A \cdot B \vee A \cdot C$
T10. Second distributive law	$a \vee b \cdot c \equiv (a \vee b) \cdot (a \vee c)$	$A \vee B \cdot C \equiv (A \vee B) \cdot (A \vee C)$
T11. Contraposition Law	$\bar{a} \supset b \equiv \bar{b} \supset a$ $a \supset b \equiv \bar{b} \supset \bar{a}$	$-A \supset B \equiv -B \supset A$ $\bar{A} \to B \equiv \bar{B} \to A$ $A \supset B \equiv -B \supset -A$
T12. Dissolution of Equivalence	$(a \equiv b) \equiv (a \supset b) \cdot (b \supset a)$	$(A \equiv B) \equiv (A \supset B) \cdot (B \supset A)$ $(A \overset{\equiv}{\equiv} B) \equiv (A \rightleftharpoons B) \cdot (-A \rightleftharpoons -B)$
T13. *Reductio ad absurdum*	$(a \supset \bar{a}) \supset \bar{a}$	$(A \supset \bar{A}) \supset \bar{A}$ $(A \to \bar{A}) \to \bar{A}$

two kinds of phenomena which is expressed in the form of a true proposition, while a statement concerning the interphenomena separately receives the value 'indeterminate'.

Similar solutions for other problems involving causal anomalies can be developed, such as the problem of the energy barrier traversed by particles, and one can also formulate the relation of complementarity for two non-commutative quantities. These results indicate that the three-valued logic offers us a form of language which allows us to speak of the quantum world without undesirable consequences.

Let us try to summarize this analysis of the problem of unobserved objects. The alternative between corpuscular theories and wave theories, which marked the two first phases of conceptions of the nature of light, has been replaced, as a result of the discoveries of M. de Broglie, by a conjunction: the word 'or' has been replaced by the word 'and'. At the same time, this conjunction, which indicates the third phase of the historical development of investigations of the nature of light, has been extended in such a way as to encompass the nature of matter. Yet, a logical analysis shows that this word 'and', indicating the conjunction, does not belong to the language of physical objects; it refers to a duality of descriptions and hence belongs to the metalanguage, a language which deals with properties of linguistic systems with the help of which we describe the physical world. One can even pass from the duality of descriptions to a plurality: a class of equivalent descriptions can be constructed. It is in the form of this class that modern logic treats the problem of unobserved objects; and the choice of descriptions is presented as a choice among extension rules of language.

Although this plurality of descriptions is already applicable to classical physics, it is not of much importance there; among the equivalent descriptions there exists a normal system, which is convenient to employ, forgetting all the others. By contrast, in quantum physics, the class of equivalent descriptions does not include a normal system. Each description entails causal anomalies if it is exhaustive, that is to say, if it attributes a well defined state to the interphenomena. To avoid the anomalies, restrictive descriptions can be constructed which exclude statements about interphenomena from the domain of true assertions. The three-valued logic offers us an adequate form of this kind of description and allows us to speak of the interphenomena in an indirect way, such that the relations between phenomena and interphenomena are expressed by true assertions, while it is not possible separately to derive statements concerning the interphenomena.

Does this result show that the true logic of quanta is three-valued? I do not believe that one can speak of the truth of a logic. A system of logic is empty, that is to say, it has no empirical content. Logic expresses the form of a language, but does not formulate any physical laws. Nevertheless, one can consider the consequences for the language of the choice of logic; and, to the extent that these consequences depend at the same time on physical laws, they reflect properties of the physical world. It is thus the combination of logic and physics which indicates the structure of the reality which concerns the physicist. A quantum physics under the form of a two-valued logic exists; but it includes causal anomalies, while a quantum physics which employs a three-valued logic does not possess any such anomalies. This conclusion, which is the result of all the experiments encompassed by the wave mechanics, expresses the strange structure of the microcosm, a structure which has so worried the physicist, but which he has learned to master by means of an ingenious system of equations and of experimental apparatus. Here is the general result established by the work of the physicists; it remains only for us to accept it, although we are well aware that it is here again a question of certitude 'humanly speaking'.

3. THE DIRECTION OF TIME IN CLASSICAL PHYSICS

In order to understand the contribution of quantum mechanics to the problem of time, one must begin by studying this problem in classical physics. We shall see that, on the one hand, wave mechanics includes a development destructive of the concept of time, while, on the other hand, wave mechanics has given this concept a new foundation, a foundation that classical mechanics could not furnish.

The concept of time comprises metrical properties and topological properties. The metrical properties have been treated, in our day, by the theory of relativity; I do not need to speak here of these well-known discoveries which have profoundly changed our ideas of the simultaneity of spatially separated events. I wish to speak about the topological properties, and I would like to show that classical physics furnishes us with powerful instruments appropriate for analyzing the topological aspect of time, an aspect which determines so completely our conception of the physical world as well as the form of our psychological experiences.

The study of the topological problem of time requires that we carefully distinguish between *the order* and *the direction* of time. The order of time corresponds to the order of points on a straight line; this one-dimensional extension is ordered without possessing a direction. In other words, the points of a line are ordered with respect to the relation 'between', but one cannot say whether the line is extended from right to left or from left to right. In order to assign a direction to the line, one must use other means; for example, one can choose one point and specify that it is situated to the left of a certain other point. By contrast, if one is given three points, the line itself will determine that which is situated between the other two. In the same way, the order of events in time concerns relations expressed with the help of the word 'between', while the direction of time assigns a sense to temporal lines, a unique sense which manifests itself in the flux of time, in the conception that time goes from past to future.

It is well known that classical mechanics cannot furnish us with the *direction* of time. The differential equations of classical mechanics are of the second order; thus, given a solution, the replacement of the variable t by the variable $-t$ leads to another solution. A ball thrown into the air, a planet moving around the Sun, these are *reversible* phenomena: the motion can occur in one or the other direction. Thus, the observation of motion considered as a mechanical phenomenon gives us no information about the direction of the motion.

This result being well known, it has often been forgotten that classical mechanics can very well give us precision as regards the *order* of time. If a ball goes from point A, via point B, to point C, mechanics lets us consider this motion as a transition from C via B to A, but we are required to keep B between C and A. It is for this reason that classical mechanics presents us with a temporally ordered physical world. In the framework of the theory of relativity, this order has become the source of the causal theory of time, according to which the concept of time originates from the causal order and is based on the following definition: an event A precedes an event C if a signal sent from A arrives at C. We know that this definition, because of the limiting character of the speed of light, leaves undetermined the order of certain events situated on space-like world-lines, for which $ds^2 < 0$. But apart from this restriction, this definition determines a temporal order of the physical world which corresponds to the order of our psychological experience.

The causal time of the theory of relativity is ordered, but it does not have a direction. The Lorentz transformation is invariant with respect to time reversal: If t is replaced by $-t$, and t' by $-t'$, one obtains a new Lorentz transformation, describing a motion in the opposite direction. This is why relativity, despite its reduction of time to causality, has contributed nothing to the problem of the direction of time.

A definition of the direction of time requires us to distinguish between cause and effect, that is to say, to add to the concept of causal connection a criterion which gives meaning to direction. Such a criterion is furnished by the second law of thermodynamics, by the concept of entropy. Let us examine the significance of this concept for the direction of time.

At the time of classical thermodynamics, there was no doubt that one had found, in the concept of entropy, an instrument which permits the introduction of the direction of time in physical equations. This optimistic opinion ran into serious difficulty when L. Boltzmann gave the statistical formulation of the second law of thermodynamics. The difficulties came to be well known in the objection concerning the reversal of the motions of molecules: if it were possible to reverse the individual motions of all the molecules, the gas would pass from a state of higher entropy to a state of lower entropy. Such a situation cannot be excluded if entropy is interpreted as a statistical property, because the individual motions of the molecules are reversible. Moreover, such a situation ought to occur, in the history of an isolated system, as often as the corresponding situation of the passage from lower entropy to higher entropy. This result follows from certain theorems established by J. Loschmidt and H. Poincaré; it means that in the graph of entropy of an isolated system, there are, from time to time fluctuations, such that the number of peaks and of valleys is infinite and their quotient converges to unity.

It is very interesting to study the publications dealing with the objection of time reversal toward the end of the last century. Boltzmann assures us that the inverse transitions occur very rarely in the history of an isolated system; it follows that one would be justified in saying that if a system is observed to be in a low entropy state, then one could conclude that it will soon pass into a state of higher entropy. Unfortunately, the same conclusion follows as regards the preceding state; it can be shown that it is very probable that the system arises from a state of higher entropy. The inference is applicable, symmetrically, to the future and to the past; it is thus impossible for us to define a direction of time using the entropy of an isolated system.

Let us put this result in more precise terms. One must distinguish between an *inference from time to entropy* and an *inference from entropy to time*. The first pertains to the following question: suppose that the system is observed in a state A of lower entropy, what will be the value of the entropy in a subsequent state B? The answer is that it is extremely probable that this value is high. This result allows us to predict the future state of the system. The second inference pertains to the question: suppose that we have observed two states A and B of the system of which A has a lower entropy and B a higher entropy; which of the two states preceded the other? The answer is that it is just as probable that A preceded B as it is that B preceded A. This result follows from the symmetry of the first inference, regarding which we concluded that a state of lower entropy would be preceded, in all probability, by a state of higher entropy. We arrive at the result that the inference from time to entropy is admissible, as it is commonly acknowledged; but that the inference from entropy to time should be rejected and that it is impossible for us to define a direction of time for an isolated system.

There exists an isolated system which is of particular interest to us: the universe. This system, without doubt, is completely isolated. Thus our universe does not have any direction of time; this conclusion cannot be rejected if it is admissible to speak of the entropy of the universe.

But, is it admissible to speak of it? One can doubt it. If the universe is spatially infinite, the concept of entropy is not applicable to it. We would, however, have no need to study this question. The direction of time is something which concerns the ensemble of our immediate environs; if we have to have recourse to distant nebulas, hidden in the profound depths of the universe, in order to resolve the question of the direction of time, we would not be able to obtain an interpretation of the experience of habitual everyday time. We have to examine anew the inferences which the physicist makes when he passes from entropy to time.

In fact, when he observes a given entropy state, the physicist refuses to conclude that this state has been preceded by another state having a higher entropy. Suppose that you are shown two pictures cut from a film, one of them showing a house in good order, the other showing the same house after the effects of an explosion; you would know well in what direction the film was taken. You would say that it is very probable that the orderly state, of lower entropy, preceded the state of disorder, of higher entropy, thus rejecting the conclusion of Boltzmann according to which the probability of this affirmation is but 50%, the same as for the inverse order of time. How does one explain this refusal to accept the results of physical theory?

One runs up against difficulties when one wants to render precise the meaning of the word 'probable', since the latter is used so often in daily language. A probability denotes a frequency, and a frequency refers to a class of events, or, more exactly, to a sequence of events, that is, to an ordered class. Differences in the meaning of the word 'probable' are often explained by the differences among the sequences to which the word refers. Let us examine the question from this point of view.

The probability calculated by Boltzmann refers to the history of an isolated system, and thus to a *time ensemble*. By contrast, the probability used in the example of pictures of houses is not of this kind. In the history of a house, there are not enough changes of state to warrant constructing a frequency; an explosion happens once, and the history of the house is finished. If a frequency is to be reckoned, it concerns a plurality of houses; that is, it concerns a *space ensemble*. The probability that we use to infer a direction of time does not refer to a time ensemble, but to a space ensemble.

Let us study this ensemble. There are many physical systems which originate from larger systems and which subsequently get separated from the latter; once separated, they remain pretty much isolated for a long enough period. These systems, which I shall call *branch systems*, begin by being in an ordered state of lower entropy, and they proceed toward a state of disorder, of higher entropy. The ordered state is not presented here as a result of chance in the history of the isolated system, but as a result of a mutual action between the system and its surroundings; the larger system proceeds toward a state of higher entropy, while a system of low entropy is created in the form of a part of the larger system.

The universe has plenty of branch systems of this sort. Milk is poured into coffee and thus a branch system of low entropy is created; the sun's rays heat a rock which, together with the snow around it, represents a system of low entropy tending toward temperature equalization, etc. It is the ensemble of such branch systems which offers the possibility of explaining the form of probability used in the inference from entropy to time: instead of saying that it is probable that the ordered system is the unlikely product of a process of separation in the history of the isolated system, we say that it is probable that the system had not been isolated in the past and is a product of outside intervention. For example, if we find a cigarette burning on an ashtray, we do not believe that this state was preceded by a state where the cigarette was all ashes, a state from which the present state had developed by an improbable separation of the chemical components of the ashes. We would prefer to suppose that somebody had lighted the cigarette several minutes ago, and thus that the present improbable state is the result of outside intervention. We would say that this explanation is more probable than the preceding one. This second occurrence of the word 'probable' refers not to a frequency in the time ensemble, but to a frequency computed in the space ensemble of branch systems.

The probability calculus provides methods for treating these two kinds of probability.

We construct a probability lattice:

$$
\begin{array}{llllll}
x_{11} & x_{12} & x_{13} & \cdots & x_{1i} & \cdots \rightarrow p \\
x_{21} & x_{22} & x_{23} & \cdots & x_{2i} & \cdots \rightarrow p \\
\cdots & \cdots & \cdots & \cdots & \cdots \\
x_{k1} & x_{k2} & x_{k3} & \cdots & x_{ki} & \cdots \rightarrow p \\
\cdots & \cdots & \cdots & \cdots & \cdots \\
\downarrow & \downarrow & \downarrow & & \downarrow \\
p_1 & p_2 & p_3 & & p_i & \rightarrow p
\end{array}
\tag{11}
$$

Each horizontal line represents the history of a branch system, and thus a time ensemble. By contrast, the columns represent space ensembles. Let us consider the probability of a lower entropy state: for each line, this probability p has a very small value. It is quite different with the columns: in the first column, the probability p_1 of a state of lower entropy is quite high because in the column one is concerned with states of separation, of initial states of branch systems. Little by little, the probability values $p_2, p_3, \ldots$ diminish and converge eventually to the probability p.

Using a notation developed elsewhere,[6] I write

$$P(B^{ki})^i = p \tag{12}$$

for the probability of a state B in a horizontal line, and

$$P(B^{ki})^k = p_i \tag{13}$$

for the probability of B in a column. The repeated index thus indicates the direction in which the frequency is determined. The convergent lattice (11) is characterized by the relation

$$\lim_{i \to \infty} P(B^{ki})^k = P(B^{ki})^i \tag{14}$$

In the theory of probability lattices it is shown that such a lattice which describes a mixing process is characterized by certain special relations which cannot be deduced from the axioms of the probability calculus and which have to be considered as definitions of types of lattices. From the standpoint of physics, this means that the type of lattice constitutes a form of empirical hypothesis; that is, that only experience can justify the choice of lattice used as a description of physical phenomena. What is in question here, in particular, is a property which I have called *lattice invariance*[7] and which can be formulated, in a simplified form, as follows:

$$P(A^{ki}, B^{k,i+n})^i = P(A^{ki}, B^{k,i+n})^k \quad (n > 0) \tag{15}$$

These expressions denote relative probabilities. On the left-hand side, we have the horizontal probability, that for which a state A of lower entropy is followed by a state B of higher entropy. On the right-hand side, we have a vertical probability of the following form: in the column $i + n$ one chooses all the elements $x_{k,i+n}$ (here k is variable, $i + n$ is constant) for which the element x_{ki} of column i is found in a state A; and in the partial sequence thus defined, one counts the elements belonging to state B. If this frequency corresponds to the frequency determined in a line, indicated on the left-hand side, the condition of lattice invariance is fulfilled.

83

It can be shown that if a lattice composed of horizontal series with diminishing memory (like Markoff series), if it satisfies the condition of lattice invariance, it is always convergent and thus characterized by the relation (14). In other words, whatever the distribution of a state B in the first column may be, the later columns have a tendency to reproduce the B-distribution existing in the horizontal lines.

Physically, this means that the space ensemble has a tendency to reproduce the time ensemble. Since this statistical tendency does not follow from the axioms of the probability calculus, it expresses a natural law which is verified by an immense number of observations and which is applicable to all the ensembles of physical systems, be they assemblies of molecules, or macroscopic systems composed of a large number of molecules, or stars. Let me speak here of a *statistical isotropy*, applying to the statistical conditions a term which, in optics, indicates an equivalence of directions. It is the statistical isotropy which yields an explanation of the phenomena which express the direction of time.

For it is the statistical isotropy which determines that the branch systems all proceed in the same direction. Let us suppose that half of these systems proceed in the inverse direction, starting from a state of high entropy and proceeding toward a state of lower entropy. It follows that the lattice composed of all the branch systems could not be isotropic: the last columns, like the first ones, would include a large number of states of lower entropy, a result which contradicts the distribution of similar states in the horizontal lines. The relation (14) would not thus be fulfilled. We conclude that this supposition is false. Statistical isotropy expresses that property of the universe to which we owe *the parallelism of the increase of entropy*: the totality of branch systems defines a single direction of time.

Once this result is established, it is easy to account for the inference from entropy to time. If the frequency is determined vertically, that is, in a column, one finds only a small probability that a state of lower entropy be preceded, for the same system and thus on a horizontal line, by a state of higher entropy; at least, this result applies to the initial columns of the lattice. We have thus succeeded in finding the answer to the objection of the reversal of molecular speeds. The symmetry of the probabilities in regard to the future and to the past exists only for the time ensemble; it no longer exists for the space ensemble. And the direction of time can be defined in terms of entropy because the aggregate of branch systems has an asymmetry with respect to time, an asymmetry which comes from the parallelism of the increase of entropy.

These ideas, it seems to me, furnish the solution to the problem of the direction of time in classical physics. Whereas classical mechanics gives us only the order of time, statistical thermodynamics distinguishes between the two directions of this order and shows that they are qualitatively statistically different. The irreversibility of the course of macroscopic events is compatible with the reversibility of the course of elementary events.

You might ask me: just how does this solution differ from that given by Boltzmann? Let me add a few words about this. Boltzmann was well aware that the entropy curve of an isolated system cannot define a direction. He thus restricted the direction of time to an ascendant part of the entropy curve of the universe, and he remarked that the human inhabitants of the universe considered as positive time the direction for which their part of the curve is ascending. He was well aware that this conception entailed the possibility of different directions of time for different cosmic epochs.

I have nothing to say against this last conclusion; indeed, I believe that a direction of time can be defined only for one cosmic epoch, and I reject completely all attempts to treat cosmic time as a whole. Thus, it seems to me that although the existence of a rise in the entropy curve of the universe is a necessary condition for a direction of time, it is not a sufficient condition, for it. One must add the hypothesis of branch systems governed by the laws of statistical isotropy. As soon as we add this hypothesis to the ideas of Boltzmann, we come to understand the direction of time; we are able to account for numerous observed phenomena in our immediate neighborhood which manifest the direction of time. Without this hypothesis, which is not a logical consequence of the fact that the entropy curve rises, we would not be able to explain the existence of such a direction.

I would like very much to present several other results which are deducible from the ideas just developed, and which concern something new: the statistics of macroscopic phenomena. In statistics of this sort, one counts as elements not the states of molecules but the states of macroscopic systems; and one can define a macroscopic entropy for them which resembles in many ways the thermodynamical entropy. To give an example, one can take the process of shuffling in a game of cards; there are many natural processes which exhibit similar qualities. A statistics of traffic accidents, for example, can be treated from this point of view, if you agree to call a traffic accident a natural event.

There exists a consequence which follows from the hypothesis of branch systems and which accounts for a certain difference between the positive direction and the negative direction of time. If we observe an improbable coincidence, we conclude that the two events are the products of a common cause. For example, if two electric lamps are simultaneously extinguished, one concludes that a fuse has blown out, or that the current has been interrupted for the whole street; it would be too improbable to suppose that the two lamps burnt out at the same moment. There is also an effect common to the two events; it becomes dark in the room and one stops reading one's newspaper. But this common effect cannot serve as an explanation of the improbable coincidence; it is only the common cause which can offer an explanation; this is what I call *the principle of the common cause*.

This idea can be expressed in another way: the total cause of a partial effect can be inferred, but the total effect of a partial cause cannot be inferred. This is why it is easy to register the past, whereas it is very difficult to predict the future. For example, one can deduce yesterday's barometric pressure from the mark registered on the paper of a barograph, but if one wishes to predict tomorrow's pressure, one must know some meteorological data concerning a much more extended region. A definition of the direction of time can be based on these ideas.[8]

What is at stake is the familiar enough difference between the positive direction and the negative direction of time, a difference intimately tied to the distinction between causality and finality, between explanation with the help of causes and explanation with the help of goals. It is not easy to incorporate it into the general statistical principle formulated in the second law of thermodynamics. Nevertheless, by employing the hypothesis of branch systems, it can be shown that the principle of the common cause, which governs a major part of our inferences and reveals our belief in a direction of time, follows from the second law and represents an application of the principle of isotropy to macroscopic statistics. The proof makes use of the probability lattice already discussed. It then follows that the complete abandonment of the principle of final causality, an abandonment which is characteristic of modern science, does not represent an arbitrary decision to accept a certain way of thinking, but is demanded by the second law of thermodynamics, which requires us to look always for a *causal* explanation.

Finally, these ideas can be related to a new branch of applied mathematics, to information theory, which has been developed in the works of C. Shannon, W. Weaver, N. Wiener, and J. von Neumann. We can use the measure of information introduced in these works to define the direction of time in terms of a statistics of macroscopic events. By combining the concept of information with the principle of the common cause, one can explain the fact that the past can be registered, but not the future. One thus arrives at a theory of registering instruments and one can show that the increase of information indicates the flux of time, the same direction (as that) characterized by thermodynamics by means of the increase of entropy.

There is a mutual relation between information and entropy: information signifies negative entropy, and entropy can be interpreted as a measure of ignorance rather than of information. A high entropy value indicates an extensive class of possible molecular arrangements; and the inverse relation between information and entropy expresses but a well-known law of traditional logic, the law of the inverse relation between comprehension and extension.

This consideration requires that we add a remark concerning registering instruments. In these instruments, the increase of information indicates the positive direction of time; thus this direction is indicated here by a diminution of entropy. This is possible because registering instruments are not isolated systems. They are pieces of apparatus which accumulate order taken from their surroundings; in fact, their order reflects the order of the space ensemble of branch systems and informs us about acts of intervention in the past, these being causal processes registered on the dial of the instrument.

Why am I presenting you with these ideas in the course of an analysis of the logical foundations of quantum theory? Surely, quantum phenomena are not macroscopic. But it is not the quantum phenomena themselves which constitute the aspect of the physical world of our experience; it is the consequences of these phenomena for macroscopic objects which create the system of relations governing the world of our experience. And these consequences, as regards the structure of time, proceed through the intermediary of the statistics of measurements and thus of the statistics of registered information.

4. THE DIRECTION OF TIME IN QUANTUM PHYSICS

In classical physics, the problem of the direction of time presents itself in the form of a paradox: while the course of molecular events is reversible, that of macroscopic events is not reversible. The solution of the paradox is given by the statistical conception of macroscopic matter, a conception which offers us the possibility of constructing a specific distinction between the positive direction and the negative direction of time.

People have often hoped to find a different solution for quantum physics. People have thought of the possibility that elementary phenomena might be of an irreversible nature; the abandonment of determinism, it was believed, could be associated with a change of this kind because an indeterministic mechanics would be exempt from strict laws which link the future to the past in the manner of a unique correspondence. I would like to examine this problem. But before entering into the details, let us say immediately that the result will be negative, that the elementary quantum phenomena have been shown to be just as reversible as the molecular motions of classical mechanics. Moreover, certain recent developments show that the problem of time in small dimensions is much more complicated than that of the molecules of classical mechanics. These latter at least left us with the concept of the order of time; but it seems that the particles of quantum mechanics do not even admit of an order of time. For this reason, the statistical theory of time is indispensable to quantum physics as well as to classical physics. On the other hand we shall see that the combination of quantum phenomena with statistical theory brings us to some surprising consequences which present the concept of time in a different light.

Classical mechanics does not furnish a direction of time because its differential equations are of the second order. If we replace t by $-t$ in a solution of these equations, we arrive at another solution. But, things are different with Schrödinger's equation because this equation is of the first order with respect to time.

Let us write Schrödinger's equation in the form:

$$H_{op}\psi(q, t) = c\frac{\partial\psi(q, t)}{\partial t} , \ c = \frac{ih}{2\pi} \tag{16}$$

In fact, if $\psi(q, t)$ is a solution of this equation, then $\psi(q, -t)$ is not a solution, because this function satisfies the equation

$$H_{op}\psi(q, -t) = -c\frac{\partial\psi(q, -t)}{\partial t} \tag{17}$$

By contrast, it is the complex conjugate $\psi^*(q, -t)$ which satisfies (16), while the complex conjugate $\psi^*(q, t)$ satisfies (17). This is easily seen if one puts

$$\psi(q, t) = \phi(q)\, e^{2\pi i\nu t} \tag{18}$$

where $\phi(q)$ is a complex function which does not depend on t. We thus have

$$\psi^*(q, -t) = \phi^*(q)\, e^{\pi i\nu t} \tag{19}$$

and this function satisfies (16). But the functions $\phi(q)$ and $\phi^*(q)$ differ from each other; one can thus distinguish between them physically for example by calculating the distributions of a quantity which does not commute with q. Can we conclude from this that Schrödinger's equation defines a direction of time?

This conclusion would not be valid. While observations can distinguish between ϕ and ϕ^*, they cannot distinguish between $\psi(q, t)$ and $\psi(q, -t)$, that is to say between (18) and the function

$$\psi(q, -t) = \phi(q) e^{-2\pi i \nu t} \qquad (20)$$

which possesses the same factor $\phi(q)$ as does $\psi(q, t)$. As (20) satisfies equation (17), a distinction between $\psi(q, t)$ and $\psi(q, -t)$ would be possible only if we could discriminate between the forms (16) and (17). But such a possibility exists only on the condition that we knew the direction of time. The sign of the second member of Schrödinger's equation indicates the direction of time, it is true; but this sign presupposes the direction of time, and if we reverse this direction, we pass from the form (16) to the form (17) without encountering any contradiction with experience.

To summarize, any determination of the direction of time with the help of Schrödinger's equation would constitute circular reasoning: it furnishes nothing other than the direction that has already been introduced by other means.

To put it another way, in quantum physics we depend on the methods of classical physics if we wish to ascertain the direction of time. By applying these familiar methods, we arrive at a sign of the second member of Schrödinger's equation and there is no other method.

The course of elementary quantum events is reversible; the equations of quantum theory do not offer us the possibility of defining the positive direction of time. Here is the first result of our analysis. A second result can be deduced, which seems even more destructive for the concept of time.

The identity of a physical object in the course of time is a relation which must be clearly distinguished from that of logical identity. Everything is identical with itself; this tautological principle of logic cannot give us any information about a physical object, because a physical object is composed of a sequence of successive states. Each state is logically identical with itself, but if we consider the entire sequence of states as constituting a physical object, we are employing a concept of *physical identity* which can be called *genidentity*, making use of a term introduced by K. Lewin.

As regards macroscopic objects, the application of this concept offers no difficulty. One easily distinguishes one person from another and even one egg from another, because the eggs can be marked. In the first section, I have pointed out the difficulties which arise for molecules, and I have discussed indirect methods for defining the genidentity of molecules. The results of this discussion can be summarized as follows: in classical physics, there is a natural genidentity of molecules, although it is not possible to determine, with the help of observations, the individuality of each molecule among the others. In the Bose or Fermi statistics, one can speak of the genidentity of molecules; but this genidentity is arbitrary and can be introduced in different ways. This means that, if two molecules collide, one can identify them after the collision as one wishes; it is not possible to distinguish experimentally a physical continuity and a kind of 'crossing over'. Genidentity has thus become a matter of arbitrary definition. Nevertheless, there remains a genidentity in the following sense: the number of particles is constant. In this sense there remains a residue of individualism: although one cannot distinguish the individual molecules one from the other, one can at least count them.

A deviation from this principle has been observed for photons. But photons are not ordinary matter; they have no rest mass. The deviation would be more serious if one could no longer count electrons. This conclusion, which represents the end of individualism, has been drawn by E. Stückelberg and R. Feynman[9] in recent works. According to these authors, the world line of an electron can curve in such a way as to return toward the past; in certain periods of its existence, the electron would thus be displaced backward in time. This part of its motion, however, admits of a second interpretation: one can speak of a positron moving forward in time. Here then are two equivalent descriptions of the same phenomenon; one is just as true as the other, and there is no experiment that could discriminate between the two interpretations.

In the discussion of the problem of unobserved objects, we have spoken of the theory of equivalent descriptions, and we have shown that each exhaustive description of the interphenomena, in quantum physics, entails certain causal anomalies. The results of Stückelberg and Feynman offer a new illustration of this principle of anomaly. The path of an electron, in fact, belongs to the interphenomena; the same is the case for the question of genidentity. Observable phenomena give us no information concerning genidentity; if we speak of an individual (particle) which remains identical with itself during a certain time-period, we extend our language in such a way as to include statements concerning unobserved properties. According to Stückelberg and Feynman, this extension of language can be constructed in two different ways with respect to the direction of time: either we have one individual particle moving, in part, in negative time, or we have two individual particles moving in positive time. The first interpretation includes the causal anomaly of a particle going contrary to the flux of time; but the second interpretation has other anomalies of which we must now speak.

Feynman proposed his interpretation, in particular, with a view toward interpreting pair-production. In the presence of a gamma ray, an electron-positron pair is sometimes produced from nothing, the positron soon meets another electron, and both of them coalesce in such a way as to leave only another gamma ray. We thus have three individual particles, two electrons and a positron, and their history is quite strange from the point of view of causality, in that it includes the production of a pair from nothing, and then the annihilation of a pair. This anomaly is eliminated, according to Feynman, if one replaces the three individual particles by a single particle, by an electron which in one phase of its existence moves backwards in time. Obviously, one anomaly can be replaced by another, and one can choose, according to one's taste.

This duality of descriptions is extremely interesting from the point of view of the logical analysis of time. It indicates the fact that even the order of time is not an invariant property in the class of equivalent descriptions. In the first interpretation, the electron coming from event A and the positron coming from event B constitute two causal lines directed toward the event C, where the two particles coalesce; in the second interpretation, these lines constitute only parts of a single line ACB. For this second interpretation, C is thus 'causally between' A and B; for the first, it is not. In other words, the order of time is ACB, or if you wish, BCA for one of the interpretations; for the other, the order is either AC and BC, or CA and CB. It is thus a question, not only of change in direction, but also of a change in the order of time.

Classical mechanics was not able to give a direction of time, but it did give us an order. We would only have to assign a direction to a single causal line and a direction would be assigned to every line. Things are different with quantum mechanics; the direction of one causal line can be reversed without reversing the others. Thus one cannot construct a coherent order which permits the definition of a direction of time. One would arrive at contradictory results: in following the path ACB, one would find that A precedes B; in following another path, one would find that B precedes A.

It seems that in the quantum domain, the concept of time loses its direct significance. The variable t has always played a doubtful role in quantum mechanics. It acquires meaning only as one passes to macroscopic observations. If this is true, then not only the direction of time, but even its order will be a product of statistics. Time would emerge from the chaos of elementary events just as thread comes out of a skein with the help of a spinning wheel. One must await the development of the theory of the atomic nucleus before one can speak definitively of that which we can anticipate today only as a possibility.

As regards Feynman's ideas, another remark presents itself. The life of a positron is quite short; this is why electrons traveling backward in time do not play a great role in the statistics of electrons. If things were not this way, it would be doubtful that an ordered time would result from atomic chaos. Perhaps there would then be closed causal lines in the macrocosm. One could conclude that the existence of a linear time is related to the difference between negative and positive electricity, to the fact that the electricity called negative prevails over that called positive, as far as the number of free particles is concerned. According to the conceptions of Dirac, this superiority of negative electricity derives from the fact that all the negative energy states are occupied, while those of positive energy are generally free; the positron exists only in the form of a hole left open when an electron leaves its place in the sea of invisible matter. These are pictures; but they express ideas which possess equivalents in mathematical equations. A solution of these problems, which greatly occupies the thought of physicists, would be capable of clarifying the problem of time.

However, it is the relations of the time of the macrocosm which determine the appearance of the time of our immediate experience. In this respect, quantum theory presents us with a quite interesting conclusion, which resolves certain difficulties which arise in the analysis of the time of classical physics.

The measurement of a quantum entity is a process which projects quantum uncertainty into the macrocosm. This results from the fact that a measurement represents an act of detachment; the elementary phenomenon, the arrival of a particle, directs the macroscopic phenomena into a unique form, and these phenomena, dial readings, would be wholly different if the elementary phenomenon were different. This is why the measurement process can give us information about the elementary phenomenon. And this is why it is impossible for us to predict the result of the measurement if it was preceded by the measurement of a quantity which does not commute with it; that is excluded by Heisenberg's relation.

Hence there exist certain macroscopic phenomena which cannot be predicted, but which can be registered. Let us suppose that consecutive alternating measurements of two non-commutative quantities are made. We will have a sequence of macroscopic events which cannot be predicted, but which can be registered. This sequence offers us a very neat distinction between the past and the future: the past is determined, but the future is not.

This thesis needs to be explained. The quantum uncertainty, in so far as it concerns only the elementary phenomena, has a dual aspect: it turns toward the past in the same way as it turns toward the future. If we have nothing but measurements made at a certain moment, we can calculate neither the past values nor the future values of these quantities. If the situation is not the same in our example, it is because we have registered the results of past measurements. Hence we owe our knowledge of the past, not to quantum methods, but to macroscopic methods, to the methods involved in using registering apparatus.

But notice the surprising conclusion coming from this consideration. The analysis of classical physics has shown us that one can register the past, but not the future. The combination of this result with the uncertainty of Heisenberg brings us to the consequence that one can know the past, but one can not predict the future.

In order to understand the meaning of this result, let us return for a moment to the determinism of Laplace. For an intelligence superior to that of man, says Laplace, "nothing would be uncertain, and the future like the past would be present to his eyes." It is not everyday experiences which gives birth to this thesis; on the contrary, it is the opinion of the man in the street that only the past is determined, while the future is undetermined. Laplace drew his thesis from classical physics, and if science has spoken, common sense has to keep quiet. Nevertheless, modern science has switched sides; it has put itself on the side of common sense and furnishes us in a precise way the difference between the past and the future that the physics of Laplace could not recognize.

It is true that the physics of Boltzmann, if we add to it the hypothesis of branch systems, yields a certain structural difference between past and future, a difference which is expressed in the inferences directed, either toward past facts, or toward facts of the future. We have discussed the fact that the partial effect admits of an inference to the total cause, while the partial cause, in general, does not admit of an inference to the total effect. But this difference, although it permits us to distinguish between past and future, was not associated with a difference of determination: although one cannot register the future, one can predict it, by basing the prediction on the totality of causes. Hence the future can not be called undetermined; or, at least, if one is to consider it as undetermined because a prediction is never absolutely certain, one will have to apply the same conception to the past, which also can not be deduced with absolute certainty. Insofar as the question of certainty is concerned, there exists a symmetry between past and future, as long as one remains in classical physics.

It is no longer the same in quantum physics. It is true that there exist past facts which can no longer be known because they have not been registered;

and there are, future facts which can be predicted quite well, such as the movements of the planets, which are exempt from quantum uncertainty. But notice the difference: there exist future facts which are impossible to predict, while there are not any facts of the past which are impossible to know. In principle, they can always be registered. And I would like to suppose that the number of future events which depend, by way of detachment relations, on unforeseeable quantum phenomena is greater than is commonly believed. I would not be surprised if it were possible to show that a great number of human actions are of this kind.

The distinction between the indeterminism of the future and the determination of the past has found, in the end, an expression in the laws of physics. Here is the important result which comes out of the union of classical statistics with the uncertainty relation of quantum physics. The consequences for the time of our experience, for the time of every day, are evident. The concept of 'becoming' acquires a meaning in physics: the present which separates the future from the past is the moment in which that which was undetermined becomes determined, and 'becoming' means the same thing as 'becoming determined'.

There remains one question for us to discuss. What is the relation between the time of physics and the time of our experience? Why is the flux of psychological time identical with the direction of increasing entropy?

The answer is simple: man is a part of nature and his memory is a registering instrument subject to the laws of information theory. The increase of information defines the direction of subjective time. The experiences of yesterday are registered in our memory, those of tomorrow are not, and they cannot be registered before 'tomorrow' has become 'today'. The time of our experience is the time which is manifested by a registering instrument. It is not the privilege of man to define a flux of time; each registering instrument does the same thing. What we call the direction of time, the direction of becoming, is a relation between a registering instrument and its surroundings: and the statistical isotropy of the universe guarantees that this relation is the same for all instruments of this kind, including the human memory.

Let us add a word concerning the term 'now'. Symbolic logic teaches us that it is necessary to distinguish between the individual sign and the class of signs, called 'symbol'. With regard to many words, the individual sign can be neglected; the word 'house', for example, has the same meaning in all its instances. It is different for some words, like 'now', 'here', 'me', whose meaning changes with the individual signs. Allow me to speak, in this case, of *reflexive signs*, and of a *token-reflexive sign*, in using the word 'sign' in the sense of 'individual sign'.[10]

An act of thought is an event and thus defines a position in time. If our experiences always occur in the framework of a 'now', this means that each act of thought defines a reference point. We cannot avoid the 'now' because the attempt to avoid it would involve an act of thought and thus would define a 'now'. A thought without a reference point does not exist, because thought itself defines it. Grammar expresses this fact by the rule that each proposition must contain a verb, that is to say a reflexive sign indicating the time of the event of which one speaks; for the tense of a verb has a reflexive meaning.

Token-reflexiveness applies also to the concepts 'determined' and 'undetermined'. The word 'determination' denotes a relation between two states A and B; the state A determines, or does not determine, the state B. It is meaningless to say that the state B, considered separately, is determined. If we say that the past is determined, or that the future is undetermined, it is implied that this is relative to the present situation; it is relative to the 'now' that the past is determined and that the future is not. These words, and many others, thus have a token-reflexiveness. It is true nevertheless that these words express an objective relation; for it is a physical fact that, if A is the state defined by the act of speaking, then a state preceding A is determined with respect to A, while a state which follows A is not.

There would be much to add, but it is time to stop; and since the subject of this article is limited to physics, I allow myself to forego going further into the discussion of the problems of subjective time. It was my intention to show that quantum physics has nothing to fear from logic; that a logical analysis of the this portion of physics can be given without sacrificing either precision or rigor; and that an excursion into the domain of speculative philosophy is not needed in order to understand quantum theory. By contrast, I believe that science requires us to construct a scientific philosophy, and that such a philosophy can exist only by an intimate cooperation with physics.

NOTES

[1] P. S. Laplace, *Essai philosophique sur les probabilités* (Gauthier-Villars, Paris, 1921) p. 3.

[2] See my book, *The Theory of Probability* [1949f], p. 156.

[3] I use the word 'reasonable' to suggest specifically 'conforming to common sense'. A meaningful proposition is thus a proposition which has a meaning, yet it might not be at all reasonable.

⁴ David Bohm, 'A Suggested Interpretation of Quantum Theory in Terms of "Hidden" Variables' *Phys. Rev.* 85, 166 and 180 (1952). See also L. de Broglie, 'Remarques sur la théorie de l'onde pilote', *C.R. Acad.* 233, 641 (1951).

⁵ The bibliography on multi-valued logics has been cited in my book *Philosophic Foundations of Quantum Mechanics* [1944b], pp. 147–148. See this book for a detailed exposition.

⁶ See *The Theory of Probability*, *op. cit*; and my article 'Les fondements logiques du calcul des probabilités', [1937b].

⁷ *Ibid.*, Equation (60).

⁸ I thus take up again several ideas developed in a previous publication, 'Die Kausal-struktur der Welt und der Unterschied von Vergangenheit und Zukunft', [1925d; this volume, Chap. 57 – Ed.]. The inferences of which I have spoken are formulated there with the help of forkings of world-lines. However, I have lately changed my opinions about the mathematical relations governing these forkings.

⁹ E. C. G. Stückelberg, *Helv. Phys. Acta* 14, 588 (1941); and 15, 23 (1942); R. P. Feyn-man, *Phys. Rev.* 76, 149 (1949).

¹⁰ I suggest the word 'token-reflexive' in my book *Elements of Symbolic Logic* [1947c], p. 284.

Logical and Philosophical Foundations of the Special Theory of Relativity*

Adolf Grünbaum [+]

Department of Philosophy, Lehigh University, Bethlehem, Pennsylvania

(Received January 21, 1955)

It is argued that the correct understanding and teaching of the special theory of relativity are still impaired by a number of specific, widespread misconceptions. To remove these, the relevant definitional, experimental, and philosophical foundations of the kinematics of the theory are made explicit by reference to the following issues: (i) the relativity of simultaneity, which arises, in the first instance, *within a single* Galilean frame, (ii) the principle of the constancy of the speed of light, which rests on three *distinct* experimental facts as well as on an important convention, made possible by a fourth experimental result, (iii) the compatibility of the invariance of the speed of light with the *absence* of a retardation on the part of moving clocks, and the role of that optical principle in the deduction of the Lorentz transformations, (iv) the merits of the *ad hoc* charge against the Lorentz-Fitzgerald contraction hypothesis; the philosophical sources of the difference between Lorentz' and Einstein's interpretations of the transformation equations.

1. INTRODUCTION

MANY uninitiated students obtain the very erroneous impression that the Michelson-Morley experiment furnishes a sufficient experimental basis for the enunciation of the principle that the speed of light is the same constant, independent of direction, position, or time, in all inertial systems, a principle which we shall hereafter call "the light principle" for brevity. This belief is then often coupled with the further mistaken supposition that, upon a suitable choice of zeros of time, the light principle alone permits the deduction of the Lorentz transformations. And since these equations, in turn, contain the well-known physical and philosophical innovations of relativistic kinematics, the contraction of a moving rod, the clock retardation, and the relativity of simultaneity, as between different reference frames, are then held to derive their essential experimental warrant from the Michelson-Morley experiment of 1886, thereby presumably rendering other, subsequent experimental confirmation of them redundant.

Quite understandably, therefore, these misconceptions give rise to puzzlement when it is learned that it was not until the Ives-Stilwell experiment of 1938 on the transverse Doppler effect that the clock retardation received its first *empirical* sanction and that only in 1932 did the Kennedy-Thorndike experiment confirm the equality, as between different inertial systems, of the to-and-fro (round-trip) times required by light to traverse a closed path of given length. In some cases, the student's intellectual conscience had already been uneasy concerning the meaning of the light principle. For he had unfortunately inferred that merely because two different Galilean observers give *the same formal description* of the behavior of a single identical light pulse, emitted upon the coincidence of their origins, the respective expanding spherical forms of this disturbance which they will "observe" in their respective systems are *constituted* by the *same* point-events or the *same* configuration of photons. And matters could then only have become more troubled for him when he heard the philosophical remark that since only *judgments* but not *sense-data* can contradict one another, each Galilean observer can and simply does "*see*" the same light pulse spreading spherically.

These particular misunderstandings are merely those most easily avoidable by the presentation of the logical structure of the kinematics of special relativity to be given in this paper. Among the more subtle misconceptions which I hope to dispel is the view that the relativity of simultaneity *first* arises as between different inertial systems. The analysis below is designed to distinguish the multiple philosophical, definitional, and experimental ingredients of the theory, thereby also giving the reader a ready

* Written during the tenure of a Faculty Fellowship from the Fund for the Advancement of Education.

grasp of the particular respects in which it would require modification in response to the discovery of certain specific kinds of disconfirming evidence, while other separable portions would be left intact. My treatment of several of the issues is greatly indebted to two outstanding works on the philosophy of relativity by Hans Reichenbach, which are not available in English.[1]

2. THE RELATIVITY OF SIMULTANEITY

During the reign of the absolutistic Newtonian theory, space and time were held to have a structure independent of the existence of matter or the occurrence of physical events. On the basis of this conception of space and time as indifferent, merely accidental *"containers"* of physical things and events, it could plausibly be maintained that one can speak significantly of *two* events or things being distinct while yet having precisely the same qualitative constitution. For it was the occupancy of different points and/or instants in already structured container space or time which was presumed to confer a difference in identity upon such things and events. And thus philosophers were able to regard receptacle space and time as the basis for *individuation*.

But with the advent of Riemann, Poincaré, and Mach, it became clear that empty space and time are *amorphous* and that if they are to have a topology and a metric, such structure will derive its existence and character from relations between *bodies* and *physical events*. It is *not* an independently structured receptacle space or time, therefore, which endows events with their particularity. Instead, the existence of a structure of space or time presupposes that, to begin with, there exist in the universe recognizable qualitative differences which confer individuality upon physical events *independently of any system of coordinates* and prior to the construction of any such system. Thus, the theory of relativity assumes that the identity of events and things is *sui generis* and that these objects first define the points and instants which we call their loci. As against the absolutistic container-theory, this conception is *"relational"* in the sense of

regarding space and time as systems of relations between physical events and things.

Accordingly, the temporal order among non-coincident physical events must derive from *their* properties and relations. Hence we seek the attributes, *if any*, in virtue of which such events sustain *unambiguously* relations of earlier, later, and simultaneous. Let us find the basis of the temporal order within a class of events sufficiently comprehensive to include events in different galaxies. It would clearly not be feasible even to *attempt* to establish an unambiguous temporal order in this class of events by means of the readings, at the respective loci of these events, furnished by clocks previously synchronized via the transport of a master clock. And even if the required transport were feasible, the attempt would fail to yield an unambiguous temporal order, since the resulting readings would depend on the path and mode of transport of the master clock.[2] Consequently, a different criterion of time order is needed.

Within our class of events, consider those pairs which have the property of being the termini of actual *or* physically *possible* causal chains. If in such pairs one and only one of the two events does or could constitute the emission of a pulse of radiation (or of a material particle) while the other is or could be the corresponding absorption or reception, then we *define* the emission-event as sustaining the relation "earlier than" to the absorption event.[3] This relation and also its converse "later than" are *asymmetric*, so that *neither* relation would be sustained by a

---

[1] H. Reichenbach, *Axiomatik der relativistischen Raum-Zeit-Lehre*, (Vieweg, 1924), Vol. 72 of *Die Wissenschaft* [hereafter this work will be cited as "AR"]; *Philosophie der Raum-Zeit-Lehre* (de Gruyter, 1928).

[2] See AR, Secs. 22–23.
[3] It might be objected at once that this definition is vitiated by circularity on the grounds that the sole basis for characterizing *one* of the two termini of a causal chain as *the* emission-terminus is that it is the *earlier* one of the two. I have shown elsewhere [See "Carnap's Views on the Foundations of Geometry" in P. A. Schilpp (ed.), *The Philosophy of Rudolf Carnap*, Library of Living Philosophers, Vol. X (Tudor, to be published)] that this circularity can be avoided by a detailed logical construction in which recourse is had to considerations of statistical thermodynamics. See also, A. Grünbaum, "Time and Entropy," *American Scientist* (October 1955). It turns out that time's arrow is defined *not* by the entropic behavior of the universe as a whole but rather by the direction of entropy increase of the majority of (quasi-) closed systems in the space ensembles of such systems. Of two causally connected events, the *earlier* one is therefore characterized by being *simultaneous* with the *lower* of two sets of entropy states of these closed systems. The concepts of simultaneity and closed system employed in this definition of "earlier" are defined, in turn, *without* presupposing time's arrow.

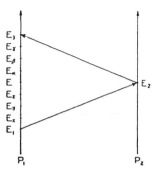

FIG. 1. World-line diagram.

pair of events in which *each* member can be the emission-event for a causal influence reaching the other, as in the case of Newtonian instantaneous action-at-a-distance.

If the physical basis for relations of earlier and later among events lies in the possibility of their being the appropriate termini of influence chains, we ask: are the physically possible causal chains of nature such as to define a *unique* set of temporal relations in which every pair of events has an unambiguous place? Consider the various relations of temporal order between a single event E_2, at a given point P_2, on the one hand, and each of a *set* of different events at a different point P_1, on the other. Since the motion of the indicator of a dial clock is a causal chain, the presence of such a clock at P_1 can define the *local* temporal order in the class of events at that point. Let $E_1, E_x, E_y, E_z, E, E_\alpha, E_\beta, E_\gamma, E_3$ be ordered members of this class,—the direction from left to right being the positive time direction—, such that t_1 is the time of E_1, t_3 the time of E_3 and $\frac{1}{2}(t_2+t_1)$ the time of E. And suppose further that among these events at P_1, the event E_1 coincides with the emission of a light ray whose arrival at P_2 coincides with E_2 and whose instantaneous reflection at P_2 issues in its return to P_1 so as to coincide with event E_3 there (see the *world-line* diagram, Fig. 1). The question before us concerns the time relations of the various events at P_1 to E_2 at P_2 which we can now ascertain in virtue of our definition of "earlier." Letting "$<$" be short for "earlier than," we know that

$$E_1 < E_2 < E_3,$$

and that

$$E_1 < E < E_3,$$

the latter relation also holding, of course, if "E" is replaced by any one of the symbols "E_x," "E_y," "E_z," "E_α," "E_β," "E_γ." But these combined relations do *not* furnish any information concerning the time relations *of* E_2 to any one of the events at P_1 lying within the *open* interval between E_1 and E_3. The existence of the latter relations will therefore depend upon the physical possibility that there be direct causal chains which could *link* E_2 to the events in the open interval at P_1.

Newtonian physics asserts that there *can be* such chains. To be sure, it would admit that a light signal emitted at P_1 will always require a *positive* time interval $t_3 - t_1$ to return to P_1 and that therefore no *electromagnetic* chain will furnish the required link. But it would go on to invoke its second law of motion, which asserts that, no matter what the velocity, the ratio of the force to the acceleration is constant and that particles can therefore be brought to arbitrarily high speeds in excess of the speed c of light by appropriately large forces acting for a sufficient time. Thus, in the Newtonian world, there can be influence chains whose *emission* would coincide with E_x, E_y, or E_z at P_1 and whose *arrival* at P_2 would coincide with E_2 but there are none for which the converse holds. E_2 can then significantly be said to be later than each of the former events. And although no signal emanating from E_α, E_β, or E_γ could reach P_2 soon enough to coincide there with E_2, suitable influence chains emanating from the latter event can, however, reach the former events and thus exhibit them to be later than E_2. There will be only *one* event E in the time interval $t_3 - t_1$ at P_1 which is such that it can be *both* the emission-terminus *and* the reception-terminus of a certain kind of influence chain of which E_2 can be a similarly dual type of terminus. It follows from our definition of temporal order that E is *neither earlier nor later* than E_2, and E thus uniquely divides the events at P_1 into those which are earlier than E_2 and those which are later. E is therefore the only event at P_1 which can justifiably be called "simultaneous" with E_2. This concept of simultaneity finds mathematical expression in Newton's law of universal gravitation and in his third law of motion, which affirm the existence of instantaneous action-at-a-distance *inter*-actions, i.e., of two-way causal chains whose emission coincides with their return and which are therefore said to travel at infinite speed (barring Wheeler-Feynman interactions and generally assuming that time is serial rather than cyclic).

We see that in a world of arbitrarily fast causal chains, the concept of absolute simultaneity would have a perfectly clear physical meaning in the temporal description of nature given by a *relational* theory of time. Contrariwise, as we shall now show in detail, Einstein's *denial* of the existence of indefinitely rapid causal chains via his affirmation of the limiting role of light propagation in the class of influence chains deprives the concept of absolute simultaneity of its physical meaning even *within* a *single* inertial system. And it will be seen that the relativity of simultaneity arising from Einstein's claim does *not* depend for its validity upon there being *disagreement* among different Galilean observers as to the simultaneity of pairs of noncoinciding events.

Although there were experiments by W. Kaufmann and others during the years 1902 to 1906 on the deflection of electrons (β rays) in electric and magnetic fields which yielded a mass-variation with velocity incompatible with Newtonian dynamics, these experiments were unable to decide between the nonrelativistic formulas of Abraham's dynamics, which allow particle velocities exceeding the velocity of light,[4] and the dynamical equations issuing from Einstein's fundamental paper of 1905, which rule out causal chains faster than light. Experiments which clearly support Einstein on this point were not successfully carried out until a good many years after 1905.[5] It was reflection, initiated by a *thought*-experiment he had carried out at the age of 16, which led Einstein to postulate the limiting character of light propagation. He writes[6]: "If I pursue a beam of light

with the velocity c (velocity of light in a vacuum), I should observe such a beam of light as a spatially oscillatory electromagnetic field at rest. However, there seems to be no such thing, whether on the basis of experience or according to Maxwell's equations." Thus the very consideration which suggested that there are no preferred inertial systems for electromagnetic phenomena also indicated that a material object cannot attain the speed of light![7]

We must now define simultaneity for a universe in which the assumption of arbitrarily fast causal chains fails to hold. But since the metrical concept of velocity presupposes that we know the meaning of a transit time and since such a time, in turn, depends upon a prior criterion of clock synchronization or *simultaneity*, we must first formulate the limiting property of electromagnetic chains *without* using the concept of simultaneity of noncoincident events. We can then state the consequences of this property for the definition of simultaneity. Accordingly, we give the following *nonmetrical* formulation of Einstein's limiting postulate: no kind of causal chain (moving particles, radiation), emitted at

---

[4] See M. Abraham, Ann. Physik 10, 105 (1903).
[5] See Møller, *The Theory of Relativity* (Oxford University Press, New York, 1952), pp. 85–89, and von Laue, *Die Relativitätstheorie* (Friedrich Vieweg & Sohn, Braunschweig, Germany, Vieweg, 1952), Vol. I, p. 27.
[6] Einstein, "Autobiographical Notes" in P. A. Schilpp (ed.) *Albert Einstein, Philosopher-Scientist* (Library of Living Philosophers, 1949), p. 53.

[7] See Laue, reference 5, p. 34n. It is of both historical and logical importance to note that in an address "The Principles of Mathematical Physics" delivered in St. Louis on September 24, 1904 and published in *The Monist 15* (1905), pp. 1–24, H. Poincaré had already envisioned the construction of a new mechanics in which the velocity of light would play a limiting role. Although he concluded this paper with the remark that "as yet nothing proves that the [old] principles will not come forth from the combat victorious and intact" (p. 24), Poincaré did prophesy that "From all these results, if they are confirmed, would arise an entirely new mechanics, which would be, above all, characterized by this fact, that no velocity could surpass that of light, any more than any temperature could fall below the zero absolute, because bodies would oppose an increasing inertia to the causes, which would tend to accelerate their motion; and this inertia would become infinite when one approached the velocity of light" (p. 16). But, unlike Einstein, Poincaré *failed* to see the *import* of this new limiting postulate for the meaning of simultaneity: in the very same paper (p. 11), he speaks of spatially separated clocks as marking "the same hour at the *same physical instant*" (my italics). And he distinguishes between watches which mark "the true time" and those that mark only the "local time," a distinction which was also invoked, as we shall see, by Larmor and Lorentz. But Poincaré does note that a Galilean observer will not be able to detect whether his frame is one in which the clocks mark the allegedly spurious local time. Accordingly, he affirms "The principle of relativity, according to which the laws of physical phenomena should be the same, whether for an observer fixed, or for an observer carried along in a uniform movement of translation; so that we have not and could not have any means of discerning whether or not we are carried along in such a motion" (p. 5).

a given point P_1 together with a light pulse can reach any other point P_2 *earlier*—as judged by a *local* clock at P_2 which merely orders events there in a metrically arbitrary fashion—than this light pulse.[8]

We now apply this postulate to the special case of two points P_1 and P_2 which are *fixed in a reference frame* S and which are connected by a chain of light signals in the manner shown in our world-line diagram above. It is then clear at once that instead of E being the only event at P_1 which is neither earlier nor later than E_2 at P_2, as in the Newtonian world, on Einstein's hypothesis each one of the entire superdenumerable infinity of point-events in the *open* interval between E_1 and E_3 at P_1 fails to have a *determinate* time-relation to E_2. For none of these events at P_1 can then be said to be either earlier or later than E_2: no signal originating at any of these events can reach P_2 soon enough to coincide there with E_2, and no causal chain emitted at P_2 upon E_2's occurrence can reach P_1 prior to the occurrence of E_3. But to say that none of these events at P_1 is either earlier or later than E_2 is to say that no one of them is *objectively* any more entitled to be regarded as simultaneous with E_2 than is any of the others. It is therefore only *by definition* that some *one* of these events comes to be simultaneous with E_2. Unlike the Newtonian situation, in which there was only a *single* event E which could be significantly held to be simultaneous with E_2, the physical facts postulated by relativity require the introduction, *within* a *single* inertial frame S, of a *convention* stipulating which *particular* pair of causally *non*-connectible events will be *called* "simulta-

<hr/>

[8] It must be remembered that, in its metrical form, this postulate does not confine purely "geometric" velocities which are *not* the velocities of causal chains to the value *c*. For example, *within* any given inertial system K, special relativity allows us to subtract *vectorially* (as against by means of the Einstein velocity addition formula) a velocity $v_1 = .9c$ of body A in K and a velocity $v_2 = -.3c$ of body B in K to obtain $1.2c$ as the relative velocity of *separation* of A and B *as judged by the* K-*observer* [though *not* by an observer attached to either A or B!]. For no body or disturbance is thereby asserted to be traveling relative to any other at the velocity $1.2c$ of separation, and since the process of separation has no direction of propagation, it cannot be a causal chain. See Laue, reference 5, pp. 40–1; L. Silberstein, *The Theory of Relativity* (The Macmillan Company, New York, 1914), p. 164n; A. Sommerfeld, Notes in *The Principle of Relativity*, a collection of original memoirs, Dover reprint, p. 94; A. Grünbaum, Phil. of Sci. 22, 53 (1955); G. D. Birkhoff, *The Origin, Nature and Influence of Relativity* (The Macmillan Company, New York, 1925), p. 104.

neous." This relativity of simultaneity prevails within a single inertial system, because the simultaneity criterion of the system is relative to the *choice* of a particular numerical value between t_1 and t_3 as the temporal *name* to be *assigned* to E_2. Accordingly, depending on the particular event at P_1 that is chosen to be simultaneous with E_2, upon the occurrence of E_2 we set the clock at P_2 to read

$$t_2 = t_1 + \epsilon(t_3 - t_1), \qquad (1)$$

where ϵ has the particular value between 0 and 1 appropriate to the choice we made. If, for example, we choose ϵ such that E_y becomes simultaneous with E_2, then all of the events between E_1 and E_y become *definitionally* (i.e., not objectively) earlier than E_2 while all of the events between E_y and E_3 become *definitionally* later than E_2. Clearly, we could alternatively choose ϵ such that instead of becoming simultaneous with E_2, E_y would become earlier than E_2 or later *in the same frame* S. This freedom to decree definitionally the relations of temporal sequence merely expresses the objective indeterminateness of the time relations between causally *non*-connectible events and, of course, such freedom can be exercised only with respect to such pairs of events.

It is apparent that no fact of nature found in the objective temporal relations of physical events precludes our choosing a value of ϵ between 0 and 1 which *differs* from $\frac{1}{2}$. But it is only for the value $\frac{1}{2}$ that the velocity of light in the direction P_1P_2 becomes *equal* to the velocity in the opposite direction P_2P_1 in virtue of the then resulting equality of the respective transit times $t_2 - t_1$ and $t_3 - t_2$. A choice of $\epsilon = \frac{3}{4}$, for instance, would make the velocity in the direction P_1P_2 only $\frac{1}{3}$ as great as the return velocity. Couldn't one argue, therefore, that the choice of a value of ϵ is not a matter of convention after all, the grounds being that only the value $\epsilon = \frac{1}{2}$ accords with such physical facts as the isotropy of space and with the methodological requirement of inductive *simplicity*, which enjoins us to account for observed facts by minimum postulational commitments? And, in any case, must we not admit (at least among friends!) that the value $\frac{1}{2}$ is more "true"? The contention that either the isotropy of space or

Occam's "razor" are relevant here is profoundly in error, and its advocacy arises from a failure to understand the import of Einstein's statement that "we establish *by definition* that the 'time' required by light to travel from A to B equals the 'time' it requires to travel from B to A."[9] For, in the first place, since no statement concerning a transit-time or velocity derives its meaning from mere facts but also requires a prior *stipulation* of the criterion of clock synchronization, a choice of $\epsilon \neq \frac{1}{2}$, which renders the transit-times (velocities) of light in opposite directions *unequal*, cannot possibly conflict with such physical isotropies and symmetries as prevail *independently* of our descriptive conventions.[10] And, in the second place, the canon of simplicity which we are pledged to observe in all of the inductive sciences is *not* implemented any better by the choice $\epsilon = \frac{1}{2}$ than by any one of the other allowed fractional values. For no distinct hypothesis concerning physical facts is made by the choice of $\epsilon = \frac{1}{2}$ as against one of the other permissible values. There can therefore be no question of having propounded a hypothesis making lesser postulational commitments by that special choice. On the contrary, it is the postulated fact that light is the fastest signal which assures that *each* one of the permissible values of ϵ will be equally compatible with all possible matters of fact which are independent of how we decide to set the clock at P_2. Thus, the value $\epsilon = \frac{1}{2}$ is not simpler than the other values in the inductive sense of assuming less in order to account for our observational data but only in the *descriptive* sense of providing a *symbolically* simpler representation of these data. But this greater symbolic simplicity arising from the value $\frac{1}{2}$ expresses itself not only in the *equality* of the velocities of light in opposite directions but also shows its unique descriptive advantages by assuring that synchronism will be both a symmetric and a transitive relation upon using *different* clocks in the same system.[11]

[9] Einstein, Ann. Physik 17, Sec. 1 (1905).

[10] An example of such an independent kind of isotropy is the fact, discovered by Fizeau [see Møller, reference 5, Chap. I, Sec. 8 and Chap. II, Sec. 16] that if, in a given system, the emission of a light beam traversing a closed polygon in a given direction coincides with the emission of a beam traveling in the *opposite* direction, then the returns of these beams to their common point of emission will likewise coincide.

[11] See AR, pp. 34–35, 38–39.

These considerations now enable us to see that physical facts which are independent of descriptive conventions do *not* dictate discordant judgments by different Galilean observers concerning the simultaneity of given events.

Consider the familiar example of the two vertical bolts of lightning which strike a moving train at points A' and B', respectively, and the ground at points A and B, respectively, A and A' being to the left of B and B', respectively. If we now define simultaneity in the ground-system by the choice $\epsilon = \frac{1}{2}$, then the velocity of light traveling from A to the midpoint C of AB becomes equal to the velocity of light traveling in the *opposite* direction BC. And if the arrivals of these two oppositely directed light pulses coincide at C, then the ground observer will say that the bolt at AA' struck simultaneously with the bolt at BB'. Will the train observer's observations of the lightning flashes *compel* him to say that the bolts did *not* strike simultaneously? Decidedly not! To be sure, since the train is moving, say, to the left relatively to the ground, it will be impossible that the two horizontal pulses which meet at the midpoint C in the ground-system also meet at the midpoint C' of the train system; the flash from A' will arrive at C' *earlier* than the one from B'. But this fact will hardly require the train observer to say that the point A' was struck by lightning *earlier* than the point B', *unless* the train observer has *also* chosen ϵ to be $\frac{1}{2}$ for his own system, a choice which then commits him to say that the time required by light from the left to traverse the distance $A'C'$ is the *same* as the transit-time of light from the right for an equal distance $B'C'$.[12] But, apart from descriptive convenience, what is to prevent the train observer from choosing a value $\epsilon \neq \frac{1}{2} (0 < \epsilon < 1)$ such that he too will say that the lightning bolts struck simultaneously, just as the ground observer did? Such a choice can always be made by the train observer: the lightning-flashes at A' and B' not being themselves signal-connectible, light-pulses originating with these flashes will always meet on the moving train at some point D' different from C' but lying between A' and B'. And in

[12] If a train observer moving to the *right* relatively to the ground chooses $\epsilon = \frac{1}{2}$, he will, of course, say that point A' was struck *later* than B'.

order to define the two flashes to have occurred simultaneously, the train observer need only decide to define the ratio of the velocity of the light coming from the left to the velocity of the light coming from the right so as to be the same as the ratio of the distance $A'D'$ to the distance $D'B'$.

It follows that it is the relativity of simultaneity *within* each inertial system which allows each Galilean observer to choose his own value of ϵ *either* so as to agree with other observers on simultaneity *or* so as to disagree. If each separate Galilean frame does choose the value $\frac{1}{2}$—which Einstein assumed in the formulation of the theory for the sake of the resulting descriptive simplicity—then the relative motion of these frames indeed makes disagreement in their judgments of simultaneity unavoidable, except in regard to pairs of events lying in the planes perpendicular to the direction of their relative motion. But, in the first instance, it is the limiting character of the velocity of light and not the relative motion of inertial systems which gives rise to the relativity of simultaneity.

Moreover, this limiting property of light is evidently an objective property of the causal structure of the physical world, quite independent of man's presence in the cosmos and of his measuring activities. The resulting relativity of simultaneity is therefore expressive of a property of the casual relatedness of physical events and is *not*, in the first instance, *generated* by *our* inability to carry out those measuring *operations* that would define absolute simultaneity. Instead, the impossibility of such operations is a *consequence* of the more fundamental impossibility of the required causal relations between physical events. To be sure, operations of measurement are indispensable for *discovering* or *knowing* that particular physical events can or cannot sustain the causal relations which would define relations of temporal succession or of ordinal simultaneity between them. But the actual or physically possible causal relations in question are or are not sustained by physical events quite apart from *our* actual or hypothetical measuring operations and are *not* first *conferred* on nature by our operations. In short, it is because no relations of absolute simultaneity *exist* to be measured that measurement cannot

disclose them; it is *not* the mere failure of measurement to disclose them that *constitutes* their nonexistence, much as that failure is *evidence* for their nonexistence.

Only a philosophical obfuscation of this state of affairs can make plausible the view that the relativity of simultaneity (or, for that matter, any of the other philosophical innovations of relativity theory) lends support to the subjectivism of homocentric operationism or of phenomenalistic positivism (for further details, see the references to other papers of mine given in footnote 29).

3. THE PRINCIPLE OF THE CONSTANCY OF THE SPEED OF LIGHT ("LIGHT PRINCIPLE")

In what sense can the null-result of the Michelson-Morley experiment legitimately be held to support the claim that the speed of light is the same constant c in all inertial systems, independent of the relative velocity of the source and observer, and of direction, position, or time?[13]

The null-result of the Michelson-Morley experiment merely showed that if *within* an inertial system, light-rays are jointly emitted from a given point in *different directions* of the system and then reflected from mirrors at equal distances from that point, as measured by rigid rods, then they will return together to their common point of emission. And it must be remembered that the equality of the round-trip times of light in different directions of the system is measured here not, of course, by material clocks but by light itself (absence of an interference fringe shift).[14] The repetition of this experiment

[13] It will be recalled that, in opposition to the emisssion theory of light, the classical ether theory had already affirmed the independence of the velocity of light in the ether from that of its *source*, while affirming, however, its dependence on the observer's motion relative to the ether medium.

J. H. Rush has pointed out [Sci. American **193**, 62 (August, 1955)] that the *time* constancy of the velocity of light may have to be questioned: measurements over the past century have yielded three fairly distinct sets of values for c.

[14] Since the classically expected time difference in the second-order terms is only of the order of 10^{-15} second, allowance must be made for the absence of a corresponding accuracy in the measurement of the equality of the two arms. This is made feasible by the fact that, on the ether theory, the effect of any discrepancy in the lengths of the two arms should *vary*, on account of the earth's motion, as the apparatus is rotated. For details, see P. Bergmann, *Introduction to the Theory of Relativity* (Prentice-Hall, Inc., New York, 1946), pp. 24–26.

at different times of the year, i.e., in different inertial systems, showed only that there is no difference within *any* given inertial system in the round-trip times as between different directions within that system. But the outcome of the Michelson-Morley experiment does *not* show at all either that (a) the round-trip (or one-way) time required by light to traverse a closed (or open) path of length $2l$ (or l) has the *same numerical value* in different inertial systems, as measured by material clocks stationed in these systems,[15] or that (b) if, contrary to the arrangements in the Michelson-Morley and Kennedy-Thorndike experiments, the source of the light is *outside* the frame K in which its velocity is measured, then the velocity in K will be independent of the velocity of the source relative to K. In fact, the statement about the round-trip times made under (a) was first substantiated by the Kennedy-Thorndike experiment of 1932, as we shall see, and the assertion under (b) received its confirmation by observations on the light from double stars.[16] Yet the light principle certainly affirms both (a) and (b) and thus clearly claims *more* than is vouchsafed by the Michelson-Morley experiment. Accordingly, we can see that *in addition* to the result of the Michelson-Morley experiment, the light principle contains at least all of the following theses:

[15] It is understood that the lengths $2l$ in the different frames are *each* ratios of the path to the *same* unit rod, which is transported from system to system in order to effect these measurements and which remains equal to unity *by definition* in the course of this transport.

[16] See Tolman, *Relativity, Thermodynamics and Cosmology* (Oxford University Press, London, 1934), pp. 16-17 and Laue, reference 5, p. 25. Professor H. Feigl has kindly called my attention to a very recent paper by P. Moon and D. E. Spencer [J. Opt. Soc. Am. 43, 635 (1953)] in which this interpretation of the data furnished by double stars is contested. These authors contend that if light is postulated to travel in a Riemannian space of constant positive curvature ($R=5$ light years), then the data on binaries admit of being interpreted as according with the Ritz emission theory. On Ritz's assumption, the velocity of light in free space is always c with respect to the *source*, but, contrary to the relativistic light principle, its value in a frame K depends on the velocity of the source relative to K.

Moon and Spencer do not explain how their hypothesis would account for the findings of R. Tomaschek [Ann. Physik 73, 105 (1924)], who repeated the Michelson-Morley experiment with stellar light and, contrary to the expectations of the Ritz theory, obtained the same results as had been found by using terrestrial light. Nor do these investigators indicate on what grounds they feel entitled to supplant the relativity of simultaneity by the absolutistic theory of time integral to their hypothesis.

(i) The assertion given under (a), which undoubtedly goes beyond the null-result of the Michelson-Morley experiment. For brevity, we shall call it "the clock axiom" in order to allude to its reference to *clock-times* of travel. Although lacking experimental corroboration at the time of Einstein's enunciation of the light principle, this "clock axiom" was suggested by the fundamental assumption of special relativity that there are no preferred inertial systems.

(ii) The Einstein postulate regarding the maximal character of the velocity of light, which is the source of the relativity of simultaneity both within a given inertial system and as between different systems. This postulate *allows* but does *not* entail that we choose the same value $\epsilon=\frac{1}{2}$ for all directions within any given sytem and also for each different system.

(iii) The claim that the velocity of light in any inertial system is independent of the velocity of its source.

To see specifically how the light principle depends for its validity upon the truth of all of these constituent theses, we now consider the consequences of abandoning any one of them while preserving the remaining two. To do so, we direct our attention to the *one-way* velocity of an outgoing light pulse that traverses a distance l in system S and also to the one-way velocity of such a pulse traversing a distance l in system S'. Let T_S and $T_{S'}$ be the respective outgoing transit times of these light pulses for the distances l in S and S', the corresponding *round-trip* times in these frames being τ_S and $\tau_{S'}$. In this notation, Eq. (1) of Sec. 2 becomes

$$T_S = \epsilon \tau_S,$$

and similarly for S'. Hence, the relevant one-way velocities of light in these systems are, respectively, given by

$$v_S = (l/T_S) = (l/\epsilon\tau_S),$$

and

$$v_{S'} = (l/\epsilon\tau_{S'}).$$

Our problem is to determine, one by one, the consequences of assuming that of the three above ingredients of the light principle, only two can be invoked at-a-time in an attempt to assure that if $v_S = c$ in virtue of the observed value of τ_S and of a choice of $\epsilon = \frac{1}{2}$ for S, then $v_{S'}$ will also have the value c.

(i) In the absence of the clock axiom, it can well be that $\tau_S \neq \tau_{S'}$. In that case, the relations $v_S = (l/\frac{1}{2}\tau_S) = c$ and $v_{S'} = (l/\epsilon\tau_{S'})$ tell us that it will *not* be possible to make $v_{S'} = c$ by a choice of $\epsilon = \frac{1}{2}$ in S'. To be sure, it may still be possible then to choose an appropriately different value of ϵ for S' so as to make $v_{S'} = c$. But, much as the latter choice of ϵ might yield the value c for the outgoing *one-way* velocity of light in S', it would also inescapably entail a value *different* from c for the *return*-velocity. And such a result would not be in keeping with the light principle.

(ii) If, on the other hand, we guarantee that $\tau_S = \tau_{S'}$ by assuming the clock axiom but disallow the freedom to choose a value of ϵ for each inertial system by withdrawing the second constituent thesis above, then $v_{S'}$ could readily be different from c. For the *illegitimacy* of choosing ϵ to be $\frac{1}{2}$ in S' would then derive from physical facts incompatible with the relativity of simultaneity which *objectively fix* the value of $T_{S'}$ as different from $\frac{1}{2}\tau_{S'}$.

(iii) The need for the third ingredient is obvious in the light of what has already been said.

It is of importance to note that even when *all* of the above constituent principles of the light principle are assumed, no fact of nature independent of our descriptive conventions would be contradicted, if we chose values of ϵ other than $\frac{1}{2}$ for each inertial system, thereby making the velocity of light different from c in both senses along each direction in all inertial systems. The assertion that the invariant velocity of light is c is therefore not, in its entirety, a purely factual assertion. But we saw that it is nonetheless a consequence of a presumed physical fact that the specification of the velocity of light involves a stipulational element which, in combination with other factual principles, *allows* us to say that the velocity of light is c.

The importance of complete clarity on the logical status of the light principle is apparent from the fact that misunderstandings of it still issue in misconceived and irrelevant attacks upon it. A very recent case in point is a paper by the noted experimental physicist H. E. Ives ["Revisions of the Lorentz Transformations," Proc. Am. Phil. Soc. **95**, 125 (1951)], who rests his proposed revisions of the Lorentz transformations on the following argumentation, whose unsoundness is evident from the analysis given above: "Einstein . . . proposed two principles: the first, 'The Principle of Relativity,' is the same in name and content as Poincaré's principle of relativity of the year before [see footnote 7 of Sec. 2 above]; the second, 'The Principle of the Constancy of the Velocity of Light' . . . is that the velocity of light *is* the same on all relatively moving bodies. Adhering at the same time to the independence of the velocity of light from that of the source, Einstein thus asked the acceptance of a paradox. . . . He also decreed a *pseudo* operational procedure for obtaining [the Lorentz transformations] . . . Distances were to be measured by rods laid end to end, distant clocks were to be set by light signals *ascribed* the velocity c. This proposed procedure . . . fell short of actually meeting the [operational] requirement. The assignment of a definite value to an unknown velocity, by fiat, without recourse to measuring instruments, is not a true physical operation, it is more properly described as a ritual. . . . Einstein . . . invoked supposed experimental fact to support his principle, . . .

"This appeal to experiment to support a logical contradiction, is however invalid. The 'experience' cited [by Einstein] is the customary laboratory measurement of the velocity of light by signals *sent out and back*, while the Lorentz transformations describe signals *sent in one direction*. No precision measurement of this sort has been performed; it would require, as compared with the experimental equipment adequate for out and back measurement, additional instruments, whose behavior under manipulation would have to be established and taken account of. It is an unwarranted assumption that such a measurement would yield the value 'c'. . . . It develops that the velocity of light measured by signals sent in one direction is *not* 'the universal constant c.' The 'principle' of the constancy of the velocity of light is not merely 'un-understandable,' it is *not* supported by 'objective matters of fact'; it is untenable, and, . . . , unnecessary.

"On any randomly chosen material platform consider a light signal proceeding from one end, at which is a clock, to a reflector at the far end, and back to the clock. Because of the independence of the velocity of light from that of matter we may expect the velocity of light in the out and back directions, in terms of divisions on the platform, and the clock at the origin, to be in general different. Call the out and back velocity c_o and c_b. . . . A point of great importance may here be noted. It is that we do not need to assign definite individual values to c_o and c_b, such for instance as calling them equal as is done in Einstein's arbitrary 'definition' of simultaneity. We carry these in our calculations as real although undetermined quantities, . . . "

4. THE EXPERIMENTAL CONFIRMATION OF THE KINEMATICS OF THE SPECIAL THEORY OF RELATIVITY

Since the Michelson-Morley experiment could hardly be regarded as empirical proof for the "clock axiom" contained in the light principle, Kennedy and Thorndike devised an experiment having a direct bearing on that axiom's relativistic denial of the existence of preferred inertial systems.[17] The apparatus used was essentially similar to that of the Michelson-Morley experiment, except for making the arms of the interferometer as *different* in length as possible in order to maximize the difference between the travel-times of the partial beams. Assume that the apparatus has a velocity relative to the ether, which could not be detected under the particular conditions of the Michelson-Morley experiment in virtue of a *bona fide* Lorentz-Fitzgerald contraction. Then, if the period of the light source does not itself depend upon this velocity relative to the fixed ether, the difference between the travel times of the two partial beams should be a function of the diurnally and annually changing velocity of the apparatus (as well as of the length difference between the two arms) and thus ought to give rise to corresponding observable shifts in the fringe pattern.[18]

[17] See R. J. Kennedy and E. M. Thorndike, Phys. Rev. 42, 400 (1932).
[18] The proviso that the period of the light source may *not* depend upon the velocity of the apparatus postulates a constant relation between the time-metric defined by light itself and that defined by material clocks. This

Now, if the Kennedy-Thorndike experiment had yielded a positive effect instead of the null-result which it did actually yield, then it could have been cogently argued that the Michelson-Morley experiment was evidence for a *bona fide* Lorentz-Fitzgerald contraction, just as a fringe shift produced by heating one of the arms of the interferometer could be held to be evidence for the elongation of that arm. But, in view of the *de facto* null outcome of the Kennedy-Thorndike experiment, there is very good reason indeed to attribute the absence of a diurnal or annual variation in the time-difference between the two partial beams to a constancy, as between different inertial systems, in the time required by each of the partial beams to traverse its own closed path. And thus we are entitled to say that the Kennedy-Thorndike experiment has provided empirical sanction for the clock axiom.

We recall that in 1905, there was no unambiguous experimental evidence supporting Einstein's postulate that light is the fastest signal. But subsequent experiments showed that the mass and kinetic energy of accelerated particles become indefinitely large as their velocity approaches that of light.[19]

relation is presupposed in using the null outcome of the Kennedy-Thorndike *optical* experiment as corroboration of the "clock axiom," as is evident from the fact that this axiom involves a claim concerning the behavior of material clocks.
[19] See W. Gerlach, *Handbuch der Physik* (Springer-Verlag, Berlin, Germany, 1926), pp. 61 ff. and Møller, reference 5, Ch. III, Sec. 32. This interpretation has been challenged by V. Bush in a paper "The Force Between Moving Charges," J. Math. and Phys. 5, 129 (1925-1926) to which Professor Parry Moon has kindly called my attention. Bush offers a *non*-relativistic account of the experimental results obtained by Kaufmann, Bucherer and others with accelerated charged particles by assigning the reason for the variation of e/m with velocity to the charge rather than to the mass, postulating a velocity-dependent charge and an invariant mass. Significantly, Bush shares the relativistic affirmation that no electrons can exceed the velocity of light but links this fact *not* with a mass increase but rather with a postulate that "the force between charges becomes zero when their relative velocity is equal to the velocity of light" (p. 149). But, as we saw, the postulate of the limiting character of the velocity of light makes acceptance of the relativity of simultaneity unavoidable. And the latter relativity together with (a) the other constituents of relativistic kinematics and (b) very plausible conservation principles of dynamics entail the variation of *mass* with velocity. Bush does not adduce any experimental facts which support his hypothesis of charge variation while being incompatible with relativity. Neither does he remove one's *prima facie* doubts that the hypothesis of charge variation could actually be consistently incorporated in the body of well-confirmed physical principles. I therefore conclude that his citation of Miller's experiments (pp. 133, 155) as warrant for offering a non-

There is therefore impressive empirical evidence for all of the constituent theses of the light principle. It would be an error, however, to suppose that the experimental justification of the light principle suffices also to substantiate the Lorentz transformations. For these equations entail the clock retardation, whereas the light principle alone does *not*. A lucid demonstration of this fact has recently been given by H. P. Robertson.[20] He considers a linear transformation

$$T: \quad (t',x',y',z') \rightarrow (t,x,y,z)$$

between a kind of primary inertial system Σ and a "moving" inertial system S'. Upon having fixed 13 of the 16 coefficients of this transformation by various conventions, symmetry conditions and a specification of the velocity v of S' relative to Σ, Robertson is concerned with the *experimental* warrant for asserting that the remaining 3 coefficients have the values required by the Lorentz transformations. And he then shows that the Michelson-Morley and Kennedy-Thorndike experiments, which do succeed in completing the confirmation of the light principle, do *not* suffice to fix the remaining three coefficients of the transformation such that these have the values required by the Lorentz transformations. An additional experiment is needed to do so: the laboratory work of Ives and Stilwell (1938) furnished the lacking data by observations on high speed canal rays. And it was their confirmation of the transverse Doppler effect that constituted the first experimental proof of the clock retardation affirmed by the Lorentz transformations. Additional confirmation has been provided by data on the rate of disintegration of mesons.[21]

5. THE PHILOSOPHICAL ISSUE BETWEEN LORENTZ AND EINSTEIN

Many writers tell us that the Lorentz-Fitzgerald contraction hypothesis was an *ad hoc*

explanation of the result of the Michelson-Morley experiment, claiming that Einstein, on the other hand, "explained" the Lorentz-Fitzgerald contraction. Thus they charge Lorentz with having formulated a hypothesis whose addition to classical physics could serve to explain only the otherwise embarrassing outcome of the Michelson-Morley experiment while precluding, in principle, any other experimental test. The unfoundedness of this charge is already apparent from our discussion of the reasoning underlying the Kennedy-Thorndike experiment. And, as we saw, far from explaining the outcome of the Michelson-Morley experiment as a consequence of more fundamental principles, Einstein incorporated its null result as a physical *axiom* in his light principle.

It is a widespread error to suppose that Einstein nonetheless furnished an explanation for the Lorentz-Fitzgerald contraction by exhibiting the contraction of a moving rod to be a consequence of the relativity of simultaneity as between different frames on the basis of the Lorentz transformations. This error is inspired by the numerical equality of the contraction factors of these two kinds of contraction, the value being $(1-\beta^2)^{\frac{1}{2}}$ in each case. But they are different, because the Lorentz-Fitzgerald contraction is measured in the very system in which the contracted arm is *at rest*, whereas the contraction that Einstein derived from the Lorentz transformations pertains to the length measured in a system relative to which the arm is *in motion*. More explicitly, the Lorentz-Fitzgerald contraction hypothesis asserts a comparison of the actual length of the arm, as measured by the round-trip time of light, to the greater length that the travel-time of light would have revealed, if the classical ether theory were true. Thus, using light as the standard for effecting the comparison, this hypothesis affirms that in the *same* system and under the same conditions of measurement, the metrical properties of the arm are different from the ones predicted by classical ether theory. And this difference or contraction is clearly quite independent of any contraction based on comparisons of lengths in different inertial systems. On the other hand, the contraction which Einstein deduced from the Lorentz transformations is based on a compari-

relativistic hypothesis is not convincing, all the more so, since a careful recent analysis of Miller's data by Shankland, McCuskey, Leone, and Kuerti in Revs. Modern Phys. 27, 167 (1955) has shown them to be due to errors. For additional doubts concerning Miller's findings, see L. Essen's report in Nature 175, 794 (1955).

[20] H. P. Robertson, Revs. Modern Phys. 21, 378 (1949). An earlier proof of the compatibility of the light principle with the *denial* of the clock retardation was given by Reichenbach (see AR, pp. 79–83, esp. pp. 81–3), who exhibits a consistent set of coordinate transformations embodying both assertions.

[21] B. Rossi and D. B. Hall, Phys Rev. 59, 223 (1941).

son of the length of a rod, as measured from an inertial system relative to which it is in motion, to the length of that same rod, as measured in its own rest system. Unlike the Lorentz-Fitzgerald contraction, this "Einstein contraction" is a *symmetrical* relation between the measurements made in any two inertial systems and is a consequence of the intersystemic relativity of simultaneity, because it relates lengths determined from *different* inertial perspectives of measurement, instead of contrasting conflicting claims concerning the results obtained under the *same* conditions of measurement. What Einstein did explain, therefore, is this "metrogenic" contraction, a phenomenon which poses no greater logical difficulties than the differences in the angular sizes of bodies that are observed from different distances. Lest it be thought that the confusion of the two kinds of contraction just discussed is a thing of the past, I cite the following recent statements from Sir E. T. Whittaker's Tarner Lectures: "The Lorentz transformation . . . supplies at once an explanation of the Fitzgerald contraction.[22] . . . the failure of all attempts to determine the velocity of the earth by comparing the Fitzgerald contraction in rods directed parallel and perpendicular to the terrestrial motion . . . is necessitated by the Postulate of Relativity. But there is no impossibility, in principle at any rate, in observing the contraction, provided we can make use of an observation-post which is *outside* the moving system."[23] To be sure, if Whittaker's proviso of an ether-system observation post outside the moving earth be granted, then an observer at that post who interprets his data *pre*-relativistically would confirm the Lorentz-Fitzgerald contraction by finding that the "true" length of the moving arm, which he believes himself to be observing from his vantage, is smaller than the "spurious" length measured by the rod of a terrestrial observer. But the relativistic explanation of the numerically equal Einstein contraction actually involved here re-

jects, as we shall see in detail, the very conceptions which alone give meaning to (a) construing the findings of the extra-terrestrial observer as *equivalent* to a contraction *within* the moving system in the sense of Lorentz and Fitzgerald, and (b) asking the question as to *why* there is a L-F contraction *within* the moving system. And the relativistic deduction of the Einstein contraction from the Lorentz transformations can therefore have no bearing at all on *why* the Michelson-Morley experiment failed to fulfill the predictions of the classical ether theory.

We see that the locus of the philosophical difference between Lorentz and Einstein has been misplaced by the proponents of the *ad hoc* charge against Lorentz. To understand Einstein's philosophical innovation, we must take cognizance of the fact that the Lorentz-Fitzgerald contraction hypothesis was not the only addition to the ether theory made by Lorentz in order to account for the available body of experimental data. In addition, he had been driven to postulate with J. Larmor that just as rods are caused to contract in any inertial system moving relatively to the ether, so also clocks are *caused* by that very motion to modify their rates and to read a spurious "local" time (as distinct from the "true" time shown by the clocks in the ether-system). The conceptual framework of Lorentz' interpretation of the transformation equations known by his name was the absolutistic one in which the clocks and rods even in the privileged ether system merely *recorded* but did not *define* the topology and metric of container space and time. On this basis, Lorentz was led to reason somewhat as follows:

(1) Since the horizontal arm of the Michelson-Morley experiment is shorter than a rod lying alongside it but conforming to the expectations of classical optics and its container theory of space, we must infer that when a unit rod in the ether system is transported to a moving system, it can no longer be a true unit rod but becomes shorter than unity in the moving system, and similarly for clocks.

(2) The *deviation* from the classically expected behavior exhibited by rods and clocks must have a *cause* in the sense of being due to a perturbational influence. For in the absence of such

[22] E. T. Whittaker, *From Euclid to Eddington* (Cambridge University Press, London, 1949), p. 63.
[23] See reference 22, p. 64, italics supplied. The same error is repeated by Whittaker in his *A History of the Theories of Aether and Electricity* (Thomas Nelson and Sons, New York, 1953), Vol. 2, p. 37, which should be read in conjunction with his "G. F. Fitzgerald," Sci. American 189, 98 (November 1953).

a cause, the classically expected behavior would have occurred spontaneously.

Einstein left the Lorentz transformations *formally* unaltered. But the reasoning underlying his radical reinterpretation of their physical meaning was probably somewhat as follows:

As Riemann has pointed out, "in the case of a discrete manifold the criterion of length is already contained in the concept of this manifold but in a continuous manifold, it must be brought in from elsewhere."[24] This means that, contrary to the container theory, the length of a body is not a measure of an *intrinsic* amount of space between its end-points or a relation between these end-points by themselves. More specifically, as Cantor has shown, the continuity of space assures that there are just as many points between the end-points of a long segment as between those of a short one. There is thus no *intrinsic* attribute of the spaces between these pairs of end-points which would endow each of their segments with a distinct length. The relational theory of space therefore rightly maintains that the length of a body AB is an attribute of the *relation* between two *pairs* of points: the termini of AB on the one hand and those of the chosen unit rod on the other. And it is inherent in this definition of length as a ratio that the unit rod be at rest relative to AB when performing its metrical function.[25]

But, on this new conception of length, one is not entitled to infer with Lorentz that a unit rod in the ether system will no longer be unity, *as a matter of physical fact*, once it has been transported to a moving system. For the relational conception *allows* us to *call* that same rod unity in the moving system *by definition*. If Lorentz had realized that the length of this rod in the moving system can be legitimately *decreed* by definition, and similarly for the periods of material clocks, then it would have been clear to him that the ground is cut from under his distinction between "true" and "local" (i.e., spurious) lengths and times and thereby from his idea that the horizontal arm in the Michelson-Morley experiment is actually shorter than the vertical arm. Thus, a coupling of the epistemological insights of the relational theory of length with the experimental findings of optics deprives reference to a preferred ether system of all objective physical significance and makes possible the enunciation of the principle of relativity.

We can understand Einstein's philosophical departure from the *second* step in Lorentz' reasoning by giving an analysis of Lorentz' invocation of a *cause* for the contraction revealed by the Michelson-Morley experiment.

Every physical theory tells us what particular behavior of physical entities or systems it regards as "natural" in the absence of *perturbational* influences. Concurrently, it specifies the the influences or causes which it regards as responsible for any *deviations* from the assumedly

---

[24] Riemann, "Über die Hypothesen welche der Geometrie zu Grunde liegen," in *Gesammelte mathematische Werke*, ed. Dedekind and Weber (Leipzig, 1876), p. 268.

[25] B. Hoffmann has shown recently [Phys. Rev. **89**, 49–52 (1953)] that the usual affine tensor calculus of the *general* theory of relativity does not implement the conception that length is a ratio, because it fails to yield invariance of the equations of physics under a change of scale. To achieve such implementation, Hoffmann proposes an enlargement of that tensor formalism assuring its conformity to an appropriate similarity principle.
It is perfectly clear that relations or properties of relations between physical objects (which are expressed numerically as ratios) are fully as objective physically and exist just as independently of the human mind as simple properties of individual objects. Thus, the *relation* between a copper bar at rest in a system K and the unit rod in K might have the property that the copper bar has a length of 5 units in K. But the *different relation* of that bar's *projection* onto the x axis of a system S, relative to which it is moving along that axis, to the unit rod of S may then yield a length of only 4.7 S-units. It is incontestable that the *differences* among the various relations sustained by the bar do *not* render these relations subjective products of the physicist's mind, any more than they do the fact that the bar in question is a copper bar. In a futile attempt to defend a mentalistic metaphysics on the basis of relativity theory, Herbert Dingle denies this fact. Replying

to decisive critiques of his views by P. Epstein [Am. J. Phys. **10**, 1 and 205 (1942) and **11**, 228 (1943)] and M. Born [Phil. Quart. **3**, 139 (1953)], Dingle offers the following argument [*The Sources of Eddington's Philosophy* (Cambridge University Press, Cambridge, England, 1954), pp. 11–12]: "The view that physics is the description of the character of an independent external world was simply no longer tenable. . . . Every relativist will admit that if two rods, A and B, of equal length when relatively at rest, are in relative motion along their common direction, then A is longer or shorter than B, or equal to it, exactly as you please. It is therefore impossible to evade the conclusion that its length is not a property of either rod; and what is true of length is true of every other so-called physical property. Physics is therefore [*sic*!] not the investigation of the nature of the external world." Far from having demonstrated that relativity physics is subjective, Professor Dingle has merely succeeded in exhibiting his unawareness of the fact that properties of relations do not cease to be *bona fide* objective properties just because they belong to *relations* between individuals rather than directly to individuals themselves. Only such unawareness can lead to his primitive thesis that the relations of physical entities to one another cannot constitute "the character of an independent external world."

"natural" behavior. But when such deviations are observed and a theory cannot designate the perturbations to which it proposes to attribute them, its assumptions concerning the character of the "natural" or unperturbed behavior become subject to doubt. For the reliability of our conceptions as to what pattern of occurrences is "natural" is no greater than the *scope* of the evidence on which they rest. And a theory's failure to designate the perturbing causes of the nonfulfillment of its expectations therefore demands the envisionment of the possibility that (a) the "natural" behavior of things is indeed different from what the theory in question has been supposing it to be and that (b) deviations from the assumedly natural behavior transpire *without* perturbational causes. Several examples from past and present scientific controversy attest to the mistaken search for the perturbational factors which are supposed to cause deviations from the pattern which a particular theory unquestioningly and *tenaciously* affirms to be the natural one. Thus, Aristotelian critics of Galileo, assuming that Aristotle's mechanics describes the natural behavior, asked Galileo to specify the cause which prevents a body from coming to rest and maintains its speed in the same straight line in the manner of Newton's first law of motion. It was axiomatic for them that uniform motion could not continue indefinitely in the absence of net external forces. In our own time, there are those who ask: if the "new cosmology" of Bondi and Gold is true, must there not be a divine interference (perturbation) in the natural order whereby the spontaneous accretion ("creation") of matter is caused?[26] The propounders of this question do not tell us, however, on what basis they take it for granted that the constancy of the mass (-energy) content of the universe is *cosmically* the natural state of affairs. Curiously enough, other theologians ask us to regard a state of "nothingness" (whatever that is!) as the natural state of the universe and thereby endeavor to create grounds for arguing

that the mere existence and conservation of matter or energy require a divine creator and sustainer. We see that the theological concept of miracles attributes a supernatural origin to certain phenomena by the mere fiat of declaring that a certain set of limited empirical generalizations which do not allow for these phenomena define with *certainty* what is natural. As if observed events presented themselves to us with identification tags as to their "naturalness"![27]

The basis for Einstein's philosophical objection to the second step in Lorentz' reasoning is now at hand: it was an error on Lorentz' part to persist, in the face of mounting contrary evidence, in regarding the classically expected behavior as the natural behavior. It was this persistence which forced him to *explain* the observed deviations from the classical laws by postulating the operation of a physically nondesignatable ether as a perturbational cause. Having used the relational theory of length to reject the conclusion of the first step in Lorentz' reasoning, Einstein was able to see that the unexpected results of the Michelson-Morley experiment do not require any perturbational causes at all, because they are integral to the "natural" behavior of things.

The character and significance of these fundamental *philosophical* differences between Einstein's conception of the Lorentz transformations, on the one hand, and the interpretation of earlier versions of these transformations by his predecessors, Fitzgerald, Larmor, Poincaré (see footnote 7 in Sec. 2 above) and Lorentz, on the other, altogether escaped recognition by E. T. Whittaker in his very recent account of the history of the development of the special theory of relativity. Entitling his chapter on that theory "The Relativity Theory of Poincaré and Lorentz," he reaches the following unwarranted judgment concerning Einstein's contribution to the theory[28]: "In the autumn of the same year [1905], . . . Einstein published a paper which

---

[26] The logical blunder which generates this question was committed in inverted form by Herbert Dingle. One of his reasons for rejecting the cosmology of Bondi and Gold is that it would allegedly require not merely a single act of miraculous divine interference, as Biblical creation *ex nihilo* does, but a continuous series of such acts. See A. Grünbaum, Sci. American 189, 6–8 (December, 1953).

[27] For further details on this issue, see A. Grünbaum, Sci. Monthly 79, 15–16 (1954).
[28] E. T. Whittaker, *A History of the Theories of Aether and Electricity* (Thomas Nelson and Sons, New York, 1953), Vol. 2, p. 40. Whittaker himself points out (p. 36) that even at the time of his death in 1928, Lorentz reportedly still favored the concepts of "true" time and absolute simultaneity. For Lorentz' own brief statement on this point, see his *The Theory of Electrons* (Columbia University Press, New York, 1909), pp. 329, 230.

set forth the relativity theory of Poincaré and Lorentz with some amplifications, and which attracted much attention. He asserted as a fundamental principle the *constancy of the velocity of light*, . . . an assertion which at the time was widely accepted, but has been severely criticized by later writers [at this point, Whittaker refers to the paper by H. E. Ives whose logical inadequacies the reader will recall from the discussion in Sec. 3 above]. In this paper Einstein gave the modifications which must now be introduced into the formulas for aberration and the Doppler effect."

In conclusion, it should be noted that the term "relational" has been used advisedly throughout this paper in place of the more common term "operational" to characterize the relativistic conception of space and time, for I reject Bridgman's contention that we human beings are the ones who first confer properties and relations upon physical entities by our operations of measurement, and I do not think that the theory of relativity can be validly adduced in support of this homocentric form of operationism.[39] Einstein's theory asks us to conceive the topology and metric of space-time as systems of relations between physical events and things. But both these things and their relations are independent of man's presence in the cosmos. Our operations of measurement merely discover or ascertain the structure of space-time but they do *not* generate it.

[39] For details, see A. Grünbaum, Sci. Monthly 79, 228 (1954), and Am. J. Phys. 22, 499 (1954).

* Currently, he is the Andrew Mellon Professor of Philosophy, Research Professor of Psychiatry, and Chairman of the Center for Philosophy of Science at the University of Pittsburgh.

Psychology in Physical Language

BY RUDOLF CARNAP

(TRANSLATED BY GEORGE SCHICK)

1. INTRODUCTION. PHYSICAL LANGUAGE AND PROTOCOL LANGUAGE

IN WHAT FOLLOWS, we intend to explain and to establish the thesis that *every sentence of psychology may be formulated in physical language.* To express this in the material mode of speech: *all sentences of psychology describe physical occurrences, namely, the physical behavior of humans and other animals.* This is a sub-thesis of the general thesis of *physicalism* to the effect that *physical language is a universal language,* that is, a language into which every sentence may be translated. The general thesis has been discussed in an earlier article,[1] whose position shall here serve as our point of departure. Let us first briefly review some of the conclusions of the earlier study.

In meta-linguistic discussion we distinguish the customary *material mode of speech* (e.g. "The sentences of this language speak of this and that object.") from the more correct *formal mode of speech* (e.g. "The sentences of this language contain this and that word and are constructed in this and that manner.") In using the material mode of speech we run the risk of introducing confusions and pseudo-problems. If, because of its being more easily understood, we occasionally do use it in what follows, we do so only as a paraphrase of the formal mode of speech.

Of first importance for epistemological analyses are the *protocol*

This article was originally published in Volume III of *Erkenntnis* (1932/33). It is reproduced here with the kind permission of Professor Carnap.

1. Carnap, "Die Physikalische Sprache als Universalsprache der Wissenschaft," *Erkenntnis* II, 1931, pp. 432-465. [The English translation of this article by Max Black was published as a monograph under the title "The Unity of Science" (London: Kegan Paul, 1934).]

language, in which the primitive protocol sentences (in the material mode of speech: the sentences about the immediately given) of a particular person are formulated, and the *system language,* in which the sentences of the system of science are formulated. A person S *tests* (verifies) a system-sentence by deducing from it sentences of his own protocol language, and comparing these sentences with those of his actual protocol. The possibility of such a deduction of protocol sentences constitutes the *content* of a sentence. If a sentence permits no such deductions, it has no content, and is meaningless. If the same sentences may be deduced from two sentences, the latter two sentences have the same content. They say the same thing, and may be translated into one another.

To every sentence of the system language there corresponds some sentence of the physical language such that the two sentences are inter-translatable. It is the purpose of this article to show that this is the case for the sentences of psychology. Moreover, every sentence of the protocol language of some specific person is inter-translatable with some sentence of physical language, namely, with a sentence about the physical state of the person in question. The various protocol languages thus become sub-languages of the physical language. The *physical language is universal and inter-subjective.* This is the thesis of physicalism.

If the physical language, on the grounds of its universality, were adopted as the system language of science, all science would become physics. Metaphysics would be discarded as meaningless. The various domains of science would become parts of unified science. In the material mode of speech: there would, basically, be only one kind of object—physical occurrences, in whose realm law would be all-encompassing.

Physicalism ought not to be understood as requiring psychology to concern itself only with physically describable situations. The thesis, rather, is that psychology may deal with whatever it pleases, it may formulate its sentences as it pleases—these sentences will, in every case, be translatable into physical language.

We say of a sentence P that it is *translatable* (more precisely, that it is reciprocally translatable) into a sentence Q if there are rules, independent of space and time, in accordance with which Q may be deduced from P and P from Q; to use the material mode of speech, P and Q describe the same state of affairs; epistemologically speaking, every protocol sentence which confirms P also confirms Q and *vice versa.* The definition of an expression "a" by means of expressions "b," "c" . . . , represents a translation-rule with the help of which any sentence in which "a" occurs may be translated into a sentence in which "a" does not occur, but "b," "c," . . . do, and *vice versa.* The translatability of all the sentences of language L_1 into a (completely or partially) different language L_2 is assured if, for every expression of L_1, a definition is presented which directly or indirectly (i.e., with the help of other definitions) derives that expression from expressions of L_2. Our thesis thus states that a definition may be constructed for every psychological concept (i.e. expression) which directly or indirectly derives that concept from physical concepts. We are not demanding that psychology formulate each of its sentences in physical terminology. For its own purposes psychology may, as heretofore, utilize its own terminology. All that we are demanding is the production of the definitions through which psychological language is linked with physical language. We maintain that these definitions can be produced, since, implicitly, they already underlie psychological practice.

If our thesis is correct, the generalized sentences of psychology, the *laws* of psychology, are also translatable into the physical language. They are thus physical laws. Whether or not these physical laws are deducible from those holding in inorganic physics, remains, however, an open question. This question of the deducibility of the laws is completely independent of the question of the definability of concepts. We have already considered this matter in our discussion of biology.[2] As soon as one realizes that the sentences of psychology belong to the physical language, and also overcomes the emotional obstacles to the acceptance of this provable thesis, one will, indeed, incline to the conjecture, which cannot as yet be proved, that the laws of psychology are special cases of physical laws holding in inorganic physics as well. But we are not concerned with this conjecture here.

Let us permit ourselves a brief remark—apart from our principal point—concerning the emotional resistance to the thesis of physicalism. Such resistance is always exerted against any thesis when an Idol is being dethroned by it, when we are asked to discard an idea with which dignity and grandeur are associated. As a result of Copernicus' work, man lost the distinction of a central position in the universe; as a result of Darwin's, he was deprived of the dignity of a special supra-animal existence; as a result of Marx's, the factors by means of which history can be causally explained were degraded from the realm of ideas to that of material events; as a result of

---

2. "Die Physikalische Sprache," *op. cit.,* p. 449 ff., (*The Unity of Science,* p. 68 ff.).

Nietzsche's, the origins of morals were stripped of their halo; as a result of Freud's, the factors by means of which the ideas and actions of men can be causally explained were located in the darkest depths, in man's nether regions. The extent to which the sober, objective examination of these theories was obstructed by emotional opposition is well known. Now it is proposed that psychology, which has hitherto been robed in majesty as the theory of spiritual events, be degraded to the status of a part of physics. Doubtless, many will consider this an offensive presumption. Perhaps we may therefore express the request that the reader make a special effort in this case to retain the objectivity and openness of mind always requisite to the testing of a scientific thesis.

2. THE FORMS OF PSYCHOLOGICAL SENTENCES

The distinction between singular and general sentences is as important in psychology as in other sciences. A *singular psychological sentence*, e.g. "Mr. A was angry at noon yesterday" (an analogue of the physical sentence, "Yesterday at noon the temperature of the air in Vienna was 28 degrees centigrade"), is concerned with a particular person at a particular time. *General psychological sentences* have various forms, of which the following two are perhaps the most important. A sentence may describe a specific quality of a specific kind of event, e.g. "An experience of surprise always (or: always for Mr. A, or: always for people of such and such a society) has such and such a structure." A physical analogy would be: "Chalk (or: chalk of such and such a sort) always is white." The second important form is that of universal-conditional statements concerning sequences of events, that is, of causal laws. For instance, "When, under such and such circumstances, images of such and such a sort occur to a person (or: to Mr. A, or: to anyone of such and such a society), an emotion of such and such a sort always (or: frequently, or: sometimes) is aroused." A physical analogy would be: "When a solid body is heated, it usually expands."

Research is primarily directed to the discovery of general sentences. These cannot, however, be established except by means of the so-called method of induction from the available singular sentences, i.e. by means of the construction of hypotheses.

Phenomenology claims to be able to establish universal synthetic sentences which have not been obtained through induction. These sentences about psychological qualities are, allegedly, known either *a priori* or on the basis of some single illustrative case. In our view, knowledge cannot be gained by such means. We need not, however, enter upon a discussion of this issue here, since even on the view of phenomenology itself, these sentences do not belong to the domain of psychology.

In physics it sometimes seems to be the case that a general law is established on the basis of some single event. For instance, if a physicist can determine a certain physical constant, say, the heat-conductivity of a sample of some pure metal, in a single experiment, he will be convinced that, on other occasions, not only the sample examined but any similar sample of the same substance will, very probably, be characterizable by the same constant. But here too induction is applied. As a result of many previous observations the physicist is in possession of a universal sentence of a higher order which enables him in this case to follow an abbreviated method. This higher-order sentence reads roughly: "All (or: the following) physical constants of metals vary only slightly in time and from sample to sample."

The situation is analogous for certain conclusions drawn in psychology. If a psychologist has, as a result of some single experiment, determined that the simultaneous sounding of two specific notes is experienced as a dissonance by some specific person A, he infers (under favorable circumstances) the truth of the general sentence which states that the same experiment with A will, at other times, have the same result. Indeed, he will even venture—and rightly—to extend this result, with some probability, to pairs of tones with the same acoustic interval if the pitch is not too different from that of the first experiment. Here too the inference from a singular sentence to a general one is only apparent. Actually, a sentence inductively obtained from many observations is brought into service here, a sentence which, roughly, reads: "The reaction of any specific person as to the consonance or dissonance of a chord varies only very slightly with time, and only slightly on a not too large transposition of the chord." It thus remains the case that every general sentence is inductively established on the basis of a number of singular ones.

Finally, we must consider sentences about psycho-physical inter-relations, such as for instance, the connection between physical stimulus and perception. These are likewise arrived at through induction, in this case through induction in part from physical and in part from psychological singular sentences. The most important sentences of gestalt psychology belong also to this kind.

General sentences have the character of hypotheses in relation to concrete sentences, that is, the testing of a general sentence consists in testing the concrete sentences which are deducible from it. A general sentence has content insofar and only insofar as the concrete sentences deducible from it have content. Logical analysis must therefore primarily be directed towards the examination of the latter sort of sentences.

If A utters a singular psychological sentence such as "Yesterday morning B was happy," the epistemological situation differs according as A and B are or are not the same person. Consequently, we distinguish between sentences about *other minds* and sentences about *one's own mind*. As we shall presently see, this distinction cannot be made among the sentences of inter-subjective science. For the epistemological analysis of subjective, singular sentences it is, however, indispensable.

3. Sentences about Other Minds

The epistemological character of a singular sentence about other minds will now be clarified by means of an analogy with a sentence about a physical property, defined as a disposition to behave (or respond) in a specific manner under specific circumstances (or stimuli). To take an example: a substance is called "plastic" if, under the influence of deforming stresses of a specific sort and a specific magnitude, it undergoes a permanent change of shape, but remains intact.

We shall try to carry out this analogy by juxtaposing two examples. We shall be concerned with the epistemological situation of the example taken from psychology; the parallel example about the physical property is intended only to facilitate our understanding of the psychological sentence, and not to serve as a specimen of an argument from analogy. (For the sake of convenience, where the text would have been the same in both columns, it is written only once.)

A Sentence about a property of a physical substance.	*A Sentence about a condition of some other mind.*
Example: I assert the sentence P_1: "This wooden support is very firm."	Example: I assert the sentence P_1: "Mr. A is now excited."

There are two different ways in which sentence P_1 may be derived. We shall designate them as the "rational" and the "intuitive" methods. The *rational* method consists of inferring P_1 from some protocol sentence p_1 (or from several like it), more specifically, from a perception-sentence

about the shape and color of the wooden support.	about the behavior of A, e.g. about his facial expressions, his gestures, etc., or about physical effects of A's behavior, e.g. about characteristics of his handwriting.

In order to justify the conclusion, a major premise O is still required, namely the general sentence which asserts that

when I perceive a wooden support to be of this color and form, it (usually) turns out to be firm. (A sentence about the perceptual signs of firmness.)	when I perceive a person to have this facial expression and handwriting he (usually) turns out to be excited. (A sentence about the expressional or graphological signs of excitement.)

The content of P_1 does not coincide with that of p_1, but goes beyond it. This is evident from the fact that to infer P_1 from p_1 O is required. The cited relationship between P_1 and p_1 may also be seen in the fact that under certain circumstances, the inference from p_1 to P_1 may go astray. It may happen that, though p_1 occurs in a protocol, I am obliged, on the grounds of further protocols, to retract the established system sentence P_1. I would then say something like, "I made a mistake. The test has shown

that the support was not firm, even though it had such and such a form and color."	that A was not excited, even though his face had such and such an expression."

In practical matters the *intuitive* method is applied more frequently than this rational one, which presupposes theoretical knowledge and requires reflection. In accordance with the intuitive method, P_1 is obtained without the mediation of any other sentence from the identically sounding protocol sentence p_2.

"The support is firm."	"A is excited."

Consequently, one speaks in this case of *immediate perceptions*

of properties of substances, e.g., of the firmness of supports.	of other minds, e.g., of the excitement of A.

But in this case too the protocol sentence p_2 and the system sentence P_1 have different contents. The difference is generally not noted because, on the ordinary formulation, both sentences sound alike.

119

Here too we can best clarify the difference by considering the possibility of error. It may happen that, though p_2 occurs in my protocol, I am obliged, on the basis of further protocols, to retract the established system sentence P_1. I would then say "I made a mistake. Further tests have shown

that the support was not firm, although I had the intuitive impression that it was."

that A was not excited, although I had the intuitive impression that he was."

[The difference between p_2 and P_1 is the same as that between the identically sounding sentences p and P_1: "A red marble is lying on this table," of an earlier example.[3] The argument of that article shows that the inference of P_1 from p_2, if it is to be rigorous, also requires a major premise of general form, and that it is not in the least simple. Insofar as ordinary usage, for convenience's sake, assigns to both sentences the same sequence of words, the inference is, in practice, simplified to the point of triviality.]

Our problem now is: *what does sentence P_1 mean?* Such a question can only be answered by the presentation of a sentence (or of several sentences) which has (or which conjointly have) the same content as P_1. The viewpoint which will here be defended is that P_1 has the same content as a sentence P_2 which asserts the existence of a physical structure characterized by the disposition to react in a specific manner to specific physical stimuli. In our example, P_2 asserts the existence of that physical structure (microstructure)

of the wooden support that is characterized by the fact that, under a slight load, the support undergoes no noticeable distortion, and, under heavier loads, is bent in such and such a manner, but does not break.

of Mr. A's body (especially of his central nervous system) that is characterized by a high pulse and rate of breathing, which, on the application of certain stimuli, may even be made higher, by vehement and factually unsatisfactory answers to questions, by the occurrence of agitated movements on the application of certain stimuli, etc.

On my view, there is here again a thoroughgoing analogy between the examples from physics and from psychology. If, however, we were to question the experts concerning the examples from their

3. See *Erkenntnis*, Vol. II, p. 460 (*The Unity of Science*, p. 92).

respective fields, the majority of them nowadays would give us thoroughly non-analogous answers. The identity of the content of P_2

and of the content of the physical sentence P_1 would be agreed to as a matter of course by all physicists.	and of the content of the psychological sentence P_1 would be denied by almost all psychologists (the exceptions being the radical behaviorists).

The contrary view which is most frequently advocated by psychologists is that, "A sentence of the form of P_1 asserts the existence of a state of affairs not identical with the corresponding physical structure, but rather, only accompanied by it, or expressed by it. In our example:

P_1 states that the support not only has the physical structure described by P_2, but that, besides, there exists in it a certain force, namely its *firmness*.	P_1 states that Mr. A not only has a body whose physical structure (at the time in question) is described by P_2, but that—since he is a *psychophysical being*—he has, besides, a consciousness, a certain power or entity, in which that excitement is to be found.
This firmness is not identical with the physical structure, but stands in some parallel relation to it in such a manner that the firmness exists when and only when a physical structure of the characterized sort exists.	This excitement cannot, consequently, be identical with the cited structure of the body, but stands in some parallel relation (or in some relation of interaction) to it in such a manner that the excitement exists when and only when (or at least, frequently when) a physical, bodily structure of the characterized sort exists.
Because of this parallelism one may consider the described reaction to certain stimuli—which is causally dependent upon that structure—to be an *expression* of firmness.	Because of this parallelism one may consider the described reaction to certain stimuli to be an *expression* of excitement.
Firmness is thus an occult property, an obscure power which stands behind physical structure, appears in it, but itself remains unknowable."	Excitement, or the consciousness of which it is an attribute, is thus an occult property, an obscure power which stands behind physical structure, appears in it, but itself remains unknowable."

This view falls into the error of a hypostatization as a result of which a remarkable duplication occurs: besides or behind a state of affairs whose existence is empirically determinable, another, *parallel* entity is assumed, whose existence is not determinable. (Note that we are here concerned with a sentence about other minds.) But—one may now object—is there not really at least one possibility of testing this claim, namely, by means of the protocol sentence p_2 about the intuitive impression of

the firmness of the support?   the excitement of A?

The objector will point out that this sentence, after all, occurs in the protocol along with the perception sentence p_1. May not then a system sentence whose content goes beyond that of P_2 be founded on p_2? This may be answered as follows. A sentence says no more than what is testable about it. If, now, the testing of P_1 consisted in the deduction of the protocol sentence p_2, these two sentences would have the same content. But we have already seen that this is impossible.

There is no other possibility of testing P_1 except by means of protocol sentences like p_1 or like p_2. If, now, the content of P_1 goes beyond that of P_2, the component not shared by the two sentences is not testable, and is therefore meaningless. If one rejects the interpretation of P_1 in terms of P_2, P_1 becomes a metaphysical pseudo-sentence.

The various sciences today have reached very different stages in the process of their decontamination from metaphysics. Chiefly because of the efforts of Mach, Poincaré, and Einstein, physics is, by and large, practically free of metaphysics. In psychology, on the other hand, the work of arriving at a science which is to be free of metaphysics has hardly begun. The difference between the two sciences is most clearly seen in the different attitudes taken by experts in the two fields towards the position which we rejected as metaphysical and meaningless. In the case of the example from physics, most physicists would reject the position as anthropomorphic, or mythological, or metaphysical. They thereby reveal their anti-metaphysical orientation, which corresponds to our own. On the other hand, in the case of the example from psychology (though, perhaps, not when it is so crudely formulated), most psychologists would today consider the view we have been criticizing to be self-evident on intuitive grounds. In this one can see the metaphysical orientation of psychologists, to which ours is opposed.

4. Rejoinder to Four Typical Criticisms

Generalizing the conclusion of the argument which, with reference to a special case, we have been pursuing above, we arrive at the thesis that *a singular sentence about other minds always has the same content as some specific physical sentence.* Phrasing the same thesis in the material mode of speech—a sentence about other minds states that the body of the person in question is in a physical state of a certain sort. Let us now discuss several objections against this thesis of physicalism.

A. *Objection on the ground of the undeveloped state of physiology:* "Our current knowledge of physiology—especially our knowledge of the physiology of the central nervous system—is not yet sufficiently advanced to enable us to know to what class of physical conditions something like excitement corresponds. Consequently, when today we use the sentence 'A is excited,' we cannot mean by it the corresponding physical state of affairs."

Rebuttal. Sentence P_1, "A is excited" cannot, indeed, today be translated into a physical sentence P_3 of the form "such and such a physico-chemical process is now taking place in A's body" (expressed by a specification of physical state-coordinates and by chemical formulae). Our current knowledge of physiology is not adequate for this purpose. Even today, however, P_1 may be translated into another sentence about the physical condition of A's body, namely into the sentence P_2, to which we have already referred. This takes the form "A's body is now in a state which is characterized by the fact that when I perceive A's body the protocol sentence p_1 (stating my perception of A's behavior) and (or) the protocol sentence p_2 (stating my intuitive impression of A's excitement) or other, analogous, protocol sentences of such and such a sort are produced." Just as, in our example from physics, sentence P_1, "The wooden support is firm," refers to the physical structure of the wooden support—and this even though the person using the sentence may sometimes not be capable of characterizing this physical structure by specifying the distribution of the values of the physical state-coordinates, so also does the psychological sentence P_1, "A is excited," refer to the physical structure of A's body—though this structure can only be characterized by potential perceptions, impressions, dispositions to react in a specific manner, etc., and not by any specification of state-coordinates. Our ignorance of physiology can therefore affect only the mode of our characterization of the physical state of affairs in question. It in no way touches upon the principal point: that sentence P_1 refers to a physical state of affairs.

B. *Objection on the ground of analogy:* "When I myself am angry, I not only act out the behavior-pattern of an angry man, I experience a special *feeling* of anger. If, consequently, I observe someone else acting out the same behavior-pattern I may, on grounds of analogy, conclude (if not with certainty, at least with probability) that he too, besides acting as he does, now has a *feeling* of anger (which is not meant as a physical state of affairs)."

Rebuttal. Though arguments from analogy are not certain, as probability arguments they are undoubtedly admissible. By way of an example let us consider an everyday argument from analogy. I see a box of a certain shape, size, and color. I discover that it contains matches. I find another box of a similar appearance, and now, by analogy, draw the probability inference that it too contains matches. Our critic believes that the argument from analogy he presents is of the same logical form as the argument just presented. If this were the case, his conclusion would certainly be sound. But this is not the case. In our critic's argument, the conclusion is *meaningless*—a mere pseudo-sentence. For, being a sentence about other minds, not to be physically interpreted, it is in principle not testable. This was the result of our previous considerations; objection D will offer us an opportunity for discussing it again. In the non-testability of our critic's conclusion rests also the difference between his arguments and the example just cited. That the second box also contains matches may in principle be tested and confirmed by observation sentences of one's protocol. The two analogous sentences, "The first box contains matches" and "The second box contains matches" are both logically and epistemologically of the same sort. This is why the analogy holds here. The case is different with "I am angry" and "That person is angry." We consider the former of these two sentences to be meaningful and the latter (if its physical interpretation is rejected) to be meaningless. Our critic, who considers the latter as well as the former sentence to be meaningful, will believe that the person who asserts the sentence finds it testable, only in a manner altogether different from that in which the former is testable. Thus both of us agree that the latter sentence is epistemologically different from the former. The use of the same grammatical structure in these two sentences is logically illegitimate. It misleads us into believing that the two sentences are of the same logical form, and that one may be used as an analogue of the other.

If the conclusion is acknowledged to be meaningless, it remains to be explained how this pseudo-sentence was introduced into the argument. The logical analysis of concept formation and of sentences in science and (especially) in philosophy very frequently discloses pseudo-sentences. However, a pseudo-sentence rarely turns up as the conclusion of an argument from analogy with meaningful premises. This may readily be accounted for. An argument from analogy has (in a simple case) the following form. Premises: If A has the property E, it always also has the property F; A' resembles A in many respects; A' has the property E. We conclude (with probability): A' also has the property F. Now, according to semantics, if "A" and "B" are object-names, "E" and "F" property-names, and "E(A)" means that A has the property E, then a) if "E(A)" and "E(B)" are meaningful (i.e. either true or false), "A" and "B" belong to the same semantic type; b) if two names, "A" and "B," belong to the same semantic type, and "F(A)" is meaningful, then "F(B)" is also meaningful. In the case under discussion here "E(A)" and "E(A')" are meaningful, and consequently—in accordance with b)—"F(A')," the conclusion of the argument from analogy, is also meaningful. Thus if the premises of an argument from analogy are meaningful and yet the conclusion is meaningless, the formulation of the premises must be in some way logically objectionable. And this is indeed the case with the argument from analogy presented by our critic. The predicative expression "I am angry" does not adequately represent the state of affairs which is meant. It asserts that a certain property belongs to a certain entity. All that exists, however, is an experienced feeling of anger. This should have been formulated as, roughly, "now anger." On this correct formulation the possibility of an argument from analogy disappears. For now the premises read: when I (i.e. my body) display angry behavior, anger occurs; the body of another person resembles mine in many respects; the body of the other person is now displaying angry behavior. The original conclusion can now no longer be drawn, since the sentence "Anger occurs" contains no "I" which may be replaced by "the other person." If one wanted to draw the appropriate conclusion, in which no substitution is made but the form of the premises simply retained, one would arrive at the meaningful but plainly false conclusion, "Anger occurs"—which states what would be expressed in ordinary language by "I am now angry."

` C. *Objection on the ground of mental telepathy.* "The telepathic transmission of the contents of consciousness (ideas, emotions, thoughts) occurs without any determinable physical mediation. Here we have an instance of the knowledge of other minds which involves no perception of other people's bodies. Let us consider an example. I wake up suddenly one night, have a distinct sensation of fear, and know that my friend is now experiencing fear; later, I discover that at that very moment my friend was in danger of death. In this case, my knowledge of my friend's fear cannot refer to any state of his body, for I know nothing of that; my knowledge concerns itself immediately with my friend's sensation of fear."

Rebuttal. Psychologists are not yet unanimously decided on the degree to which they ought properly to credit the occurrence of cases of telepathy. This is an empirical problem which it is not our business to solve here. Let us concede the point to our critic, and assume that the occurrence of cases of telepathic transmission has been confirmed. We shall show that, even so, our earlier contentions are not affected in the least. The question before us is: what does sentence P_1, "My friend now experiences fear" mean, if I take P_1 to be a statement of telepathically derived cognition? We maintain that the meaning of P_1 is precisely the same as it would be if we used it on the grounds of some normally (rationally or intuitively) derived cognition. The occurrence of telepathy in no way alters the meaning of P_1.

Let us consider a precisely analogous situation involving the cognition of some physical event. I suddenly have the impression that a picture has fallen from the wall at my house, and this when neither I nor anyone else can in any normal way perceive that this has happened. Later, I discover that the picture has, indeed, fallen from the wall. I now express this cognition which I have obtained by clairvoyance in sentence Q, "The picture has now fallen from the wall." What is the meaning of this sentence? The meaning of Q here is clearly the same as it would be if I used it on the ground of some normally derived cognition, that is, on the ground of some cognition by direct perception of the event in question. For in both cases Q asserts that a physical event of a certain sort, a specific displacement of a specific body, has taken place.

The case is the same with telepathic cognition. We have already considered the case in which the state of some other mind is intuitively grasped, though by means of a perception of the other person's body. If a telepathic cognition of the state of some other mind occurs, it too is based on an intuitive impression, this time without a simultaneous perception. That which is cognized, however, is the same in both cases. Earlier, we remarked that P_1 does not have the same content as the protocol sentence p_2 about the (normally) intuitive impression, and that p_2 cannot support a sentence about something beside or behind the physical condition of the other person's body. Our remarks hold equally for telepathically intuitive impressions.

D. *Objection on the ground of statements by others.* "We are, to begin with, agreed that A is in a certain physical state which is manifested by behavior of a certain sort and produces in me, apart from sense-perceptions, an intuitive impression of A's anger. Beyond this, however, I can find out that A really does experience anger by questioning him. He himself will testify that he experienced anger. Knowing him to be a truthful person and a good observer, why should I not consider his statement to be true—or at least probably true?"

Rebuttal. Before I can decide whether I should accept A's statement as true, or false, or probably true—before, indeed, I can consider this question at all—I must first of all understand the statement. It must have meaning for me. And this is the case only if I can test it, if, that is, sentences of my protocol are deducible from it. If the expression is interpreted physically it is testable by means of protocol sentences such as my p_1 and p_2, that is, by sentences about specific perceptions and intuitive expressions. Since, however, our critic rejects the physical interpretation of the expression, it is in principle impossible for me to test it. Thus it is meaningless for me, and the question whether I should consider it to be true, or false, or probable, cannot even be posed.

Should unusual, brilliant patterns suddenly appear in the sky— even if they took the form of letters which seemed to compose a sentence—science could not comprehend them except by first conceiving them, describing them, and explaining them (i.e. subsuming them under general causal-sentences) as physical facts. The question whether such an arrangement of symbols constitutes a meaningful sentence must be decided without taking into consideration whether or not it appears in the sky. If this symbol-arrangement is not a meaningful sentence at other times, it cannot become one no matter how effulgent an appearance it makes in the sky. Whether a sentence is true or false is determined by empirical contingencies; but whether a sentence is or is not meaningful is determined solely by the syntax of language.

It is no different in the case of those acoustic phenomena that issue from the mouths of certain vertebrates. They are first of all facts, physical occurrences, and specifically, sound waves of a certain sort. We can, further, also interpret them as symbols. But whether or not such an arrangement of symbols is meaningful cannot depend on its occurrence as an acoustic phenomenon. If the sentence "A was angry yesterday at noon" has no meaning for me —as would be the case if (insofar as our critic rejects its physical meaning) I could not test it—it will not be rendered meaningful by the fact that a sound having the structure of this sentence came from A's own mouth.

But—it will be asked—do we not need the statements of our fellow-men for the elaboration of inter-subjective science? Would not physics, geography, and history become very meager studies if I had to restrict myself in them to occurrences which I myself had directly observed? There is no denying that they would. But there is a basic difference between a statement by A about the geography of China or about some historical event in the past on the one hand, and, on the other, a statement by A about the anger he felt yesterday. I can, in principle, test the statements of the first sort by means of perception sentences of my own protocol, sentences about my own perceptions of China, or of some map, or of historical documents. It is, however, in principle impossible for me to test the statement about anger if our critic asks me to reject the physical meaning of the sentence. If I have often had occasion to note that the geographical or historical reports that A makes can be confirmed by me, then, on the basis of an inductive probability inference, I consider myself justified in using his other statements—insofar as they are meaningful to me—in the elaboration of my scientific knowledge. It is in this way that inter-subjective science is developed. A sentence, however, which is not testable and hence not meaningful prior to its statement by A is not any the more meaningful after such a statement. If, in accordance with our position, I construe A's statement about yesterday's anger as a statement about the physical condition of A's body yesterday, this statement *may* be used for the development of inter-subjective science. For we use A's sentence as evidence (just to the extent to which we have found A to be trustworthy) in support of the attribution of a corresponding physical structure to the corresponding spatio-temporal region of our physical world. Neither do the consequences which we draw from this attribution generically differ from those that are obtained from any other physical statement. We build our expectations of future perceptions on it—in this case with respect to A's behavior, as in other cases with respect to the behavior of other physical systems.

The assertions of our fellow men contribute a great deal to extending the range of our knowledge. But they cannot bring us anything *basically* new, that is, anything which cannot also be learned in some other way. For the assertions of our fellow men are, at bottom, no different from other physical events. Physical events are different from one another as regards the extent to which they may be used as signs of other physical events. Those physical events which we call "assertions of our fellow man" rank particularly high on this scale. It is for this reason that science, quite rightly, treats these events with special consideration. However, between the contribution of these assertions to our scientific knowledge and the contributions of a barometer there is, basically, at most a difference of degree.

5. Behaviorism and "Intuitive" Psychology

The position we are advocating here coincides in its broad outlines with the psychological movement known as "behaviorism"—when, that is, its epistemological principles rather than its special methods are considered. We have not linked our exposition with a statement of behaviorism since our only concern is with epistemological foundations while behaviorism is above all else interested in a specific method of research and in specific concept formations.

The advocates of behaviorism were led to their position through their concern with animal psychology. In this domain, when the material given to observation does not include statements but only inarticulate behavior, it is most easy to arrive at the correct method of approach. This approach leads one to the correct interpretation of the statements of human experimental subjects, for it suggests that these statements are to be conceived as acts of verbalizing behavior, basically no different from other behavior.

Behaviorism is confronted with views, more influential in Germany than in the United States, which uphold the thesis that psychology's concern is not with behavior in its physical aspect, but rather, with *meaningful behavior*. For the comprehension of meaningful behavior the special method known as "intuitive understanding" ("Verstehen") is said to be required. Physics allegedly knows nothing of this method. Neither meaningful behavior considered collectively nor the individual instances of such behavior which psychology investigates can possibly—so it is maintained—be characterized in terms of physical concepts.

In intuitive psychology this view is generally linked with the view that beside physical behavior there is yet another, psychical event, which constitutes the true subject-matter of psychology, and to which intuitive understanding leads. We do not want to consider this idea any further here, since we have already thoroughly examined it.

But even after one puts this idea aside, intuitive psychology poses the following objection to physicalism.

Objection based on the occurrence of "meaningful behavior." "When psychology considers the behavior of living creatures (we disregard here the question whether it deals only with such behavior), it is interested in it as meaningful behavior. This aspect of behavior cannot, however, be grasped in terms of physical concepts, but only by means of the method of intuitive understanding. And this is why psychological sentences cannot be translated into physical sentences."

Rebuttal. Let us recall a previous example of the *physicalization* of an intuitive impression, i.e. of a qualitative designation in the protocol language.[4] We there showed that it is possible by investigating optical state-coordinates, to determine the entirety of those physical conditions which correspond to "green of this specific sort" and to subsume them under laws. The same is the case here. It simply depends on the physical nature of an act—say, of an arm-movement —whether I can intuitively understand it—as, say, a beckoning-motion—or not. Consequently, physicalization is possible here too. The class of arm-movements to which the protocol-designation "beckoning motion" corresponds can be determined, and then described in terms of physical concepts. But perhaps doubts may be raised as to whether the classification of arm-movements as intelligible or unintelligible, and, further, the classification of intelligible arm-movements as beckoning motions or others really depends, as our thesis claims, solely on the physical constitution of the arms, the rest of the body, and the environment. Such doubts are readily removed if, for instance, one thinks of films. We understand the *meaning* of the action on the movie screen. And our understanding would doubtless be the same if, instead of the film presented, another which resembled it in every physical particular were shown. Thus one can see that both our understanding of meaning and the particular forms it takes are, in effect, completely determined by the physical processes impinging on our sense-organs (in the film-example, those impinging on our optic and auditory sense-organs).

The problem of physicalization in this area, that is, the problem of the characterization of *understandable* behavior as such and of the various kinds of such behavior by means of concepts of systematized physics, is not as yet solved. But does not then our basic thesis rest on air? It states that all psychological sentences can be translated into physical sentences. One may well ask to what extent

---

4. *Erkenntnis*, Vol. II, *op. cit.*, pp. 444 ff. (*The Unity of Science*, p. 58 ff.).

such a translation is possible, given the present state of our knowledge. Even today every sentence of psychology *can* be translated into a sentence which refers to the physical behavior of living creatures. In such a physical characterization terms do indeed occur which have not yet been physicalized, i.e. reduced to the concepts of physical science. Nevertheless, the concepts used *are* physical concepts, though of a primitive sort—just as "warm" and "green" (applied to bodies) were physical concepts before one could express them in terms of physical state-coordinates (temperature and electromagnetic field, respectively).

We should like, again, to make the matter clear by using a *physical example*. Let us suppose that we have found a substance whose electrical conductivity is noticeably raised when it is irradiated by various types of electro-magnetic radiation. We do not yet, however, know the internal structure of this substance and so cannot yet explain its behavior. We want to call such a substance a "detector" for radiation of the sort involved. Let us suppose, further, that we have not yet systematically determined to what sorts of radiation the detector reacts. We now discover that the sorts of radiation to which it responds share still another characteristic, say, that they accelerate specific chemical reactions. Now suppose that we are interested in the photo-chemical effects of various sorts of radiation, but that the determination of these effects, in the case of a specific sort of radiation, is difficult and time-consuming, while the determination of the detector's reaction to it is easy and quickly accomplished; then we shall find it useful to adopt the detector as a test-instrument. With its aid we can determine for any particular sort of radiation whether or not it is likely to have the desired photo-chemical effect. This practical application will not be impeded by our ignorance of the detector's micro-structure and our inability to explain its reaction in physical terms. In spite of our ignorance, we can certainly say that the detector isolates a certain physically specified class of rays. The objection that this is not a physical class since we cannot characterize it by a specification of optical state-coordinates but only by the behavior of the detector will not stand. For to begin with, we know that if we carried out a careful empirical investigation of the electro-magnetic spectrum, we could identify the class of rays to which the detector responds. On the basis of this identification we could then physicalize the characterization of the rays in terms of detector-reactions, by substituting for it a characterization in terms of systematic physical concepts. But even our present way of characterizing the radiation in terms of the detector-test is a physical characterization, though an indirect one. It is distinguished from the direct characterization which is our goal only through being more circumstantial. There is no difference of kind between the two characterizations, only one of degree, though the difference of degree is indeed sufficiently great to give us a motive for pursuing the empirical investigations which might bring the direct physical characterization within our grasp.

Whether *the detector is organic or inorganic* is irrelevant to the epistemological issue involved. The function of the detector is basically the same whether we are dealing with a physical detector of specific sorts of radiation or with a tree-frog as a detector of certain meteorological states of affairs or (if one may believe the newspapers) with a sniffing dog as a detector of certain human diseases. People take a practical interest in meteorological forecasts. Where barometers are not available they may, consequently, use a tree-frog for the same purpose. But let us be clear about the fact that this method does not determine the state of the tree-frog's soul, but a physically specified weather condition, even if one cannot describe this condition in terms of the concepts of systematized physics. People, likewise, have a practical interest in medical diagnoses. When the directly determinable symptoms do not suffice, they may, consequently, enlist a dog's delicate sense of smell for the purpose. It is clear to the doctor that, in doing so, he is not determining the state of the dog's soul, but a physically specified condition of his patient's body. The doctor may not be able, given the present state of physiological knowledge, to characterize the diseased condition in question in terms of the concepts of systematic physics. Nonetheless, he knows that his diagnosis—whether it is based on the symptoms he himself has directly observed or on the reactions of the diagnostic dog—determines nothing and can determine nothing but the physical condition of his patient. Even apart from this, the physiologist acknowledges the need for physicalization. This would here consist in describing the bodily condition in question, i.e. defining the disease involved in purely physiological terms (thus eliminating any mention of the dog's reaction). A further task would be to trace these back to chemical terms, and these, in turn, to physical ones.

The case with *intuitive psychology* is precisely analogous. The situation here happens to be complicated for epistemological analysis (though for psychological practice it is simplified) by the fact that in the examination of an experimental subject the intuitive psychologist is both the observer and detector. The doctor here is his own diagnostic dog; which, indeed, is also often the case in medical diagnoses—in their intuitive phases. The psychologist calls the behavior of the experimental subject "understandable" or, in a special case, for instance, "a nod of affirmation," when his detector responds to it, or—in our special case—when it results in his protocols registering "A nods affirmatively." Science is not a system of experiences, but of sentences; it does not include the psychologist's experience of understanding, but rather, his protocol sentence. The utterance of the psychologist's protocol sentence is a reaction whose epistemological function is analogous to the tree-frog's climbing and to the barking of the diagnostic dog. To be sure, the psychologist far surpasses these animals in the variety of his reactions. As a result, he is certainly very valuable to the pursuit of science. But this constitutes only a difference of degree, not a difference of kind.

In the light of these considerations, two demands are to be made of the psychologist. First, we shall expect him (as we expect the doctor) to be clear about the fact that, in spite of his complicated diagnostic reaction, he establishes nothing but the existence of some specific physical condition of the experimental subject, though a condition which can be characterized only indirectly—by his own diagnostic reaction. Secondly, he must acknowledge (as the physiologist does) that it is a task of scientific research to find a way of physicalizing the indirect characterization. Psychology must determine what are the physical conditions to which people's detector-reactions correspond. When this is carried out for every reaction of this sort, i.e. for every result of intuitive understanding, psychological concept formation can be physicalized. The indirect definitions based on detector-reactions will be replaced by direct definitions in terms of the concepts of systematized physics. Psychology, like the other sciences, must and will reach the level of development at which it can replace the tree-frog by the barometer. But even in the tree-frog stage psychology already uses physical language, though of a primitive sort.

6. Physicalization in Graphology

The purpose of this section is not to justify physicalism, but only to show how psychological concepts can in fact be physicalized. To this end we shall examine a branch of psychology in which physicalization has already been undertaken with some success. In doing so we may perhaps also meet the criticism which is occasionally voiced, that the achievement of physicalization, assuming it were possible, would in any case be fruitless and uninteresting. It is held that, given sufficient information concerning the social group and the circumstances of the people involved, one might perhaps be able to specify arm-movements which are interpreted as beckoning-motions in such a way that they would be characterizable in terms of kinematic (i.e. spatio-temporal) concepts. But it is alleged that this procedure would not provide us with any further insight into anything of interest, least of all into the connections of these with other events.

Remarkably enough, physicalization can show significant success in a branch of psychology which until comparatively recent times was pursued in a purely intuitive (or at most a pseudo-rational) manner and with wholly inadequate empirical data, so that it then had no claim to scientific status. This is graphology. Theoretical graphology—we shall concern ourselves here with no other sort—investigates the law-like relationships which hold between the formal properties of a person's handwriting and those of his psychological properties that are commonly called his "character."

We must first of all explain what is meant by *character* in physical psychology. Every psychological property is marked out as a disposition to behave in a certain way. By "actual property" we shall understand a property which is defined by characteristics that can be directly observed: by "disposition" (or "dispositional concept") we shall understand a property which is defined by means of an implication (a conditional relationship, an if-then sentence). Examples of familiar dispositional concepts of physics may serve to illustrate this distinction, and will, at the same time, illustrate the distinction between occurrent and continuant properties, a distinction which is important in psychology. An example of a physical *occurrent property* is a specific degree of temperature. We define "Body K has temperature T" to mean "When a sufficiently small quantity of mercury is brought into contact with K, then . . . " When defined in this way, the concept of temperature is a dispositional concept. Now that physics has disclosed the micro-structure of matter and determined the laws of molecular motion, a different definition of temperature is used: temperature is the mean kinetic energy of molecules. Here, then, temperature is no longer a dispositional concept, but an actual property. The *occurrent properties of psychology* are logically analogous to the familiar dispositional concepts of physics. Indeed, on our view, they are themselves nothing else than physical concepts. Example: "Person X is excited" means "If, now, stimuli of such and such a sort were applied, X would react in such and such a manner" (both stimuli and reactions being physical events). Here too the aim of science is to change the form of the definition; more accurate insight into the micro-structure of the human body should enable us to replace dispositional concepts by actual properties. That this is not a utopian aim is shown by the fact that even at the present time, a more accurate knowledge of physiological macro-events has yielded us a set of actual characteristics of occurrent states (e.g. for feelings of various sorts: frequency and intensity of pulse and respiration, glandular secretion, innervation of visceral muscles, etc.). Such a change of definitions is markedly more difficult when the states which have to be delimited are not emotional, for it then presupposes a knowledge of the micro-structure of the central nervous system which far surpasses the knowledge currently available.

Physical constants, e.g. heat-conductivity, coefficient of refraction, etc. might be taken as examples of physical *continuant properties*. These too were originally defined as dispositional concepts, e.g. "A substance has a coefficient of refraction n" means "If a ray of light enters the substance, then . . . " Here again the aim of transforming the definition has already been achieved for some concepts, and is being pursued in the case of the remainder. The reference to dispositions gives way to an actual designation of the composition (in terms of atoms and electrons) of the substance in question. The *psychological continuant properties* or "character properties" (the word "character" is here being used in a broad, neutral sense—to mean more than volitional or attitudinal properties) can, at present, be defined only in the form of dispositional concepts. Example: "X is more impressionable than Y" means "If both X and Y have the same experience under the same circumstances, more intense feelings are experienced by X than by Y." In these definitions, both in the characterization of the stimuli (the statement of the circumstances) and in that of the reaction, there are names which still designate psychological occurrent properties, for which the problem of physicalization has not yet been solved. To physicalize the designations of continuant properties will be possible only when the designations of occurrent properties have been dealt with. So long as these are not completely physicalized, the physicalization of continuant properties and, as a result, that of characterology as a whole, must remain in a scientifically incomplete state, and this no matter how rich our stock of intuitive knowledge may be.

There is no sharp division between occurrent and continuant designations. Nonetheless, the difference of degree is large enough to justify their being differently labelled and differently treated, and, consequently, large enough to justify the separation of characterology

135

from psychology as a whole (considered as the theory of behavior). Graphology sets itself the task of finding in the features of a person's handwriting indications of his character and, to some extent, of his occurrent properties. The practising graphologist does not intend the rational method to replace intuition, but only to support or to correct it. It has, however, become clear that the pursuit of the task of physicalization will serve even this purpose. Along these lines graphology has already, of late, made some significant discoveries.

Since the problem of graphology is to discover the correspondences holding between the properties of a person's handwriting and those of his character, we may here divide the problem of physicalization into three parts. The physicalization of the properties of handwriting constitutes the *first part of the problem*. A certain script gives me, for instance, an intuitive impression of something full and juicy. In saying so, I do not primarily refer to characteristics of the writer, but to characteristics of his script. The problem now is to replace intuitively identified script-properties of this sort by properties of the script's shape, i.e. by properties which may be defined with the aid of geometrical concepts. That this problem can be solved is clear. We need only thoroughly investigate the system of forms which letters, words, and lines of script might possibly take in order to determine which of these forms make the intuitive impression in question on us. So, for instance, we might find that a script appears full or two-dimensional (as opposed to thin or linear) if rounded connections are more frequent than angles, the loops broader than normal, the strokes thicker, etc. This task of the physicalization of the properties of handwriting has in many cases been accomplished to a large extent.[5] We are not objecting to the retention of the intuitively derived descriptions (in terms, for instance, of "full," "delicate," "dynamic," etc.). Our requirement will be adequately met as soon as a definition in exclusively geometric terms is provided for each such description. This problem is precisely analogous to the problem, to which we have frequently referred, of identifying in quantitative terms those physical conditions which correspond to a qualitative designation—such as "green of such and such a sort"—in the protocol language.

The *second part of the problem* consists of the physicalization of the character properties referred to in graphological analyses. The traditional concepts of characterology—whose meaning is as a

---

5. *Cf.* Klages, L., *Handschrift und Character*, Leipzig, 1920. Several of our examples are taken from this book or suggested by it.

rule not clearly defined, but left to be expressed in our everyday vocabulary or by means of metaphorical language—have to be systematized and given physicalistic (behavioristic) definitions. We have already seen that such a definition refers to a disposition to behave in a certain way, and further, that the task of the construction of such definitions is difficult and presupposes the physicalization of psychological occurrent properties.

We can see that in both parts of the problem the task is one of replacing primitive, intuitive concept formations by systematic ones, of replacing the observer with a tree-frog by the observer with a barometer (in graphology, as in intuitive medical diagnoses, the observer and the tree-frog coincide).

In addition to these questions there is a third aspect of the problem to be considered: the basic empirical task of graphology. This consists of the search for the correlations which hold between the properties of handwriting and those of character. Here too, a systematization, though of a different sort, takes place. The correspondence of a specific property of handwriting to a specific property of character may, at first, be recognized intuitively—for instance, as a result of an empathetic reflection on the arm-movements which produced the script in question. The problem of systematization here is to determine the degree of correlation of the two properties by a statistical investigation of many instances of script of the type in question and the characters of the corresponding writers.

Our position now is that the further development and clarification of the concepts of psychology as a whole must take the direction we have illustrated in our examination of graphology, the direction, that is, of physicalization. But, as we have already emphasized several times, psychology is a physical science even prior to such a clarification of its concepts—a physical science whose assignment it is to describe systematically the (physical) behavior of living creatures, especially that of human beings, and to develop laws under which this behavior may be subsumed. These laws are of quite diverse sorts. A hand movement, for instance, may be examined from various aspects: first, semiotically, as a more or less conventional sign for some designated state of affairs; secondly, mimically, as an expression of the contemporaneous psychological state—the occurrent properties of the person in question; thirdly, physiognomically, as an expression of the continuant properties—the character of the person in question. In order to investigate, say, the hand movements of people (of certain groups) in their mimical and physiognomic aspects one might perhaps take motion pictures of them, and, from these, derive kinematic diagrams of the sort which engineers construct for machine parts. In this manner the shared kinematic (i.e., spatio-temporal) characteristics of the hand movements with whose perception certain intuitive protocol designations tend to be associated (e.g. "This hand movement looks rushed," ". . . grandiose," etc.) would have to be determined. It will now be clear why precisely graphology—the characterological investigation of writing movements, a very special sort of hand movements, identifiable in terms of their specific purpose—should be the only study of this sort which can as yet show any results. The reason is that writing movements themselves produce something resembling kinematic diagrams, namely, the letters on the paper. To be sure, only the track of the movements is drawn. The passage of time is not recorded —the graphologist can subsequently only infer this, imperfectly, from indirect signs. More accurate results would be demonstrable if the complete three-dimensional spatio-temporal diagram, not only its projection on the writing plane, were available. But even the conclusions to which graphology currently subscribes allay whatever misgivings there might have been that investigations directed at the physicalization of psychological concepts would prove to be uninteresting. It may not even be too rash a conjecture that interesting parallels may be found to hold between the conclusions of characterological investigations of both the involuntary and the voluntary motions of the various parts of the human body on the one hand, and on the other hand the conclusions of graphology which are already available to us. If specific properties of a person's character express themselves both in a specific form of handwriting and in a specific form of arm motion, a specific form of leg motion, specific facial features, etc., might not these various forms resemble one another? Perhaps, after having first given fruitful suggestions for the investigation of other sorts of bodily movements, graphology may, in turn, be stimulated by the results to examine script properties it had previously overlooked. These, of course, are mere conjectures; whether or not they are justifiable cannot affect the tenability of our thesis, which maintains the possibility of translating all psychological sentences into physical language. This translatability holds regardless of whether or not the concepts of psychology are physicalized. Physicalization is simply a higher-level, more rigorously systematized scientific form of concept formation. Its accomplishment is a practical problem which concerns the psychologist rather than the epistemologist.

7. SENTENCES ABOUT ONE'S OWN MIND; "INTROSPECTIVE PSYCHOLOGY"

Our argument has shown that a sentence about other minds refers to physical processes in the body of the person in question. On any other interpretation the sentence becomes untestable in principle, and thus meaningless. The situation is the same with sentences about one's own mind, though here the emotional obstacles to a physical interpretation are considerably greater. The relationship of a sentence about one's own mind to one about someone else's may most readily be seen with respect to a sentence about some *past state* of one's own mind, e.g. P_1: "I was excited yesterday." The testing of this sentence involves either a *rational* inference from protocol sentences of the form of p_1—which refer to presently perceived script, photographs, films, etc. originating with me yesterday; or it involves an *intuitive* method, e.g. utilizing the protocol sentence p_2, "I recall having been excited yesterday." The content of P_1 exceeds both that of the protocol sentence p_1 and that of the protocol sentence p_2, as is most clearly indicated by the possibility of error and disavowal where P_1 is concerned. P_1 can only be progressively better confirmed by sets of protocol sentences of the form of p_1 and p_2. The very same protocol sentences, however, also confirm the physical sentence P_2: "My body was yesterday in that physical condition which one tends to call 'excitement.'" P_1 has, consequently, the same content as the physical sentence P_2.

In the case of a sentence about the *present state* of one's own mind, *e.g.* P_1: "I now am excited" one must clearly distinguish between the system sentence P_1 and the protocol sentence p_2, which, likewise, may read "I now am excited." The difference rests in the fact that the system sentence P_1 may, under certain circumstances, be disavowed, whereas a protocol sentence, being an epistemological point of departure, cannot be rejected. The protocol sentences p_1 which rationally support P_1 have here some such form as "I feel my hands trembling," "I see my hands trembling," "I hear my voice quavering," etc. Here too, the content of P_1 exceeds that of both p_1 and p_2, in that it subsumes all the possible sentences of this sort. P_1 has the same content as the physical sentence P_2, "My body is now in that condition which, both under my own observation and that of others, exhibits such and such characteristics of excitement," the characteristics in question being those which are mentioned both in my own protocol sentences of the sort of p_1

and p_2 and in other people's protocol sentences of corresponding sorts (discussed above in our example of sentences about other minds).

The table opposite shows the analogous application of the physicalist thesis to the three cases we have discussed by exhibiting the parallelism of sentences about other minds, sentences about some past condition of one's own mind, and sentences about the present condition of one's own mind, with the physical sentence about the wooden support.

Objection from introspective psychology: "When the psychologist is not investigating other experimental subjects, but pursues self-observation, or "introspection," instead, he grasps, in a direct manner, something non-physical—and this is the proper subject-matter of psychology."

Rebuttal. We must distinguish between a question of the justification of the use of some prevalent practical method of inquiry and a question of the justification of some prevalent interpretation of the results of that method. *Every* method of inquiry is justified; disputes can arise only over the question of the purpose and fruitfulness of a given method, which is a question our problem does not involve. We may apply any method we choose; we cannot, however, interpret the obtained sentences as we choose. The meaning of a sentence, no matter how obtained, can unequivocally be determined by a logical analysis of the way in which it is derived and tested. A psychologist who adopts the method of what is called "introspection" does not thereby expose himself to criticism. Such a psychologist admits sentences of the form "I have experienced such and such events of consciousness" into his experiment-protocol and then arrives at general conclusions of his own by means of inductive generalization, the construction of hypotheses, and, finally, a comparison of his hypotheses with the conclusions of other persons. But again we must conclude, both on logical and epistemological grounds, that the singular as well as the general sentences must be interpreted physically. Let us say that psychologist A writes sentence p_2: "(I am) now excited" into his protocol. An earlier investigation[6] has shown that the view which holds that protocol sentences cannot be physically interpreted, that, on the contrary, they refer to something non-physical (something "psychical," some "experience-content," some "datum of consciousness," etc.) leads directly to the consequence that every protocol sentence is meaningful only to its author. If A's protocol sentence p_2 were not subject to a

---

6. *Erkenntnis*, Vol. II, p. 454, (*The Unity of Science*, pp. 78-79).

THE PHYSICALISTIC INTERPRETATION OF PSYCHOLOGICAL SENTENCES

	1. Sentence about the Wooden Support (As an Analogy)	2. Sentence about the State of Someone Else's Mind	3. Sentence about the State of One's Own Mind at Some Time in the Past	4. Sentence about the Present State of One's Own Mind
System sentence P_1: a) *rationally* derived from protocol sentence p_1:	"The support is firm"	"A is excited"	"I was excited yesterday"	"I am now excited"
or b) *intuitively* derived from protocol sentence p_2:	"The support has such and such a color and shape"	"A has such and such an expression"	"These letters (written by me yesterday) have such and such a shape"	"My hands are now trembling"
	"The support looks firm"	"A is excited (A looks excited)"	"Now a recollection of excitement"	"Now excited"
P_1 has the same content as the *physical sentence* P_2:	"The support is physically firm"	"A's body is physically excited"	"My body was physically excited yesterday."	"My body is now physically excited"
The physical term: is hereby defined as a disposition to react under certain circumstances in a specified way:	"physically firm"		"physically excited"	
	"Under such and such a load, such and such a distortion occurs; under such and such a load, breakage occurs"	"Under such and such circumstances, such and such gestures, expressions, actions, and words occur."		

141

physical interpretation, it could not be tested by B, and would, thus, be meaningless to B. On the previous occasion in question we showed, further, that the non-physical interpretation leads one into insoluble contradictions. Finally, we found that every protocol sentence has the same content as some physical sentence,[7] and that this physical translation does not presuppose an accurate knowledge of the physiology of the central nervous system, but is feasible even at present. Sentences about one's own mind—whether one takes these to be inter-subjective system sentences or so-called introspective protocol sentences—are thus in every case translatable into sentences of the physical language.

One may perhaps object that there is, after all, a difference between an experience and an utterance about it, and that not every experience has to be expressed in a protocol sentence. The difference referred to certainly exists, though we would formulate it differently. Sentences P_1: "A now sees red" and P_2: "A now says 'I see red'" do not have exactly the same content. Nor does P_1 justify the inference of P_2; only the conditional sentence "If this and that occurs, then P_2" may be inferred. For P_1 ascribes a physical state to A of such a kind that, under certain circumstances, it leads to the event of speaking the sentence referred to in P_2.

If we consider the method in accordance with which the conclusions of so-called introspection are generally integrated with the body of scientific knowledge, we shall note that these conclusions are, indeed, physically evaluated. It so happens that the physicalism adopted in practice is generally not acknowledged in theory. Psychologist A announces his experimental results; reader B reads in them, among others, the sentence "A was excited" (for the sake of clarity we write "A" instead of the word "I" which B in reading must replace by "A"). For B, this is a sentence about someone else's mind; nothing of its claim can be verified except that A's body *was* in such and such a physical condition at the time referred to. (We argued this point in our analysis of sentence P_1 about someone else's mind.) B himself could not, indeed, have observed this condition, but he can now indirectly infer its having existed. For, to begin with, he sees the sentence in question in a book on whose title-page A is identified as the author. Now, on the basis of a general sentence for which he has already obtained indirect evidence, B infers (with some degree of probability) that A wrote the sentences printed in this book; from this, in its turn, on the basis of

---

7. *Ibid.*, pp. 457 ff., (*The Unity of Science*, pp. 84 ff.).

a general sentence, with regard to A's reliability, for which he again has good inductive evidence, B infers that, had he observed A's body at the relevant time he would (probably) have been able to confirm the existence of the state of (physical) excitement. Since this confirmation can refer only to some physical state of A's body, the sentence in question can have only a physical meaning for B.

Generally speaking, a psychologist's spoken, written, or printed protocol sentences, when they are based on so-called introspection, are to be interpreted by the reader, and so figure in inter-subjective science, *not chiefly as scientific sentences, but as scientific facts.* The epistemological confusion of contemporary psychology stems, to a large extent, from this confusion of facts in the form of sentences with the sentences themselves considered as parts of science. (Our example of the patterns in the sky is relevant here.) The introspective statements of a psychologist are not, in principle, to be interpreted any differently from the statements of his experimental subjects, which he happens to be reporting. The only distinction the psychologist enjoys is that, when the circumstances justify it, one may accept his statements as those of an exceptionally reliable and well-trained experimental subject. Further, the statements of an experimental subject are not, in principle, to be interpreted differently from his other voluntary or involuntary movements—though his speech movements may, under favorable circumstances, be regarded as especially informative. Again, the movements of the speech organs and of the other parts of the body of an experimental subject are not, in principle, to be interpreted differently from the movements of any other animal—though the former may, under favorable circumstances, be more valuable in the construction of general sentences. The movements of an animal are not, again, in principle, to be interpreted any differently from those of a volt-meter—though under favorable circumstances, animal movements may serve scientific purposes in more ways than do the movements of a volt-meter. Finally, the movements of a volt-meter are not, in principle, to be interpreted differently from the movements of a raindrop—though the former offer more opportunities for drawing inferences to other occurrences than do the latter. In all these cases, the issue is basically the same: from a specific physical sentence, other sentences are inferred by a causal argument, i.e. with the help of general physical formulae—the so-called natural laws. The examples cited differ only in the degree of fruitfulness of their premises. Volt-meter readings will, perhaps, justify the inference of a greater number of scientifically important sentences than the behavior of some specific raindrop will; speech movements will, in a certain respect, justify more such inferences than other human bodily movements will. Now, in the case with which we are concerned here, the inference from the sign to the state of affairs signified has a quite remarkable form. In using someone's introspective statement about the state of his own mind (e.g. A's statement: "A is excited"), the statement, taken as an acoustic event, is the sign; under favorable conditions, which are frequently satisfied in scientific contexts, the state of affairs referred to is such that it can be described by a sentence ("A is excited") of the very same form as the acoustic event which functions as a sign of it. [The requisite conditions are that the person in question be considered reliable and qualified to make psychological reports, and further that the language of these reports be the same as that of the scientific system.] This identity of the form of the acoustic fact and the scientific sentence which is to be inferred from it explains why the two are so easily and so obstinately confused. The disastrous muddle into which this confusion leads us is cleared up as soon as we realize that here, as in the other cases cited, it is only a question of drawing an inference from a sign to that which it indicates.

It becomes all the more clear that so-called introspective statements cannot be given a non-physical interpretation when we consider how their use is learned. A tired child says "Now I am happy to be in bed." If we investigated how the child learned to talk about the states of his own mind we would discover that, under similar circumstances, his mother had said to him, "Now you are happy to be in bed." Thus we see that A learns to use the protocol sentence p_2 from B—who, however, interprets this series of words as constituting the system sentence P_2, a sentence, for B, about someone else's mind. Learning to talk consists of B's inducing a certain habit in A, a habit of "verbalizing" (as the behaviorists put it) in a specific manner in specific circumstances. And, indeed one tends so to direct this habit that the series of words produced by the speech movements of the child A coincides with the sentence of the intersubjective physical language which not only describes the appropriate state of A, but—and this is the essential point—describes A's state *as B perceives it,* that is, the physical state of A's body. The example of the child shows this especially clearly. The sentence, "You are happy," spoken by the mother, is a sentence about someone else's mind, and thus, according to our earlier analysis, can designate nothing but some physical state of affairs. The child is thus induced to develop the habit of responding to specific circumstances by uttering a sentence which expresses a physical state observed by some other person (or inferred by some other person from observed signs). If the child utters the same sounds again on some other occasion, no more can be inferred than that the child's body is again in that physical state.

8. SUMMARY

So-called psychological sentences—whether they are concrete sentences about other minds, or about some past condition of one's own mind, or about the present condition of one's own mind, or, finally, general sentences—are always translatable into physical language. Specifically, every psychological sentence refers to physical occurrences in the body of the person (or persons) in question. On these grounds, psychology is a part of the domain of unified science based on physics. By "physics" we wish to mean, not the system of currently known physical laws, but rather the science characterized by a mode of concept formation which traces every concept back to state-coordinates, that is, to systematic assignments of numbers to space-time points. Understanding "physics" in this way, we can rephrase our thesis—a particular thesis of physicalism—as follows: *psychology is a branch of physics.*

REMARKS BY THE AUTHOR (1957)

While I would still maintain the essential content of the main thesis of this article, I would today modify some special points. Perhaps the most important of them is the following. In the article I regarded a psychological term, say "excited," as designating a state characterized by the disposition to react to certain stimuli with overt behavior of certain kinds. This may be admissible for the psychological concepts of everyday language. But at least for those of scientific psychology, as also of other fields of science, it seems to me more in line with the actual procedure of scientists, to introduce them not as disposition concepts, but rather as theoretical concepts (sometimes called "hypothetical constructs"). This means that they are introduced as primitives by the postulates of a theory, and are connected with the terms of the observation language, which designate observable properties, by so-called rules of correspondence. This method is explained and discussed in detail in my article "The Methodological Character of Theoretical Concepts," in H. Feigl and M. Scriven, (eds., *Minnesota Studies in the Philosophy of Science, Vol. I.*

145

The main thesis of physicalism remains the same as before. It says that psychological statements, both those of everyday life and of scientific psychology, say something about the physical state of the person in question. It is different from the corresponding statements in terms of micro-physiology or micro-physics (which at the present stage of scientific development are not yet known, comp. § 4A above) by using the conceptual framework of psychology instead of those of the two other fields. To find the specific features of the correspondence will be an empirical task (comp. § 6, the third part of the procedure of physicalization). Once known, the correspondence can be expressed by empirical laws or, according to our present view, by theoretical postulates. Our present conception of physicalism, the arguments for it, and the development which led to it, are represented in the following two articles by Herbert Feigl: (1) "Physicalism, Unity of Science and the Foundations of Psychology," in: P. A. Schilpp, editor, *The Philosophy of Rudolf Carnap* (Library of Living Philosophers); see also my reply to Feigl in the same volume; (2) "The 'Mental' and the 'Physical,'" in Vol. II of *Minnesota Studies in Philosophy of Science.*

1

The Logical Analysis of Psychology

Carl G. Hempel

Author's prefatory note, 1977. The original French version of this article was published in 1935. By the time it appeared in English, I had abandoned the narrow translationist form of physicalism here set forth for a more liberal reductionist one, referred to in note 1, which presents psychological properties and states as partially characterized, but not defined, by bundles of behavioral dispositions. Since then, I have come to think that this conception requires still further broadening, and that the introduction and application of psychological terms and hypotheses is logically and methodologically analogous to the introduction and application of the terms and hypotheses of a physical theory.[*] The considerations that prompt-

ed those changes also led me long ago to abandon as untenable the verificationist construal of the "empirical meaning" of a sentence—a construal which plays such a central role in the arguments set forth in this article.

Since the article is so far from representing my present views, I was disinclined to consent to yet another republication, but I yielded to Dr. Block's plea that it offers a concise account of an early version of logical behaviorism and would thus be a useful contribution to this anthology.

In an effort to enhance the closeness of translation and the simplicity of formulation, I have made a number of small

An earlier version of this paper appeared in Ausonio Marras., ed., *Intentionality, Mind, and Language* (Urbana: University of Illinois Press, 1972), pp. 115-131, and in Herbert Feigl and Wilfrid Sellars, eds., *Readings in Philosophical Analysis* (New York: Appleton-Century-Crofts, 1949), pp. 373-384, translated from the French by W. Sellars. Reprinted, with revisions by the author, by permission of the author, Herbert Feigl, Wilfrid Sellars, and the editors of *Revue de Synthese.*

[*]My reasons are suggested in some of my more recent articles, among them "Logical Positivism and the Social Sciences," in P. Achinstein and S. F. Barker, eds., *The Legacy of Legal Posi-*
tivism (Baltimore: Johns Hopkins University Press, 1969); "Reduction: Ontological and Linguistic Facets," in S. Morgenbesser, P. Suppes, and M. White, eds., *Philosophy, Science, and Method. Essays in Honor of Ernest Nagel* (New York: St. Martin's Press, 1969); "Dispositional Explanation and the Covering-Law Model: Response to Laird Addis," in A. C. Michalos and R. S. Cohen, eds., *PSA 1974: Proceedings of the 1974 Biennial Meeting of the Philosophy of Science Association* (Dordrecht: Reidel, 1976), pp. 369-376.

changes in the text of the original English version; none of these affects the substance of the article.

I

One of the most important and most discussed problems of contemporary philosophy is that of determining how psychology should be characterized in the theory of science. This problem, which reaches beyond the limits of epistemological analysis and has engendered heated controversy in metaphysics itself, is brought to a focus by the familiar alternative, "Is psychology a natural science, or is it one of the sciences of mind and culture (*Geisteswissenschaften*)?"

The present article attempts to sketch the general lines of a new analysis of psychology, one which makes use of rigorous logical tools, and which has made possible decisive advances toward the solution of the above problem.[1] This analysis was carried out by the "Vienna Circle" (*Weiner Kreis*), the members of which (M. Schlick, R. Carnap, P. Frank, O. Neurath, F. Waismann, H. Feigl, etc.) have, during the past ten years, developed an extremely fruitful method for the epistemological examination and critique of the various sciences, based in part on the work of L. Wittgenstein.[2] We shall limit ourselves essentially to the examination of psychology as carried out by Carnap and Neurath.

The method characteristic of the studies of the Vienna Circle can be briefly defined as a *logical analysis of the language of science*. This method became possible only with the development of a subtle logical apparatus which makes use, in particular, of all the formal procedures of modern symbolic logic.[3] However, in the following account, which does not pretend to give more than a broad orientation, we shall limit ourselves to setting out the general principles of this new method, without making use of strictly formal procedures.

II

Perhaps the best way to characterize the position of the Vienna Circle as it relates to psychology, is to say that it is the exact antithesis of the current epistemological thesis that there is a fundamental difference between experimental psychology, a natural science, and introspective psychology; and in general, between the natural sciences on the one hand, and the sciences of mind and culture on the other.[4] The common content of the widely different formulations used to express this contention, which we reject, can be set down as follows. Apart from certain aspects clearly related to physiology, psychology is radically different, both in subject matter and in method, from physics in the broad sense of the term. In particular, it is impossible to deal adequately with the subject matter of psychology by means of physical methods. The subject matter of physics includes such concepts as mass, wave length, temperature, field intensity, etc. In dealing with these, physics employs its distinctive method which makes a combined use of description and causal explanation. Psychology, on the other hand, has for its subject matter notions which are, in a broad sense, mental. They are *toto genere* different from the concepts of physics, and the appropriate method for dealing with them scientifically is that of empathetic insight, called "introspection," a method which is peculiar to psychology.

One of the principal differences between the two kinds of subject matter is generally believed to consist in the fact that the objects investigated by psychology—in contradistinction to those of physics—are specifically endowed with meaning. Indeed, several proponents of this idea state that the distinctive method of psychology consists in "understanding the sense of meaningful structures" (*sinnvolle Gebilde verstehend zu erfassen*). Take, for example, the case of a man who speaks. Within the framework of physics, this process is considered to be completely explained once the movements which make up the utterance have been traced to their causes, that is to say, to certain physiological processes in the organism, and, in particular, in the central nervous system. But, it is said, this does not even broach the psychological problem. The latter begins with understanding the sense of what was said, and proceeds to integrate it into a wider context of meaning.

It is usually this latter idea which serves as a principle for the fundamental dichotomy that is introduced into the classification of the sciences. There is taken to be an *absolutely impassable gulf* between the *natural sciences* which have a subject matter devoid of meaning and the *sciences of mind and culture*, which have an intrinsically meaningful subject matter, the appropriate methodological instrument for the scientific study of which is "comprehension of meaning."

III

The position in the theory of science which we have just sketched has been attacked from several different points of view.[5] As far as psychology is concerned, one of the principal countertheses is that formulated by behaviorism, a theory born in America shortly before the war. (In Russia, Pavlov has developed similar ideas.) Its principal methodological postulate is that a scientific psychology should limit itself to the study of the bodily behavior with which man and the animals respond to changes in their physical environment, and should proscribe as nonscientific any descriptive or explanatory step which makes use of terms from introspective or "understanding" psychology, such as 'feeling', 'lived experience', 'idea', 'will', 'intention', 'goal', 'disposition', 'repression'.[6] We find in behaviorism, consequently, an attempt to construct a scientific psychology which would show by its success that even in psychology we have

to do with purely physical processes, and that therefore there can be no impassable barrier between psychology and physics. However, this manner of undertaking the critique of a scientific thesis is not completely satisfactory. It seems, indeed, that the soundness of the behavioristic thesis expounded above depends on the possibility of fulfilling the program of behavioristic psychology. But one cannot expect the question as to the scientific status of psychology to be settled by empirical research in psychology itself. To achieve this is rather an undertaking in epistemology. We turn, therefore, to the considerations advanced by members of the Vienna Circle concerning this problem.

IV

Before addressing the question whether the subject matters of physics and psychology are essentially the same or different in nature, it is necessary first to clarify the very concept of the subject matter of a science. The theoretical content of a science is to be found in statements. It is necessary, therefore, to determine whether there is a fundamental difference between the statements of psychology and those of physics. Let us therefore ask what it is that determines the content—one can equally well say the "meaning"—of a statement. When, for example, do we know the meaning of the following statement: "Today at one o'clock, the temperature of such and such a place in the physics laboratory was 23.4° centigrade"? Clearly when, and only when, we know under what conditions we would call the statement true, and under what circumstances we would call it false. (Needless to say, it is not necessary to know whether or not the statement is true.) Thus, we understand the meaning of the above statement since we know that it is true when a tube of a certain kind filled with mercury (in short, a thermometer with a centigrade scale), placed at the indicated time at the location in question, exhibits a coincidence between the level of the mercury and the mark of the scale numbered 23.4. It is also true if in the same circumstances one can observe certain coincidences on another instrument called an "alcohol thermometer"; and, again, if a galvanometer connected with a thermopile shows a certain deviation when the thermopile is placed there at the indicated time. Further, there is a long series of other possibilities which make the statement true, each of which is described by a "physical test sentence," as we will call it. The statement itself clearly affirms nothing other than this: all these physical test sentences obtain. (However, one verifies only some of these physical test sentences, and then "concludes by induction" that the others obtain as well.) The statement, therefore, is nothing but an abbreviated formulation of all those test sentences.

Before continuing the discussion, let us sum up this result as follows:

1. A statement that specifies the temperature at a selected point in space-time can be "retranslated" without change of meaning into another statement—doubtless longer—in which the word "temperature" no longer appears. That term functions solely as an abbreviation, making possible the concise and complete description of a state of affairs the expression of which would otherwise be very complicated.

2. The example equally shows that *two statements which differ in formulation* can nevertheless have the *same meaning*. A trivial example of a statement having the same meaning as the above would be: 'Today at one o'clock, at such and such a location in the laboratory, the temperature was 19.44° Réaumur."

As a matter of fact, the preceding considerations show—and let us set it down as another result—that *the meaning of a statement is established by the conditions of its verification*. In particular, two differently formulated statements have the same meaning or the same effective content when, and only when, they are both true or both false in the same conditions. Furthermore, a statement for which one can indicate absolutely no conditions which would verify it, which is in principle incapable of confrontation with test conditions, is wholly devoid of content and without meaning. In such a case we have to do, not with a statement properly speaking, but with a "pseudo-statement," that is to say, a sequence of words correctly constructed from the point of view of grammar, but without content.[7]

In view of these considerations, our problem reduces to one concerning the difference between the circumstances which verify psychological statements and those which verify the statements of physics. Let us therefore examine a statement which involves a psychological concept, for example: "Paul has a toothache." What is the specific content of this statement, that is to say, what are the circumstances in which it would be verified? It will be sufficient to indicate some test sentences which describe these circumstances.

a. Paul weeps and makes gestures of such and such kinds.

b. At the question "What is the matter?," Paul utters the words "I have a toothache."

c. Closer examination reveals a decayed tooth with exposed pulp.

d. Paul's blood pressure, digestive processes, the speed of his reactions, show such and such changes.

e. Such and such processes occur in Paul's central nervous system.

This list could be expanded considerably, but it is already sufficient to bring out the fundamental and essential point, namely, that all the circumstances which verify this psychological statement are expressed by physical test sentences. [This is true even of test condition *b*, which merely expresses the fact that in specified physical circumstances (the propagation of vibrations produced in the air by the enunciation of the words, "What is the matter?") there occurs in the body of the subject a certain physical process (speech behavior of such and such a kind).]

The statement in question, which is about someone's "pain," is therefore, just like that concerning the temperature, simply an abbreviated expression of the fact that all its test sentences are verified.⁸ (Here, too, one verifies only some of the test sentences and then infers by way of induction that the others obtain as well.) It can be retranslated without loss of content into a statement which no longer contains the term "pain," but only physical concepts. Our analysis has consequently established that a certain statement belonging to psychology has the same content as a statement belonging to physics; a result which is in direct contradiction to the thesis that there is an impassable gulf between the statements of psychology and those of physics.

The above reasoning can be applied to *any psychological statement*, even to those which concern, as is said, "deeper psychological strata" than that of our example. Thus, the assertion that Mr. Jones suffers from intense inferiority feelings of such and such kinds can be confirmed or falsified only by observing Mr. Jones' behavior in various circumstances. To this behavior belong all the bodily processes of Mr. Jones, and, in particular, his gestures, the flushing and paling of his skin, his utterances, his blood pressure, the events that occur in his central nervous system, etc. In practice, when one wishes to test statements concerning what are called the deeper layers of the psyche, one limits oneself to the observation of external bodily behavior, and, particularly, to speech movements evoked by certain physical stimuli (the asking of questions). But it is well known that experimental psychology has also developed techniques for making use of the subtler bodily states referred to above in order to confirm the

psychological discoveries made by cruder methods. The statement concerning the inferiority feelings of Mr. Jones—whether true or false—means only this: such and such happenings take place in Mr. Jones' body in such and such circumstances.

We shall call a statement which can be translated without change of meaning into the language of physics, a "physicalistic statement," whereas we shall reserve the expression "statement of physics" to those which are already formulated in the terminology of physical science. (Since every statement is in respect of content equivalent to itself, every statement of physics is also a physicalistic statement.) The result of the preceding considerations can now be summed up as follows: *All psychological statements which are meaningful, that is to say, which are in principle verifiable, are translatable into statements which do not involve psychological concepts, but only the concepts of physics. The statements of psychology are consequently physicalistic statements. Psychology is an integral part of physics.* If a distinction is drawn between psychology and the other areas of physics, it is only from the point of view of the practical aspects of research and the direction of interest, rather than a matter of principle. This logical analysis, the result of which shows a certain affinity with the fundamental ideas of behaviorism, constitutes the physicalistic conception of psychology.

V

It is customary to raise the following fundamental objection against the above conception. The physical test sentences of which you speak are absolutely incapable of formulating the intrinsic nature of a mental process; they merely describe the physical *symptoms* from which one infers, by purely psychological methods—notably that of understanding—the presence of a certain mental process.

But it is not difficult to see that the

use of the method of understanding or of other psychological procedures is bound up with the existence of certain observable physical data concerning the subject undergoing examination. There is no psychological understanding that is not tied up physically in one way or another with the person to be understood. Let us add that, for example, in the case of the statement about the inferiority complex, even the "introspective" psychologist, the psychologist who "understands," can confirm his conjecture only if the body of Mr. Jones, when placed in certain circumstances (most frequently, subjected to questioning), reacts in a specified manner (usually, by giving certain answers). Consequently, even if the statement in question had to be arrived at, *discovered*, by "empathetic understanding," the only *information* it gives us is nothing more nor less than the following: under certain circumstances, certain specific events take place in the body of Mr. Jones. It is this which constitutes the meaning of the psychological statement.

The further objection will perhaps be raised that men can feign. Thus, though a criminal at the bar may show physical symptoms of mental disorder, one would nevertheless be justified in wondering whether his mental confusion was "real" or only simulated. One must note that in the case of the simulator, only some of the conditions are fulfilled which verify the statement "This man is mentally unbalanced," those, namely, which are most accessible to direct observation. A more penetrating examination—which should in principle take into account events occurring in the central nervous system—would give a decisive answer; and this answer would in turn clearly rest on a physicalistic basis. If, at this point, one wished to push the objection to the point of admitting that a man could show *all the "symptoms"* of a mental disease without being "really" ill, we reply that it would be absurd to characterize such a man as

"really normal"; for it is obvious that by the very nature of the hypothesis we should possess no criterion in terms of which to distinguish this man from another who, while exhibiting the same bodily behavior down to the last detail, would "in addition" be "really ill." (To put the point more precisely, one can say that this hypothesis contains a *logical contradiction*, since it amounts to saying, "It is possible that a statement should be false even when the necessary and sufficient conditions of its truth are fulfilled.")

Once again we see clearly that the meaning of a psychological statement consists solely in the function of abbreviating the description of certain modes of physical response characteristic of the bodies of men or animals. An analogy suggested by O. Neurath may be of further assistance in clarifying the logical function of psychological statements.[9] The complicated statements that would describe the movements of the hands of a watch in relation to one another, and relatively to the stars, are ordinarily summed up in an assertion of the following form: "This watch runs well (runs badly, etc.)." The term "runs" is introduced here as an auxiliary defined expression which makes it possible to formulate briefly a relatively complicated system of statements. It would thus be absurd to say, for example, that the movement of the hands is only a "physical symptom" which reveals the presence of a running which is intrinsically incapable of being grasped by physical means, or to ask, if the watch should stop, what has become of the running of the watch.

It is in exactly the same way that abbreviating symbols are introduced into the language of physics, the concept of temperature discussed above being an example. The system of physical test sentences *exhausts* the meaning of the statement concerning the temperature at a place, and one should not say that these sentences merely have to do with "symptoms" of the existence of a certain temperature.

Our argument has shown that it is necessary to attribute to the characteristic concepts of psychology the same logical function as that performed by the concepts of "running" and of "temperature." They do nothing more than make possible the succinct formulation of propositions concerning the states or processes of animal or human bodies.

The introduction of new psychological concepts can contribute greatly to the progress of scientific knowledge. But it is accompanied by a danger, that, namely, of making an excessive and, consequently, improper use of new concepts, which may result in questions and answers devoid of sense. This is frequently the case in metaphysics, notably with respect to the notions which we formulated in section II. Terms which are abbreviating symbols are imagined to designate a special class of "psychological objects," and thus one is led to ask questions about the "essence" of these objects, and how they differ from "physical objects." The time-worn problem concerning the relation between mental and physical events is also based on this confusion concerning the logical function of psychological concepts. Our argument, therefore, enables us to see that *the psycho-physical problem is a pseudo-problem*, the formulation of which is based on an inadmissible use of scientific concepts; it is of the same logical nature as the question, suggested by the example above, concerning the relation of the running of the watch to the movement of the hands.[10]

VI

In order to bring out the exact status of the fundamental idea of the physicalistic conception of psychology (or logical behaviorism), we shall contrast it with certain theses of psychological behaviorism and of classical materialism, which give the appearance of being closely related to it.[11]

1. Logical behaviorism claims neither that minds, feelings, inferiority complexes, voluntary actions, etc., do not exist, nor that their existence is in the least doubtful. It insists that the very question as to whether these psychological constructs really exist is already a pseudoproblem, since these notions in their "legitimate use" appear only as abbreviations in physicalistic statements. Above all, one should not interpret the position sketched in this paper as amounting to the view that we can know only the "physical side" of psychological processes, and that the question whether there are mental phenomena behind the physical processes falls beyond the scope of science and must be left either to faith or to the conviction of each individual. On the contrary, the logical analyses originating in the Vienna Circle, one of whose consequences is the physicalistic conception of psychology, teach us that every meaningful question is, in principle, capable of a scientific answer. Furthermore, these analyses show that what, in the case of the mind-body problem, is considered as an object of belief, is absolutely incapable of being expressed by a factual proposition. In other words, there can be no question here of an "article of faith." Nothing can be an object of faith which cannot, in principle, be an object of knowledge.

2. The thesis here developed, though related in certain ways to the fundamental idea of behaviorism, does not demand, as does the latter, that psychological research restrict itself methodologically to the study of the responses organisms make to certain stimuli. It by no means offers a theory belonging to the domain of psychology, but rather a logical theory about the statements of scientific psychology. Its position is that the latter are without exception physicalistic statements, by whatever means they may have been obtained.

Consequently, it seeks to show that if in psychology only physicalistic statements are made, this is not a limitation because it is logically *impossible* to do otherwise.

3. In order for logical behaviorism to be valid, it is not necessary that we be able to describe the physical state of a human body which is referred to by a certain psychological statement—for example, one dealing with someone's feeling of pain—down to the most minute details of the phenomena of the central nervous system. No more does it presuppose a knowledge of all the physical laws governing human or animal bodily processes; nor *a fortiori* is the existence of rigorously deterministic laws relating to these processes a necessary condition of the truth of the behavioristic thesis. At no point does the above argument rest on such a concrete presupposition.

VII

In concluding, I should like to indicate briefly the clarification brought to the problem of the division of the sciences into totally different areas, by the method of the logical analysis of scientific statements, applied above to the special case of the place of psychology among the sciences. The considerations we have advanced can be extended to the domain of sociology, taken in the broad sense as the science of historical, cultural, and economic processes. In this way one arrives at the result that every sociological assertion which is meaningful, that is to say, in principle verifiable, "has as its subject matter nothing else than the states, processes and behavior of groups or of individuals (human or animal), and their responses to one another and to their environment,"[12] and consequently that every sociological statement is a physicalistic statement. This view is characterized by Neurath as the thesis of "social behaviorism," which he adds to that of "individual behaviorism" which we have expounded

above. Furthermore, it can be shown[13] that every statement of what are called the "sciences of mind and culture" is a sociological statement in the above sense, provided it has genuine content. Thus one arrives at the "thesis of the unity of science":

The division of science into different areas rests exclusively on differences in research procedures and direction of interest; *one must not regard it as a matter of· principle. On the contrary, all the branches of science are in principle of one and the same nature; they are branches of the unitary science, physics.*

VIII

The method of logical analysis which we have attempted to explicate by clarifying, as an example, the statements of psychology, leads, as we have been able to show only too briefly for the sciences of mind and culture, to a "physicalism" based on logic (Neurath): *Every statement of the above-mentioned disciplines, and, in general, of empirical science as a whole, which is not merely a meaningless sequence of words, is translatable, without change of content, into a statement containing only physicalistic terms, and consequently is a physicalistic statement.*

This thesis frequently encounters strong opposition arising from the idea that such analyses violently and considerably reduce the richness of the life of mind or spirit, as though the aim of the discussion were purely and simply to eliminate vast and important areas of experience. Such a conception comes from a false interpretation of physicalism, the main elements of which we have already examined in section VII above. As a matter of fact, nothing can be more remote from a philosophy which has the methodological attitude we have characterized than the making of decisions, on its own authority, concerning the truth or falsity of particular scientific statements, or the desire to eliminate any matters of fact whatsoever. *The subject matter of this philosophy is limited to the form of scientific statements, and the deductive relationships obtaining between them.* It is led by its analyses to the thesis of physicalism, and establishes on purely logical grounds that a certain class of venerable philosophical "problems" consists of pseudo-problems. It is certainly to the advantage of the progress of scientific knowledge that these imitation jewels in the coffer of scientific problems be known for what they are, and that the intellectual powers which have till now been devoted to a class of meaningless questions which are by their very nature insoluble, become available for the formulation and study of new and fruitful problems. That the method of logical analysis stimulates research along these lines is shown by the numerous publications of the Vienna Circle and those who sympathize with its general point of view (H. Reichenbach, W. Dubislav, and others).

In the attitude of those who are so bitterly opposed to physicalism, an essential role is played by certain psychological factors relating to individuals and groups. Thus the contrast between the constructs (*Gebilde*) developed by the psychologist, and those developed by the physicist, or, again, the question as to the nature of the specific subject matter of psychology and the cultural sciences (which present the appearance of a search for the essence and unique laws of "objective spirit") is usually accompanied by a strong emotional coloring which has come into being during the long historical development of a "philosophical conception of the world," which was considerably less scientific than normative and intuitive. These emotional factors are still deeply rooted in the picture by which our epoch represents the world to itself. They are protected by certain affective dispositions which surround them like a rampart, and for all these reasons appear to us to have genuine content

—something which a more penetrating analysis shows to be impossible.

A psychological and sociological study of the causes for the appearance of these "concomitant factors" of the metaphysical type would take us beyond the limits of this study,[14] but without tracing it back to its origins, it is possible to say that if the logical analyses sketched above are correct, the fact that they necessitate at least a partial break with traditional philosophical ideas which are deeply dyed with emotion can certainly not justify an opposition to physicalism—at least if one acknowledges that philosophy is to be something more than the expression of an individual vision of the world, that it aims at being a science.

Notes

1. I now consider the type of physicalism outlined in this paper as too restrictive; the thesis that all statements of empirical science are *translatable*, without loss of theoretical content, into the language of physics, should be replaced by the weaker assertion that all statements of empirical science are *reducible* to sentences in the language of physics, in the sense that for every empirical hypothesis, including, of course, those of psychology, it is possible to formulate certain test conditions in terms of physical concepts which refer to more or less directly observable physical attributes. But those test conditions are not asserted to exhaust the theoretical content of the given hypothesis in all cases. For a more detailed development of this thesis, cf. R. Carnap, "Logical Foundations of the Unity of Science," reprinted in A. Marras, ed., *Intentionality, Mind, and Language* (Urbana: Univ. of Illinois Press, 1972).

2. *Tractatus Logico-Philosophicus* (London, 1922).

3. A recent presentation of symbolic logic, based on the fundamental work of Whitehead and Russell, *Principia Mathematica,* is to be found in R. Carnap, *Abriss der Logistik* (Vienna: Springer, 1929; vol. 2 of the series *Schriften zur Wissenschaftlichen Weltauffassung*). It includes an extensive bibliography, as well as references to other logistic systems.

4. The following are some of the principal publications of the Vienna Circle on the nature of psychology as a science: R. Carnap, *Scheinprobleme in der Philosophie. Das Fremdpsychische und des Realismusstreit* (Leipzig: Meiner, 1928); *Der Logische Aufbau der Welt* (Leipzig: Meiner, 1928) [English trans.: *Logical Structure of the World* (Berkeley: Univ. of California Press, 1967)]; "Die Physikalische Sprache als Universalsprache der Wissenschaft," *Erkenntnis*, 2 (1931-32), 432-465 [English trans.: *The Unity of Science* (London: Kegan Paul, 1934)]; "Psychologie in physikalischer Sprache," *Erkenntnis*, 3 (1932-33), 107-142 [English trans.: "Psychology in Physical Language," in A. J. Ayer, ed., *Logical Positivism* (New York: Free Press, 1959)]; "Ueber Protokollsaetze," *Erkenntnis*, 3 (1932-33), 215-228; O. Neurath, "Protokollsaetze," *Erkenntnis*, 3 (1932-33), 204-214 [English trans.: "Protocol Sentences," in *Logical Positivism*]; *Einheitswissenschaft und Psychologie* (Vienna: Springer, 1933; vol. 1 of the series *Einheitswissenschaft*). See also the publications mentioned in the notes below.

5. P. Oppenheim, for example, in his book *Die Natuerliche Ordnung der Wissenschaften* (Jena: Fischer, 1926), opposes the view that there are fundamental differences between any of the different areas of science. On the analysis of "understanding," cf. M. Schlick, "Erleben, Erkennen, Metaphysik," *Kantstudien*, 31 (1926), 146.

6. For further details see the statement of one of the founders of behaviorism: J. B. Watson, *Behaviorism* (New York: Norton, 1930); also A. A. Roback, *Behaviorism and Psychology* (Cambridge, Mass.: Univ. Bookstore, 1923); and A. P. Weiss, *A Theoretical Basis of Human Behavior*, 2nd ed. rev. (Columbus, Ohio: Adams, 1929); see also the work by Koehler cited in note 11 below.

7. Space is lacking for further discussion of the logical form of test sentences (recently called "protocol sentences" by Neurath and Carnap). On this question see Wittgenstein, *Tractatus Logico-Philosophicus*, as well as the articles by Neurath and Carnap which have appeared in *Erkenntnis* (above, note 4).

8. Two critical comments, 1977: (a) This reference to verification involves a conceptual confusion. The thesis which the preceding considerations were intended to establish was clearly that the statement "Paul has a tooth-ache" is, in effect, an abbreviated expression of all its test sentences; not that it expresses the claim (let alone the "fact") that all those test sentences have actually been tested and verified. (b) Strictly speaking, none of the test sentences just mentioned is implied by the statement "Paul has a toothache": the latter may be true and yet any or all of those test sentences may be false. Hence, the preceding considerations fail to show that the given psychological statement can be "translated" into sentences which, in purely physical terms, describe macro-behavioral manifestations of pain. This failure of the arguments outlined in the text does not preclude the possibility, however, that sentences ascribing pain or other psychological characteristics to an individual might be "translatable," in a suitable sense, into physical sentences ascribing associated physical micro-states or micro-events to the nervous system or to the entire body of the individual in question.

9. "Soziologie im Physikalismus," *Erkenntnis*, 2 (1931-32), 393-431, particularly p. 411 [English trans.: "Sociology and Physicalism," in A. J. Ayer, ed., *Logical Positivism*].

10. Carnap, *Der Logische Aufbau der Welt*, pp. 231-236; id. *Scheinprobleme in der Philosophie*. See also note 4 above.

11. A careful discussion of the ideas of so-called "internal" behaviorism is to be found in *Psychologische Probleme* by W. Koehler (Berlin: Springer, 1933). See particularly the first two chapters.

12. R. Carnap, "Die Physikalische Sprache als Universalsprache," p. 451. See also: O. Neurath, *Empirische Soziologie* (Vienna: Springer, 1931; the fourth monograph in the series *Schriften zur wissenschaftlichen Weltauffassung*).

13. See R. Carnap, *Der Logische Aufbau der Welt*, pp. 22-34 and 185-211, as well as the works cited in the preceding note.

14. O. Neurath has made interesting contributions along these lines in *Empirische Soziologie* and in "Soziologie im Physikalismus" (see above, note 9), as has R. Carnap in his article "Ueberwindung der Metaphysik durch logische Analyse der Sprache," *Erkenntnis*, 2 (1931-32), 219-241 [English trans.: "The Elimination of Metaphysics through Logical Analysis of Language," in A. J. Ayer, ed., *Logical Positivism*].

21. ON THE RELATION BETWEEN PSYCHOLOGICAL AND PHYSICAL CONCEPTS*

I

Recent philosophy has not been lacking in attempts to free the Cartesian problem of the relation between mind and body from its metaphysical obscurities, by refusing to pose it in terms of mental and physical substances; beginning, instead, with the harmless question as to how, in general, we have come by our physical and psychological concepts. That this is actually the correct way to approach the solution of the problem, I have no doubt. Indeed, I am convinced that the problem will already be solved the moment we become completely clear as to the rules in accordance with which we employ the words 'mental' and 'physical'. For we shall then grasp the proper meaning of all physical and psychological propositions, and in doing so will know in what relation the propositions of physics stand to those of psychology.

When Descartes sought to define his 'corporeal substance' by specifying the attribute '*extensio*' as its characteristic mark, he took the first step in a direction which must be followed to the end before one can hope to form a clear idea of the properties which belong to all 'physical' concepts, and to these alone. '*Extensio*' refers, of course, to *spatial* extension; and it is indeed possible confidently to assert that an analysis of the concept of spatial extension yields without further ado a definition of the concept 'physical'.

The problem, however, is by no means so simple that it suffices to say 'whatever is spatially extended is physical', for there are words which make sense when combined with the predicate 'spatially extended', and which nevertheless refer to 'mental' states; such words, for example, as 'visual image', 'tactual sensation', 'pain', etc. Consequently, the difference we are seeking

---

* 'De la Relation entre les Notions Psychologiques et les Notions Physiques', *Revue de Synthèse* 10 (1935) 5–26.
Translated by Wilfrid Sellars. Reprinted with the kind permission of the translator and the publisher from *Readings in Philosophical Analysis*, edited by H. Feigl and W. Sellars, Appleton-Century-Crofts, Inc., New York 1949. The German original 'Über die Beziehung zwischen den psychologischen und den physikalischen Begriffen' was first published in Moritz Schlick, *Gesammelte Aufsätze 1926–1936*, Wien 1938, pp. 267–287.

can be found along the above lines only if the word 'extended' has different meanings in its psychological and physical usages.

Is this the case? Do I have the same thing in mind, or something different, when I say of a pain that it spreads over a certain area, as compared to when I ascribe a certain spatial extensity to a physical object, for example my hand? Is the visual image of the moon 'extended' in the same sense as the moon itself? Do my visual impressions on looking at a book have extension in the same sense as the tactual impressions I obtain by holding it in my hand?

The answering of these questions is the first step in the process of clarifying our concepts, nay the second — for the first and more difficult step is to ask these questions at all. This step was not taken by Descartes nor by those who follow him — the possibility not even occurring to them that the word 'extensio' is used in more than one sense. It would therefore not be correct to describe their use of this word by saying that they took it to have the *same* meaning in significantly different cases. They didn't even see that there were different cases. Berkeley alone was a famous exception. He posed the third of our three questions. The first two couldn't be raised in his system, since a by no means inconsiderable part of his philosophy consisted exactly in a proof that these questions do not exist. For him there is no other kind of extension than that which can be attributed to the representations of sight and touch; indeed, in Berkeley's philosophy it is already a mistake to speak of these as 'representations', since there is nothing which is copied by them and is their original. Kant, who philosophized so much later than Berkeley, believed he had nothing to learn from him, and didn't succeed in raising our questions. He invariably speaks, as did Descartes before him, of extension, of Space, and omits any investigation as to whether it may not be necessary to distinguish between several space-concepts; first, between the physical and the psychological, and under the latter, between visual-space, tactual-space, etc. This neglect had unfortunate consequences for Kant's philosophy of geometry, and through this, for his system as a whole. Physical space, the space of nature, is for him also psychological space, since nature is for him 'mere appearance', that is, mere 'idea', and this is a psychological term.

It is possible to regard Kant's distinction between 'outer' and 'inner' sense as an attempt to draw a boundary between the physical and the mental. His doctrine that Space, the form of intuition for outer sense, is lacking in the case of inner sense is indeed reminiscent of Descartes, as well as a forerunner of recent attempts to characterize the mental as simply the non-spatial. It is said that even where the mental has to do with the spatial (in ideas and perception), it is itself non-spatial. The idea of a red triangle is itself neither

159

red nor triangular, nor is the perception of an extended object itself extended.

This assertion owes its appearance of plausibility to the fact that the words 'perception' and 'idea' are ambiguous. By them one can refer either to the content, that which is given (*une donnée actuelle*), or to the event, the act of perception, which is characterized as a 'mental process' and concerning which there is indeed no question of 'extension'. (We leave unraised the question as to the justifiability of this distinction between content and act, and limit ourselves to pointing out that surely it first occurred to us to speak of an act of perception – and, later, of imaging – only after we had gained the knowledge that the occurrence of 'contents' is somehow dependent on processes in the sense organs, and, furthermore that these processes are physical.) One can certainly not say of the contents of perception – at least in the case of sight and touch – that they are 'non-spatial'; rather they are beyond doubt extended. Indeed it is from them that we first derive this concept.

Nevertheless, we do not mean the same by 'extension' in psychological and physical contexts. In order to make the difference clear it is best to examine exactly those cases where it is most difficult to distinguish psychological from physical space. We asked above if, for example, a pain is extended in the same sense as is a physical object, say, my hand. But what about the case where the pains are in my hand itself, where my whole hand aches? Do we not have here a mental datum the spatial extension of which is identical with that of the physical object which is 'my hand'?

The answer is, 'absolutely not!' Pain has its own space just as visual sensations have theirs and as do sensations of touch. The fact that several sensations of pain can occur *simultaneously* is sufficient to require us to speak of a 'pain-space'. Every arrangement of simultaneous items is a side-by-side (as opposed to a sequence) and it is customary to call such facts 'spatial'. It is experience which first brings about the coördination of the several spaces of visual and tactual sensations, feelings of pain, etc.

This can be made to stand out most clearly by conceiving of a man who lives in complete darkness and complete absence of motion. He would be acquainted with neither visual nor tactual sensations, but he could very well have 'pain throughout his hand' (even though he would not use these words). Should he be freed from his cell, he would slowly form the customary spatial notions and on the basis of the observation of certain coexistences and sequences of events would gradually learn to interpret these pains as pains of the 'hand', that is to say, of the five-fingered visual and tactual object which is connected with his body by another bodily structure, the 'arm'. For he would observe that his pains depend in a definite way on what befalls a physical object which he calls 'my hand', which is visible in the visual field and touchable in the tactual field. Thus, a wounding or movement of this object would increase the pain, while other processes (medical treatment) would diminish them. In this way, the pain-space would be coördinated with the sight-space. Since experience alone teaches us that the several kinds of extension always appear together, the conclusion is to be drawn that there are several 'spaces' rather than only one. If the world were otherwise than it actually is, if, for example, the person concerned always felt a pain when a certain object, for example the candlestick on the table, was violently disturbed, and, should the candlestick move, perceived a sensation akin to the kinaesthetic sensation which normally accompanies the movement of his hand, such experience could lead him to coördinate the space of the 'hand-ache' with that of the candlestick (and if, for example, the candlestick had five branches, its extension would correspond to that of the five fingers). He could thus meaningfully say, 'I have a pain in the candlestick'. (Similar and as yet unpublished considerations have been advanced by Ludwig Wittgenstein in another connection.) Thus, it is possible to conceive of experiences which would result in the localization of the *same* handaches in quite different physical spaces. If follows that mental pain-space and physical space are entirely different things.

The difference is obvious in extreme cases. Let us compare, referring back to our second example, the extension of the moon with that of the visual image of the moon. The diameter of the moon, a physical magnitude, can be given in miles; the diameter of the visual image, on the other hand, is not even a 'size'. (Needless to say, the visual image must not be confused with the retinal image, which has physical magnitude, and, consequently, a diameter which can be specified in units of measure). The extensity of the visual image is frequently assigned an angular measure. The latter is, indeed, a physical magnitude, but it does not make one of the visual image itself. Rather, such a method of assigning a measure can be justified only by means of a definitional coördination, which, however, is not practical for many purposes. Thus, if one compares the visual image of the moon at the zenith, with that of the moon at the horizon, the angle is the same in both cases; nevertheless, as is well known, we call the extent of the mental visual image of the moon greater in the second case than in the first. Whatever is meant by the 'extension' or 'size' of a mental image, it is in any case something quite different from the extension or size of a physical object.

II

In what, then, does the difference consist which is to lead us to a definition of the 'physical'?

Here we shall apply the method which seems to me the sole method of true philosophy: We shall turn our attention to the way in which propositions about physical objects are *verified*. That which is common to all the methods by which such propositions are verified, must then be that which is characteristic of the physical. All propositions are tested with respect to their truth or falsity by the performance of certain operations, and to give an account of the meaning of the propositions consists in specifying these operations. Of what sort, then, are these operations in the case of propositions in which physical terms appear? In other words, in what does the process of determining physical properties consist?

Physical properties are *measurable* properties. They are defined by the methods of measurement [1]. It will suffice if we limit our discussion of these methods to the *scientific* methods of physics. There are, of course, pre-scientific ways of noting the presence of physical properties which continue to play a dominating role in everyday life, but there is no difference in principle between the procedures of everyday life and those of research. Since, however, the methods of science stand out more clearly, we shall limit ourselves to these. In everyday life, also, physical concepts arise only where measurements of one kind or another have taken place (even if by the thoroughly crude methods of pacing, touch, visual estimation, etc.), that is, *quantitative* determinations have been achieved. Every measurement springs from a counting, and can in the last analysis always be traced to a numbering of 'coincidences', where by a coincidence is to be understood the spatial coming together of two previously separated singularities of the visual or tactual fields (marks, pointers, etc.). This characteristic of measurement whereby spatial extension is, as it were, mastered by division into discrete parts has often been pointed out. It is this way of determining the spatial which is the *physical* way.

Why exactly do we make use of this procedure?

The only correct answer is, because of its objectivity, that is, because of its inter-sensual and inter-subjective validity. What this means can be easily clarified by an example. If I move the tips of my index fingers toward one another, there occurs in the visual field an event which is called 'meeting of the finger tips', and another event in the tactual field which I call 'contact of the finger tips'. These two events, each of which is a discrete and distinguishable element in its field, always occur simultaneously. This is a fundamental empirical relation between them. Every time that a coincidence occurs in the field of touch, one also occurs in the visual field (as least under favourable circumstances of an exactly specifiable sort, for example, illumination, position of the eyes, etc.). This relationship is independent of the particular sense in question; it is intersensual. We also learn from experience that it is inter-subjective. That is to say, all other people who are present affirm (again under given, readily specifiable circumstances) that the same number of homologous coincidences occur in their visual and tactual fields. Thus, not only the different several senses, but also the different subjects agree in their testimony concerning the occurrence of coincidences. The order of these coincidences is nothing other than physical space-order (properly, space-time-order); it is an *objective* order (for by this word we bring together the two ideas of inter-sensual and inter-subjective).

In general, objectivity obtains only for those physical propositions which are tested by means of coincidences, and not for propositions which are concerned with qualities of colour or sound, feelings such as sadness or joy, with memories and the like, in short, 'psychological' propositions.

The meaning of all physical propositions thus consists in the fact that they formulate either coincidences or laws relating to coincidences; and these are spatio-temporal determinations. One may be tempted to say that this makes sense only if the coinciding items are specified, and that the propositions are incomplete without this addition. But closer examination shows that such specifications (which indeed must be made) refer us back to propositions concerning other coincidences. (Here we find the justification for the thesis, elaborated particularly by A. S. Eddington, that physics as a whole is to be understood as geometry. 'Geometry' in this connection clearly refers to an empirical science, rather than a purely formal mathematical discipline.) Even explication by means of ostensive gestures, which alone, in the last analysis, relates our concepts to the world, and makes them signs of objects in nature, is readily seen to consist in the bringing about of coincidences (for example, of a pointing finger with the object singled out). The fact that the spatial description of atomic processes does not occur in modern quantum theory does not alter the fact that all physical laws are verified by the occurrence of coincidences; for this holds also of the laws in which magnitudes relating to atoms appear. These magnitudes also have meaning only by their relation to physical space determinations.

According to what we have said above, the essential feature of physical concepts is that they are arrived at by selecting out of the infinite variety of events a special class, namely these 'coincidences', and describing their interrelationships with the help of numbers. Physical magnitudes are identical with the number-combinations which are thus arrived at. The question which we are seeking to answer (in principle) can therefore be put as follows: What is the relation of these coincidences to all other events, for example to the occurrence of a pain, to the change of a colour, to a feeling of pleasure, to the emergence of a memory, and so forth?

III

It is usually claimed that the physicist simply and deliberately avoids reference to whatever is not a matter of space-time determinations. He ignores, it is said, the 'qualitative' and limits himself to describing the quantitative relationships to be found in the world. This usually develops into the charge that physics is 'one-sided', that it plays a narrowly circumscribed role in our knowledge of reality; that it gives us only a fragment which must be supplemented, an empty space-time hull which must be filled with content. This content, it is urged, is the psychological. Psychology would therefore confront physics as an autonomous discipline. Indeed, we often hear the opinion that not even physics and psychology together exhaust the modes of describing the world, and that there remains a place where metaphysics is privileged to lay down the law.

To the assertion of the one-sidedness and limitations of the methods of physics, there stands in sharp opposition the claim that an absolutely complete description of the world is possible by the use of physical methods; that every event in the world can be described in the language of physics, and therefore specifically, that every psychological proposition can be translated into an expression in which physical concepts alone occur. This claim — which is referred to (in somewhat inelegant terminology) as the thesis of 'physicalism' — is correct, if the physical language is not only objective, which we have already seen, but is in addition the *only* objective language; or, more accurately, if translatability into the physical language is a necessary condition of objectivity. This seems indeed to be the case. All experience up to now points to the conclusion that only physical concepts and concepts which are reducible to physical concepts fulfil the requirement of objectivity, which is, of course, essential to a language, for without it the language could not serve as a means by which different subjects could arrive at an understanding.

I therefore hold the thesis of physicalism to be correct[2] but — and this can hardly be overemphasized — it is correct only on the basis of specific *experiences*. The thesis is therefore a factual one, an empirical proposition, as is, say, the proposition that England is an island, or the assertion that conservation of energy obtains in nature. The thesis is therefore not a philosophical discovery. The philosopher as such is not interested in facts of experience as such, for each fact is only one of indefinitely many possible facts. Rather he is interested in the *possibility* of facts. Since, in my opinion, his task is that of determining the meaning of propositions, and since a proposition has meaning only when it formulates a *possible* state-of-affairs (whether or not the state-of-affairs actually exists is irrelevant), it is one and the same thing to say that the philosopher is concerned with the meaning of propositions, and to say that he deals with the possibility of facts.

That the world is exactly as it is, that matters stand exactly as experience shows they do, is — in a readily intelligible sense — a contingent fact; and it is in exactly the same sense a contingent fact that the physical language is an inter-subjective universal language. (Even one of the most ardent exponents of 'physicalism', Carnap, explains it as a stroke of good luck[3].) As far as we are concerned, it follows directly from this that the word 'physicalism' in no way designates a 'philosophical movement'. This is an admonition to us to evaluate and make use of the facts which the term brings to mind no differently than any other empirical matter of fact; to treat them, namely, as a paradigm, as one possibility among others. It is exactly by picturing other possible states-of-affairs from which the one that is actually realized stands out as against a background, that we shall first come to understand the latter correctly, and to grasp the role actually played by physical concepts, as well as their relation to psychological concepts.

IV

What, then, are the data of experience on which the objectivity and universality of the physical language rests? They consist in the fact that between the 'coincidences' and all other events, there can be found systematic relationships such that to every difference in any of the other events, there corresponds a determinate difference in the coincidences, so that, in principle, the world contains no variation nor constancy which does not go hand in hand with a variation or constancy in the domain of coincidences. If this is the case, then clearly the entire world of experience is uniquely determined by these coincidences; when these are known, so is it. It is from this that stems the universal character of the physical language. Two examples may suffice to illustrate. For the first we choose the relationship that exists between the psychological and the physical concepts of colour. Physically, a colour is defined by a frequency, a number of vibrations per second. This number, as is well known, is arrived at by the familiar procedure of counting the interference fringes of the light or measuring a spectrum, and from the resulting figures along with other measurements read off the apparatus, calculating the 'frequency'. That is to say, one observes the coincidence of a spectral line or of an interference fringe with certain marks on the measuring apparatus. Now experience shows that these coincidences always occur at the same places, and in accordance with the same general laws, whenever the light has visually the same colour. For monochromatic light of an absolutely specific shade of red, I always get exactly the same frequency. Consequently, if I know that a source of light is emitting rays of this frequency, then I know what colour I will see when it meets my eye. Thus, to designate the colour, it is sufficient to give the frequency. Indeed, this physical designation is actually far more accurate than the corresponding colour word (for example, 'Bordeaux-red') used by the psychologist.

But is the correspondence of the frequency with the colour as seen truly unambiguous? Do I always see the same colour when I look at a source emitting the same frequency? Obviously not, for if my eye is tired, or has previously been affected by light of another colour, or if my nervous system is under the influence of santonin, then I have different colour impressions although objectively the radiation is the same. Doesn't experience refute the 'thesis of physicalism'? No, for experience teaches that in all these cases in which, in spite of the identity of the frequency, I see a different colour, *other* physical changes are detectable, namely those which concern the state of my organism, in particular my nervous system. The investigation of my nervous system, which is naturally a physical investigation, making use of the method of coincidences, shows (as far as our experience goes) that every difference in colour quality goes hand in hand with a difference in the optical segment of the nervous system.

But without concerning ourselves as to whether a physiological investigation of the nervous system will be carried to completion, or is even a technical possibility, we find other physically describable processes which can be used in place of neural events to achieve an unambiguous correspondence between sense-quality and coincidence system, namely the physical behaviour of the individual – in particular the reactions (speech, writing, etc.) by which he reports on his sensations when he is asked about their qualities. It will be supposed that the reason these reactions are as satisfactory for the purpose as the above-mentioned neural processes is because they in their turn can be unambiguously correlated with these processes (by virtue of the causal connection between them). But this is irrelevant to our purpose. What concerns us is solely the fact that it is possible unambiguously to coördinate quality of sensation with coincidence systems.

Every change of colour quality thus corresponds to a change in the system of coincidences; but this is a matter, not of those coincidences alone which are involved in the measuring of the frequency of the light, but also of other coincidences, observable in the body of the perceiver, the belonging of which to the sum-total of coincidences is a matter of empirical fact. With the taking into account of all relevant coincidences, the coördination of physical concepts with the qualities becomes completely unambiguous, as 'physicalism' asserts.

One cannot reproach the physicist with the intentional overlooking of all qualities, for it is just not true that he overlooks them. On the contrary, every difference is for him an occasion and a hint to search for a difference of coincidences. If, for example, I were to say that I see blue under circumstances in which one is expected to have a sensation of yellow (say, at the place of the sodium line of the spectrum), the physicist would not rest until he had 'explained' this unexpected fact, that is, until he had discovered physical peculiarities in my body, in other words, abnormal measurements shown by certain coincidences, which appear in this case and in no other. The world of qualities is thus of highest importance for him. He in no way forgets it, but on the contrary only regards his quantitative system as a satisfactory description of nature if the manifold of the world of qualities is represented in it by a corresponding multiplicity of numbers.

For our second example, let us take the question as to how the mental datum which is a feeling of grief is expressed and communicated. A feeling of this kind is neither localized, nor do we ascribe it a spatial extent, and its structure is essentially different from that of a sense quality. To be sure, grief is for the most part evoked by external events, that is to say, by events which occur outside the body of the griever, and which can be described in physical terms (for example, someone's death, or the news of a death). But the difference between this case and the preceding consists in the fact that no one believes that there exists a one-to-one correspondence between the quality of the feeling of grief and these external events. Rather, the dependence of the feeling on the state of the subject is so obvious that everybody looks to the body of the griever himself for the coincidences which are here principally in question. Once again we do not need to consider the events in the nervous system – which are for the most part unknown – for it is sufficient to pay attention to his expression, his utterances, his whole deportment. In these processes – which are describable in terms of coincidences – we have the facts by which feelings are expressible in the physical language.

Let it not be thought that the physicist must leave something out of his description, that there is something which he cannot formulate, which it remains, say, for the poet to express. For even the poet can only perceive someone's grief in terms of bodily behaviour, and only in terms of bodily behaviour can he make it intuitive for the listener. Indeed, the better a psychologist he is, the more he is a master of poetic language, the less he will make use of psychological terms to describe the grief. Instead he will attempt to achieve his purpose in an apparently indirect way by describing how the griever walks, his expression, how he holds his head, the weary movements of his hand, or by repeating his broken words, – occurrences, in short, which can also be described by the physicist, although he would make use of other symbols.

V

How exactly do we build our 'psychological' concepts? Whereas the physical language gives formulation to events in their extensive spatial-temporal relationships, the psychologist brings them together from quite a different point of view, namely, in accordance with their 'intensive similarity'. Thus, each of a large number of different but resembling properties which occur in experience, is called by the common name 'green', another manifold is called 'yellow', and so on. Both of the manifolds exhibit such a resemblance to one another as well as to certain other qualities, that they are grouped under the common term 'colour'. In addition, there are other elements which differ from these, but resemble each other and therefore receive a common name, as for example, 'sound', 'pleasure-feeling', 'anger', 'odour', 'pain', 'uneasiness', etc. Furthermore, there are families of events which are called 'change of colour', 'intensification of sound', 'decrease in brightness', 'dying away of a feeling', 'visual motion', 'tactual motion', and so on. With these there naturally belong the classes of events, 'visual coincidence' and 'tactual coincidence'.

We must therefore include the latter in the list of 'psychological' concepts. If this strikes one as paradoxical or seems to contradict our earlier statements, then he is far removed from an understanding of the relation between physical and psychological concepts. It would be clearly a mistake to say: 'The coincidences are of a much more complicated nature'. If, for example, I dream that I am playing billiards, I see the balls come together in such a way that at certain points on their surfaces there occur coincidences which cannot, however (in this case), be used to construct a physical or objective space. For they are only dream-events. One cannot fit them into the same structure with the corresponding events of an actual game. They obey different laws. The 'physical space' that one might construct with their aid, would be an unreal physical space, whereas the visual coincidences of a dream as mental events have naturally the same reality as the facts of waking life. But they do not have the intersubjectivity which distinguishes the coincidences observed in real life. Indeed, the difference from an actual billiard game consists exactly in the fact that the coincidences of the dream are not suited to the construction of an inter-subjective space, whereas the coincidences of normal life fit in a direct and easy way into the system of physical space and natural law. Thus, it is not the coincidences as such, which constitute the 'physical world', rather it is their incorporation into a certain system (the system of objective space) which makes possible the formation of physical concepts. The adjectives 'physical' and 'mental' formulate only two different representational modes by which the data of experience are ordered; they are different ways of describing reality. That in which one counts ordered coincidences in inter-subjective space, is the physical; whereas that which operates by the grouping of intensive properties is a psychological description.

The so-called 'psycho-physical problem' arises from the mixed employment of both modes of representation in one and the same sentence. Words are put side by side which, when correctly used, really belong to different languages. This gives rise to no difficulties in everyday life, because there language isn't pushed to the critical point. This occurs first in philosophical reflection on the propositions of science. Here the physicist must needs assure us that, for example, the sentence, 'The leaf is green' merely means that a certain spatial object reflects rays of a certain frequency only: while the psychologist must needs insist that the sentence says something about the quality of a perceptual content. The different 'mind-body theories' are only outgrowths of subsequent puzzled attempts to make these interpretations accord with one another. Such theories speak for the most part of a duality of percept and object, inner-world, outer-world, etc., where it is actually only a matter of two linguistic groupings of the events of the world. The circumstance that the physical language as a matter of experience seems to suffice for a complete description of the world, has, as history teaches, not made easy the understanding of the true situation, but has favoured the growth of a materialistic metaphysics, which is as much a hindrance to the clarification of the problem as any other metaphysics.

VI

In our world, the physical language has the character of objectivity and universality, which the psychological language seems to lack. It is possible to conceive that matters were turned around – that the formation of psychological concepts was inter-sensual and inter-subjective, while no universal agreement could be achieved in the case of assertions concerning coincidences. Such a world would bear no resemblance to the actual world, but one could nevertheless picture it to oneself – as consisting, for example, of a finite number of discrete qualities (classifiable in various resemblance-classes) the simultaneous or successive occurrence of which was shown by experience to be governed by certain laws, but which were never clearly distinguished from one another by clear-cut boundaries. Naturally, in this world, the means of communication, the linguistic symbols, would be constructed of material entirely different from our words, and the individuals who speak with one another would not possess spatial bodies of the sort to which we are accustomed, – but all this is not impossible.

The reason for the fact that exactly the physical language, the language of spatial coincidences, is for us an inter-subjective means of communication, lies naturally in the fact that it is by spatial relationships that individuals are both distinguished from and yet bound up with one another. Putting it somewhat differently: The external world is a spatial world. Indeed, the word 'external' serves to designate a spatial relation; and it is easy to see that the opposition between 'I' and 'external world' is as a matter of fact only the difference between 'one's own' body and other physical objects. But the clarification of such complicated concepts as 'I' or even 'consciousness' lies beyond the scope of this paper. We content ourselves here with the examination of the employment of certain simple psychological and physical terms. It is a preliminary task which prevents the emergence of those difficulties which hide behind the words 'psycho-physical problem'.

VII

We have emphasized that the circumstances on which rests the universality of the physical language, that is to say, the 'thesis of physicalism', are of an empirical rather than a logical character. They are, however, of such a pervasive sort, and we are so thoroughly accustomed to them, that it is by no means easy to form an idea as to how the world would look if only these decisive relationships did not obtain, though everything else remained the same. It would be a world enormously different from the actual world.

In it there would be no uniform one-to-one correspondence between coincidences and qualities. Perhaps we can imagine this most easily if we consider *feelings*. I can, for example, imagine that my feeling of grief corresponded in no way to any bodily condition. If, for example, I laughed, skipped around, sang and told witty stories, no one would be able to conclude from this that I was gay, rather this behaviour would be as compatible with a sorrowful as with a cheerful mood. Above all, and this is a significant point, it would have to be impossible for me to communicate my state of feeling under interrogation. I must not be able, even if I desired, to give information concerning my feelings. (It is extremely difficult to express oneself accurately on such considerations; in our case, the correct formulation would be: in the changed world it would be a law of nature relating to my will, that there was no such thing as a wish to give expression to a feeling.) For if I could say something concerning my feelings, then there would be spatially describable processes, namely, speech movements and speech sounds — by reference to which the feeling qualities could be unambiguously described, and that would contradict our hypothesis. There must be no uniform relation between any kind of external events and the occurrence of my feelings, for otherwise someone could describe my feeling-state as 'that which one has on the death of a friend or relative'. Only if my feelings occurred entirely without connection with my sense-perceptions, would it be impossible to designate that which in the actual world we call 'grief' by a word belonging to inter-subjective language which anyone can understand. It would be impossible to give a definition for such a word.

In the described case there would be a world of feeling which could not be talked about in the physical language. To be sure, all that I could communicate would be expressible in this language. It would be the sole inter-subjective language (in contrast with the possibility suggested in the preceding Section), but it would no longer be universal, for in addition to it there would be a private language in which I could reflect about the world of feeling.

Similar considerations arise in connection with the 'sense qualities'. It is, for example, possible that although all visual coincidences continue as before, they should be accompanied by entirely different perceptual contents from those to which we are accustomed, and, indeed, in a fully irregular way. For example, in the case of the observation of optical spectra, the lines might preserve their exact position, but appear in varying colours, so that the location of the D-line of sodium appeared first as yellow, then as red, then green, etc., without my being able to discover any rule by which the appearance of a specific colour was bound up with determinate external conditions capable of being specified by means of coincidences. In this case, while I could always order the colours in classes and assign them symbols, these symbols would not belong to an objective language; they would have only a private use.

With the aid of these symbols I could formulate such regularities as might very well be present and discoverable in the domain of the qualities. Here are a few examples of such possibilities:

1. At every moment, the entire visual field has only one colour – with different intensity at different places – but undergoes a temporal variation such that the various colours appear in their spectral order: red, yellow, green, blue, etc.

2. We see the world as red when we are in a cheerful mood; as blue, on the other hand, when we are in an unpleasant mood. These feelings – in accordance with our assumption – must be in no way bound up with bodily events.

3. I have the ability to bring about 'arbitrary' changes of quality; I can act in this domain. This, however, can only be allowed on the assumption that the motive for such activity always lies in the qualities themselves, and never in the coincidences. These would not, if I may so express myself, influence my will in so far as it was concerned with qualities; nor, on the other hand, could my will be influenced (if we are to be consistent with our assumptions) by the qualities in so far as it was concerned with coincidences (actions in the external world).

4. If I feel warm, the colour qualities change in one direction of the spectrum, if I feel cold, in the other – here as well, needless to say, warmth and coldness must be independent of coincidences. – etc., etc.

In circumstances such as those described, and in a thousand others more or less fantastic, there would be no possibility of assigning words for the colour-qualities in an inter-subjective language. We would as a matter of course think of language *qua* means of communication as something which belongs only to the domain of coincidences. We wouldn't even conceive of an alternative possibility, for it wouldn't even occur to us that there could be a connection between coincidences and changes of quality, – just as now many a physicalist may think that there couldn't fail to be such a connection.

The notion of worlds which differ from actuality in the ways we have indicated perhaps makes by no means inconsiderable demands on our imagination; the laws of such a world – and with them the conditions of our own existence – would strike us as extremely strange and would have an entirely different form. But is imagination a privilege of the poet alone? May we not assume it in the philosopher?

VIII

What could be said about such a non-physicalistic world as we have pictured in several examples? First of all perhaps this, that we should hardly speak of it as *one* world but rather as two different domains, one physical, public, and common, and one private, psychological and suited only to monologue. The latter would be to such an extent mine alone, that I couldn't even arrive at the thought of communicating facts concerning it to others. The two worlds would run on side by side. Yet they would not be lacking in all connection. There would be certain relations between the spatial characteristics of the two, for the coincidences would in any case mark the boundaries of the qualities.

By means of a comparison of the constructed example with the actual world we first learn to understand and evaluate the structure of the latter. It is, as far as experience tells us, so constructed that it is fully describable by means of the spatio-temporal conceptual apparatus of physics; this implies the existence in the world of a certain determinate mode of interconnection. The instant we think away this property of the world, reality falls apart into several domains; it ceases to be a *universe*.

We have therefore to do with an empirical fact of far-reaching significance. But only with an empirical fact. We can be saved from attaching too much weight to this fact by noting that we can conceive of different degrees of the separation of the domain of qualities from that of the coincidences, so that a gradual transition from the actual world to our so completely different imaginary world is conceivable. For example, qualities in general might be strictly bound up with coincidences, with the exception, for example, of a limited domain of colours, let us say, shades of green, for which all our earlier assumptions would be true. In this case, the private domain excluded from physics would be of extremely limited scope. We can, however, think of it as broadened to any desired degree, first to include all visual, then all acoustic qualities, etc., so that the validity of the physicalistic assertion would be ever more restricted.

Moreover, we can think of the worlds of sight, sound, smell, etc., as related to one another in certain uniform ways or not, as we choose. In the latter case we are led to conceive of as many mutually independent domains as there are kinds of quality. Needless to say there is here no question of metaphysical pluralism any more than it would be a metaphysical dualism to contrast the world of qualities, uniformly interrelated in accordance with empirical laws, with the world of coincidences. Rather we would have to do with an empirical, contingent division of the world, just as it is an empirical contingent fact that we have exactly the number of sense-organs we do, neither more nor less.

If, as a matter of fact, the physical language is characterized by complete universality, the setting down of this circumstance is in no way the assertion of a metaphysical 'monism'. But one could hardly go wrong with the assumption that it is exactly this empirical fact which impressed the great system builders of the monistic tradition, particularly Spinoza and Leibniz, even though it was impossible at their time to find the correct way of expressing it.

Here, however, we are getting off the main track of our remarks. Our aim has been so to loosen up thought by the consideration of various logical possibilities, as to dispel the traditional associations which have so often hindered the understanding of the relation between physical and psychological propositions.

NOTES

[1] Bridgman's book, *The Logic of Modern Physics*, New York 1927, carries this thought through for physics as a whole.
[2] Cf. my *Allgemeine Erkenntnislehre*, 2nd ed., p. 271 [Engl. by A. E. Blumberg, *General Theory of Knowledge*, Wien and New York, 1974, p. 296].
[3] Cf. *Erkenntnis* 2 (1931) p. 445 ['Die physikalische Sprache als Universalsprache der Wissenschaft'].

The Mind-Body Problem
in the Development of Logical Empiricism

by Herbert FEIGL

The cluster of puzzles and perplexities that constitute the
the Mind-Body-Problem of modern philosophy owes its origin
to a great variety of motives and considerations. The central
issue, however, may justly be located in the disputes between
Dualism and Monism. The dualistic doctrines have a twofold
root: Firstly, there are the mythological, animistic, theological and religious-moral contentions as to the sharp distinction,
if not actual separability, of the mental and the physical. The
deep-rooted and culturally fairly widespread wishful belief
in some form of survival after bodily death, as well as the
exaltation of the spirit and the deprecation of the flesh in so
many eastern and western religions and moral codes may be
regarded as the emotional root of dualism. The other,
scientific, root of dualism may be found in the rise of science,
most prominently beginning with the 17th century, although
at least adumbrated in ancient thought. The striking success
of the method of the physical sciences was, at least historically,
contingent upon a clear cut division of the physical and the
mental, and the relegation of the latter to the limbo of a sort
of secondary or epiphenomenal existence. But the development of modern psycho-physics and psycho-physiology
from the 19th century on, culminating in presentday neurophysiology, Gestalt-psychology, psycho-somatic medicine and
cybernetics, has revived the interest in monistic interpretations.
One discrepant tendency may of course be seen in the dualistic
claims of the researchers in the still highly questionable fields
of Parapsychology (extrasensory perception, psychokinesis,

176

etc.). Another and very different kind opposition comes from philosophers of various schools who either on the basis of their metaphysical commitments or simply in the name of clear thinking insist that the physical and the mental are *toto genere* and irreconcilably distinct and different so that any monistic attempts at their identification must be rejected on purely logical grounds.

This is not the place to review even in outline the history of dualistic and monistic arguments and systems from Descartes and Spinoza down to our time. Two notable conclusions seem to emerge from a study of this history:

(1). The clarification of the badly tangled issues requires as an indispensable first step the discrimination between the factual and the logical questions involved in the mind-body-problem. The factual questions depend for their solution on the progress of scientific research, such as in psycho-physiology. The philosopher qua logical analyst has no business either imaginatively to anticipate or dogmatically to endorse hypotheses that can be established only by painstaking empirical investigations. Since the philosopher is concerned with the analysis of meanings, he can at best examine the consistency of various hypotheses and clarify their precise content by an examination of their logical implications.

(2) It is evident that different thinkers have been impressed with different aspects of the very complex problem of the relations of the mental and the physical. Descartes was puzzled with the question how something of the nature of a non-spatial substance (thinking) could be causally related with a spatial substance (matter). Some philosophers of the 19th and 20th century tried to tackle another "spatial" problem: the location of sense data. Still others have tried to account for the difference of the mental and the physical in terms of the distinctions of the qualitative and the quantitative or of content and structure. Some were intrigued with the "private" character of consciousness and the "public" character of behavior and of neurophysiological processes. Others again, found in the "meaningful", "intentional", "referential", nature of mental states an insuperable obstacle in the attempted identification with "blind" brain-states. Similar objections

arose out of the considerations of "purpose", "free-choice", "reason" on the mental side as juxtaposed with "mechanism", "determinism", "cause" on the physical side. Normative and critical predications (like "corect", and "incorrect", "success" and "failure", responsible" and "irresponsible", "justified" and "unjustified", (morally) "right" and "wrong", etc., seem to apply meaningfully only to minds, mental states, attitudes or functions but not to physical things, processes or events.

This list of juxtapositions, which could easily be expanded, may serve as a reminder that any present-day-advocate of monism (in the sense of an identity-theory) is confronted with a considerable task. Recent naturalistic philosophical and psychological movements, such as positivism, pragmatism, neo-realism, behaviorism and some phases of analytic philosophy, have in one way or another attempted various resolutions of the puzzles posed by the apparent incompatibilities of the essential features of the mental and the physical. A good many of the traditional questions in the total complex of the problem have fairly generally been recognized as pseudo-problems, arising out of conceptual confusions. This may be asserted with assurance in the case of the free choice vs. determinism perplexity. Almost equally definite seem to me the clarifications of the problems of spatial localization, of emergent novelty and of teleology. The proper view of the referential, normative and critical functions of "mind" or "reason" depends on an adequate formulation of rule-guided behavior. Although a good deal of work along these lines is still required, it is evident even now that some of these questions pertain not so much to the distinction of the mental and the physical, but rather to that of logical structure to psychological (or behavioral) fact. Common to all these issues however is the irrepressible and most controversial question: In which sense is the identification of the mental and the physical to be understood? It is interesting to note that Logical Empiricism in the 25 years of its career since its beginnings in the Vienna Circle has in succession embraced three different monistic views and has temporarily countenanced also a more agnostic (parallelistic) form of dualism. In recent years Logical Empiricists have prepared a return to their first monistic position, however reformulated in a more cautious and therefore more auspicious manner. In connection with the very brief review of the four previous positions that I am now going to present, it must be kept in mind that the affinities these positions display with the more traditional metaphysical doctrines are, on the wole, more of the nature of historical analogies than genuine identities of theoretical import. Logical empiricists have from the beginning disclaimed any intention of pronouncing ontological truths. Their sole concern has been the analysis of language and meaning. It was precisely on the basis of such reflections that ontologies of *all* sorts were declared as devoid of factual meaning. The metaphysicians, understandably hurt in their pride and unconvinced by the negativism of the positivists, kept reading into the logical analysis of the latter all the traditional tenets and categories. As already admitted, the flavor of the traditional monisms (or of parallelism) was there, but only historically-culturally speaking. The first position, for example, can easily be regarded as a double-aspect, or double knowledge view of the type held by critical realism. This was Schlick's (84) outlook before the formation of the Vienna Circle, i. e. before the impact of the ideas of Carnap and Wittgenstein. [1] However, even anticipating the later emphases of logical positivism, Schlick regarded the difference of the mental and the physical as a difference between two conceptual systems, of which the physical, as a matter of fundamental empirical fact, is universally applicable whereas the psychological pertains only to a small part of the total realm of reality. This early point of view is therefore more appropriately characterized as a "double-language" theory.

With the first phase of logical positivism, most markedly represented by Carnap's *"Der Logische Aufbau der Welt"*, the rational reconstruction of empirical knowledge was pursued

---

[1] This widely held position may be traced back to Spinoza, and is represented in various metaphysical versions also by Leibniz, (in a certain sense also by Kant), Schopenhauer, Fechner, Clifford, Riehl, Paulsen, the American monistic critical realists, especially R. W. Sellars, D. Drake, C. A. Strong; by one phase of B. Russell's thought; by R. Ruyer in France; by the psychologists Ebbinghaus, M. Prince, Warren; the Gestalt psychologists, especially Köhler and Koffka; by L. T. Troland, E. G. Boring, C. K. Ogden, and others.

on a phenomenalistic basis. It is therefore not surprising that metaphysicians misinterpreted this approach as a revival of a Berkeleyan subjective idealism. While Carnap explicitly disavowed any claims regarding the ultimate reality-problems of the mental and the physical, he shared of course with Berkeley, Hume, Condillac, Mill, Mach and Avenarius the conviction that there is no ontological mind-body problem that could be legitimately formulated. The only genuine problem, Carnap claimed, was one of logical analysis, i. e. the question of the formal relations between the concepts that describe the data of first-person-experience, the concepts of physics and those of (behavioristic) psychology. The "basic situation" of the mind-body-relation was identified with the parallelism of data that a person would experience if he were to observe by means of some "cerebroscope", his own cerebral processes alongside with the stream of images or feelings which "correspond" to those brain processes. But the internal difficulties of a strictly phenomenalistic reconstruction were soon recognized. The translatability of statements concerning physical objects into statements concerning phenomenal data could no longer be held to obtain in the sense of mutual deducibility. And the absurdities of a metaphysical solipsism were parallelled by the absurdities of a phenomenal language that was doomed to be "private", "soliloquistic", "incommunicable".

The second phase of logical positivism arose largely out of a reaction against the phenomenalism (experientialism) of the first phase. Under the influence of O. Neurath's and K. Popper's critical suggestions, Carnap (10, 116, 118, 119, 119a) formulated his *physicalism*. It was easy again for metaphysically minded opponents to misconstrue this position as a variant of ontological materialism. But Carnap's aim was, just as in the previous phase, merely that of an analysis of language. He outlined a logical reconstruction of factual knowledge on the basis of an intersubjective (physicalistic) thing-language. This position, though independently arrived at, was generally akin to the methodological behaviorism that had been formulated even somewhat earlier but with much less formal precision by E. A. Singer (90) and K. S. Lashley (155). It is important to distinguish two phases in the development of physicalism. The first phase was rather rash in its claim of the translatability of the statements of physics and those of psychology into those of the thing-language. Availing ourselves of the material idiom (realistic language), this radical and crude form of physicalism may be said to amount to an identification of mental states with overt behavior. Early behaviorism (especially that of J. B. Watson) has been rightly accused of just this fallacious reduction. This view was essentially revised and corrected in the later formulations (119, 119a). Strict translatability depends of course on explicit definitions. But no explicit definitions that would serve the purpose could plausibly be constructed. The concepts of physics and psychology could perhaps be *introduced* by means of test-condition-test result conditionals but not in any way be regarded as synonymous with concepts of the thing-language (or purely logical compounds thereof). Carnap (119) advanced his reduction sentences as a possible formulation of those conditionals. While it has become increasingly doubtful that this formulation is logically adequate, the underlying and related ideas of confirmability and degree of confirmation are now quite generally accepted. No statement of physics nor of (intersubjective) psychology can be considered as completely and directly verifiable (or refutable) by the observations as formulated in the protocol-statements of the thing-language. The protocol-statements confer only a degree of confirmation upon the statements in the scientific languages of physics and psychology.

Reichenbach's version of scientific empiricism (64, 176) had for many years opposed the narrow verifiability criterion of the Viennese positivists. His emphasis on probability and induction, led him to advocate a more inclusive confirmability criterion, amounting approximately to the same delimitation of factual meaning as Carnap's criterion (in the second phase of physicalism). Reichenbach's account of the mind-body problem, based on his empirical realism, represents in many ways a position similar to that of Schlick in his early realistic approach. Before we turn to a fuller discussion of this view we must briefly mention a more agnostic position which arose out of a reaction against the earlier, rather immature arguments in favor of mind-body identity.

Felix Kaufmann (34), and similarly also Norman Jacobs (150), generally in sympathy with the principles of Logical Empiricism, insisted that strict identity would have to be tantamount to *logical* equivalence of phenomenal (introspective) descriptions of mental states with the descriptions of the "correlated" neurophysiological processes. But it seems obvious, so Kaufmann argued essentially, that the investigations of psycho-physiology are of a factual-empirical character. Which mental state is correlated with which neural processes can be determined only by experimental investigations. The statement of the correlation is therefore synthetic and the "equivalence" of the two descriptions thus can at best be only of (universal) *empirical* character. Reading this conclusion again in terms of traditional metaphysics it may be taken as a formulation of dualistic parallelism. Wolfgang Köhler in one of his later works,[1] and other thinkers trying to be cautious in such delicate matters, have essentially retreated to this obviously safer (because less daring) position. If anyone (like, e. g., E. G. Boring[2]) wanted to account for the parallelism by means of a supposed more fundamental identity, he usually availed himself of the principle of parsimony.

The principle of parsimony itself needs careful analysis. Occam's razor has really, as it were, *three* blades. The simplicity it advocates may be the descriptive or purely formal (or logico-mathematical) expediency that distinguishes, e. g., the heliocentric from the geocentric description of the planetary system. It may be the factual (or inductive) simplicity that arises from a reduction of the number of independent empirical hypotheses. This is probably the purport of Newton's *regula philosophandi*. But finally, Occam's razor may be used to cut away metaphysical entities. In what follows I shall contend that this third blade, the confirmability criterion of Carnap and Reichenbach, if properly applied, removes the metaphysical surplus, without cutting into the flesh of knowledge. I shall contend also that this new point of view involves (1) a fundamental revision of phenomenalistic

---

[1] *The Place of Values in a World of Facts*, Liveright, New York, 1936.
[2] *The Physical Dimensions of Consciousness*, Appleton, New York, 1933.

positivism and radical operationism (and behaviorism); (2) a re-instatement of a clarified critical realism on the basis of pure semantics and pure pragmatics; (3) a return to a reinterpreted identity (or double-language) view of mind and body.

(1) The slogan of Vienna Logical Positivism: "The meaning of a statement is the method of its verification"; (190) and the slogan of Bridgman's operationism (5): "A concept is synonymous with the set of operations" [which determine its applications] were excellent preventives of the transcendent type of metaphysical speculations. They have had a most salutary purifying effect. Logical empiricism in its later development, however, had to replace these radical principles by more conservative ones. Als already indicated, the meaning of scientific statements cannot in general be identified with their confirming evidence. This is obvious in all those cases in which the evidence must in principle be indirect. Historical statements concerning past events, predictions of future events; existential hypotheses concerning radiations, subatomic processes in physics; genes, filterpassing viruses in biology; unconscious motivations in psychology; etc., are only some of the more striking types of assertions whose meanings (i. e. the states of affairs to which they refer) cannot be identified with the states of affairs that can conceivably serve as evidence for them. For a more specific but very simple example we may refer to the concept of the temperature of a body. As ordinary and scientific common-sense (untouched by ultra-positivistic reductionism) would put it, thermometer (or pyrometer) readings, spectroscopic findings, and other types of measurement merely indicate something about the body in question, namely the intensity of heat which is a state of that body. No matter whether this heat intensity is construed in terms of classical (macro-) thermodynamics or in terms of statistical (micro-or molecular) thermodynamics, it is in any case only *evidenced by but not identical with* those indications. Similarly for psychology: The overt symptoms and behavior that indicate an emotion, like e. g., anxiety, are confirmable and measurable in terms of skin temperature, endocrine secretions, psychogalvanic

reflexes, verbal responses, etc., but must not be confused with the emotion itself. Generally, the "theoretical constructs" of the sciences cannot be identified with (i. e. explicitly defined in terms of) concepts which apply to the directly perceptible facts as they are manifest in the contexts of ordinary observation or of experimental operations.

(2) The required correction and emendation of the phenomenalistic phase of positivism and operationism can best be achieved by means of a reconstruction in terms of pure semantics and pure pragmatics. Semantics as developed primarily by Tarski and Carnap enables us in a precise way to speak, in a metalanguage, about the relation of designation that holds between the symbols of a given language (the object language) and the objects, properties, relations and states of affairs they symbolize. The required metalanguage must of course have a sufficiently rich vocabulary to allow for this. It is in the field of pure pragmatics (thus far only sketched in outline by Wilfrid Sellars) that the rules and the scope of the metalanguage are determined. The pragmatic prerequisites of a workable scientific language extend far beyond the conditions that must be fulfilled for the sake of logical consistency and for the purposes of deductive inference. They also include the condition of confirmability, with all that this implies: a set of proper names (or co-ordinates) and of predicates only some of which correspond to directly confrontable items of immediate experience; a set of relationships that connect the directly verifiable with the only (indirectly) confirmable predicates and statements. With such a reconstruction a distinction necessarily neglected by phenomenalism can be reinstated. It is the important distinction between the evidential basis and the factual reference of terms and statements. In acknowledging this distinction we retain the empiricist conditions for meaningfulness and for factual adequacy: Only if our terms are nomologically related to terms that designate items or aspects of what is directly observable can they be factually meaningful; and only if statements are supported at least by incomplete and/or indirect evidence can they be justifiably asserted. But in the recognition of the incompleteness and indirectness of the verification

of practically all scientific statements we implicitly allow for a genuinely critical realism. This new version of realism is free from the objectionable metaphysical elements in the older forms of realism. Much of the perplexities in the time-honored reality-problems arose out of confusion of the intuitive, experiential idea of reality with the cognitive, objective concept of reality. The agonies that attend all attempts to solve the "problem of transcendence" can be avoided once it is realized that this is a pseudo-problem. The solution that had been sought involved plainly an inconsistency: The non-given was to be proved just as real as the given. But if by "real" one means *given*, then obviously the wish for a demonstration is doomed because of the self-contradiction. If however one wishes to connect with the word "real" not an ineffable but a cognitively expressible significance then the usage of this term in common life and in science may profitably be taken as a standard. "Real" and "unreal" are of course ambiguous and often emotively tinged words. But in the context of the traditional realism-phenomenalism controversy it is clear that the distinction connoted by these terms cannot be intended to achieve a division among things, events or processes. Once anything is at all classified under one of these three headings it is *co ipso* considered real. Dreams and delusions are (even according to commonsense) real enough as occurrent events. What is not real are the referents (designata) of certain terms or assertions that we sometimes formulate on the basis of certain *interpretations* of dream or delusion-experiences.

The realistic correction of positivism consists in the identification of meaning with factual reference. This conforms well with customary usage according to which a statement *means* a state of affairs; and is *true* if that state of affairs is fulfilled ("is real", "exists"). This is the obvious grammar of "meaning", "truth" and "reality". Metaphysical problems cannot arise as long as we combine those definitions with the empiricist requirement that in order to be meaning*ful*, a statement must in principle be confirmable. The confirmation rules which formulate the connections between the evidential basis and the factual referents of statements are the metalinguistic correlate of those laws without which inference of specific unobserved or unobservable states of affairs would be impossible. Just which network of laws and existential assumptions will most adequately and parsimoniously serve for a comprehensive and predictively fruitful organization of the data can of course not be settled in any a priori fashion. Nevertheless, only within the frame of a language that makes such a network possible can we legitimately assign probabilities to hypotheses on the basis of relevant evidence. The ("realistic") frame itself however cannot be justified by considerations of inductive probability. The adoption of this frame can be vindicated only by its fruitfulness for the purposes which it helps to fulfill. Like other principles which rationalists mistake for synthetic *a priori* presuppositions this is, from the viewpoint of logical reconstruction, a basic convention, capable only of pragmatic but not of cognitive justification. [1]

(3) We are now ready to develop the implications of the just outlined clarified empirical realism for the mind-body problem. There are three demonstrably mistaken reductions by means of which monistic solutions have been attempted. There is firstly the crude and simple-minded identification of the stimulus-aspects with the mental qualities. Obviously we cannot say that a color sensation is identical with the radication (of a certain intensity and frequency-pattern) which, under certain conditions merely elicits that sensation. Secondly, in our critique of phenomenalism we have also refuted the identification of physical bodies with complexes or configurations of elements of direct perception. Thirdly, the behavioristic identification of mental states with the responses (including linguistic utterances) of organisms is equally fallacious. It is of course granted that the confirmation of objective statements concerning "physical" bodies is possible only on the basis of the evidence of direct experience. Similarly, intersubjectively meaningful statements concerning mental states are confirmable only on the basis of behavioral evidence. If we are to avoid the errors of phenomenalist reduction and

---

[1] The realism of pure semantics and pragmatics is outlined in (134, 194, 195) and an analysis of the problem of justification may be found in (133).

quite generally of the negativism of orthodox positivism then all the relationships mentioned are not identities, but—at best—lawful (causal) connections between distinguishable states or events. The equivalence of statements about each pair of states or events can therefore be only of the empirical type. The precipitous assertion of a logical equivalence was of course based on the phenomenalistic claims of the explicit definability of the entities in one realm in terms of the entities of the corresponding other realm. This, as we have tried to point out, was completely unwarranted.

Curiously enough, the same sort of critique has been applied also to the identification of mental states with processes inside the organism, i. e. neurophysiological processes. It seemed quite incredible how a color sensation, a remembrance of things past, an act of thought concerning mathematical relations, or a feeling of indignation, could in any sense whatsoever be "the same" as some brain-process or other. Here again it was urged that the relation can be no other than, at best, that of a lawful correspondence or parallelism of simultaneous events. The many arguments in favor of this view are well known. One of the more important among these arguments contends that the attributes of mental states and events and the attributes of the corresponding neurophysiological processes are so different that the respective predicates characterizing each of the two types of processes can stand only in the relation of general (empirical) equivalence but never in that of a logical equivalence. Hume argued that statements of specific causal relations are synthetic *a posteriori* because alternatives are always conceivable without self-contradiction. Similarly, it is contended that a brain process which a future neurophysiology might characterize as of a definite type could conceivably be associated with a phenomenologically described immediate experience of a type radically different from that with which, as a matter of empirical regularity, it is actually associated (say, a sentiment of nostalgia). Eddington once argued that even the most detailed physiological and physical knowledge of the behavior and the nervous processes in the human organisms occurring on some November 11th at 11 a. m. in London could not possibly indicate to a Martian super-scientist unfamiliar with human history and unendowed with human sentiments that these events "mean" a commemoration of the armistice. This fascinating argument however, rests on two fallacies. Firstly, such a utopian knowledge of the neurophysiological processes would enable the Martian to derive the actual and potential verbal behavior of the Londoners; it would also enable him to reconstruct the physical account of the origin of the ritual (two minutes silence, etc.) and thus to know, in principle, everything that can be known about those events in an inter-subjective manner. Secondly, this can be achieved even if the Martian, because of the differences or limitations in his repertoire of emotions, cannot empathize, let alone share, the sentiments in question. A congenitally blind man, equipped with modern-physical devices, could investigate not only the physics of colored surfaces, of light radiations reflected by them, etc., but also the (behavioristic) psychology of color sensation, discrimination and perception (on the part of subjects equipped with eyesight). Similarly, a Martian could *know* all *about* human feelings and emotions without having *knowledge of* them, i. e. without directly experiencing them or being *acquainted* wiht them by intuition or imagination.

Quite generally, one of the difficulties that are so frequently adduced in the critique of the identity-theory of mind and body rests on a confusion of *acquaintance* with *knowledge*. No one denies that the *image* of a brain, as perceived by a surgeon or as pictured in terms of an atomic model has totally different properties from a melody-as-heard or a sentiment-of-elation-as-actually-lived-through. But images or other directly experienced acts or data are not in and by themselves concepts. Knowledge proper is always conceptual. This insight is an important point of agreement between such otherwise divergent recent philosophers as Poincaré, Bergson, James, Dewey, Russell. Eddington, R. W. Sellars, C. I. Lewis, Schlick, Wittgenstein and Carnap. What then is meant by "conceptual knowledge"? What is meant by "concept"? The best answer we can give today rests on a repudiation of psychologism and upon the results of pure semiotic. Concepts are symbols whose meaning is constituted by the syntactical, semantical and pragmatic rules which determine the relations of those symbols to one another, to their designata and to their evidential basis. The crucial question then concerns the conditions of the identity of concepts. What is the criterion for identity? We can safely follow Leibniz' *principium identatis indiscernibilium*, here as elsewhere. If two terms are defined by the same set of rules, they are merely different symbols for the same meaning, they are the same concept. Such synonymity however may arise in various ways. The most obvious and trivial case is that of explicit definition in which we arbitrarily stipulate the unrestricted mutual substitibility of symbols. More interesting and more relevant for our problem is the case of epistemic (or "systemic") synonymity. We may determine certain meanings uniquely by different definitions of the type known as "definite descriptions" (Russell). Thus two explorers may unwittingly have observed the same mountain from different directions, and only after comparing notes come to realize that it was really identically the same mountain. This is a systemic identity in that it can be established only if the system of empirical geometry and optics is presupposed. Quite analogously, the identity of the morning star with the evening star (ever since Frege a much used example in logical analyses), is based on the recognition that one and the same trunk of world-lines (the four-dimensional representation of the planet Venus) is the object of reference of the two designations, referring to alternative segments of that trunk. Only within the system of Kepler's kinematics and of ordinary geometrical optics can this identity be explicated and warranted. This and the preceding example concerned the identity of things (continuants), or more precisely speaking, the identity of the designation of a name with the designatum (descriptum) of a description or else the identity of the descripta of two descriptions of thinglike entities.[1] But quite similar considerations hold for concepts (predicates of various levels). The identity of the concept of "electric current" defined by various definite descriptions such as those based on the magnetic, chemical or thermal indications can be defended against empiricist or operationist doubts only after a full fledged system of electrodynamics enables us to *deduce* those various effects from a unitary

---

[1] Even these first two examples, could be analyzed in terms of individual-concepts (unit-classes) instead of things, whose identity is under examination.

theory of electricity, magnetism, electrolysis and heat. Those doubts could of course never be removed with finality. We not only admit but would even emphasize the empirical or inductive basis which underlies all such identifications in the realm of factual knowledge. The only kind of identification that can be proved with finality is found in the purely formal sciences. Despite the fundamental difference between the situation in empirical knowledge and that in pure mathematics, there is an instructive structural analogy here. Two different infinite series, for example, may be used for the definition (unique description) of one and the same number, as e. g. in the case of π. But such mathematical proofs of identity also presuppose a frame of concepts and postulates. Only within such a frame can we assert meaningfully and demonstrate validly the identity of the object of two descriptions. (A perfectly obvious illustration is the arithmetical identity of 2^3 with $\sqrt{64}$). The frame of arithmetic, i. e. the postulate system of Peano in the Frege-Russell interpretation, is logically or analytically valid. The situation is radically different in empirical geometry. For example the identity of two points or line-segments characterized in different ways depends upon the factual adequacy of the geometrical postulates. The same holds, *a fortiori*, for the identifications in the natural sciences. Returning to an illustration previously introduced, the identification of the temperature of a gas with the mean kinetic energy of its molecules depends of course upon the truth of the molecular-statistical theory of heat. But if the truth of the theory is assumed, the strict identity of reference becomes a matter of logical deduction. Temperature as a macro-concept refers to the state of a body which is only more fully characterized by the theoretical description of its micro-structure. Once the theory is adopted it would make no sense to speak of the temperature as something distinct and different from that set of micro-conditions. Only the pictorial connotations of the word "temperature" that remind us of thermometers or of the directly felt heat of a body seem to make the corresponding concept merely "parallel" to that of. molecular thermodynamics.

The logical principle that underlies our argument is, as indicated before, simply a variant of Leibniz' principle of

identity. The meaning of a concept is determined, not by
its pictorial connotations, but by the system of rules which
implicitly defines that meaning. If two terms, no matter
what words or symbols they are and no matter what pictorial
appeals they may convey, are mutually substitutible for each
other because they fulfill precisely the same functions in a
system of rules, then they have the same meaning, they are
the same concept.

The application of these considerations to the mind-body
problem must by now be fairly obvious. Relative to the
"molar" (or macro-) account given by behavioristic psychology, the neurophysiological account is a micro-description of
the very same events and processes. The pictorial connotation
of the two accounts are of course different, since the images
attaching to the behavioristic terms represent stimulus-response
situations, while the images connected with the neurophysiological language are apt to represent observations of nervous
tissues. The notoriously greatest difficulty however arises here
from the pictorial connotations of the mentalistic terms that
owe their introduction to a third avenue of approach to the
same processes —, *introspection*. The qualities of direct
awareness, the facts of stimuli and responses, the directly
observable data of the neurophysiologist are of course not to
be identified with one another. We have already warned
against the fallacies involved here. But we contend that the
designata of the mentalistic language are identical with the
descripta of the behavioristic language, and that both are
identitical with the designata of the neurophysiological
language. Utilizing the distinction suggested before, we may
say that the factual reference of some of the terms in each of
these different languages (or vocabularies) may be the same
while only their evidential bases differ. A state of mind,
conceived as an event in the spatio-temporal-causal structure
of the world may thus be characterized by concepts that are
evidentially anchored in quite heterogeneous areas. It is this
anchoring that gives the concepts their particular place in one
or the other vocabulary. But if we are sure not to confuse
their factual reference with their evidential base we may
rightly say that they have the same meaning. This holds
unless we countenance in principle unconfirmable assertions
or unless the facts of psychology themselves force upon us an

interactionistic dualism. The last proviso indicates the systemic nature of the proposed identifications. On the whole, I should think, the available evidence points with remarkable consistency in the direction of a system of psychology, psychophysics and psychophysiology wich provides for the monistic solution here outlined. But this is the empirical, the factual issue which philosophical analysis cannot decide and should not prejudge. We can do no more than clarify the logical structure of the problem and remove unfounded objections to the identity theory which perhaps owing to a failure of nerve seems to have been temporarily eclipsed by a return to parallelism—if not even interactionism. The view we are proposing here should not be construed as a metaphysical doctrine. It again has merely some historical affinities with certain forms of epiphenomenalistic materialism, panpsychism, or the double-aspect or double-knowledge theories. If a label is wanted, then perhaps "double-language-theory" is still the least misleading I can suggest. Within the conceptual system which fulfills the intersubjective confirmability condition and is at the same time the simplest account compatible with the accumulated facts of psychology, the terms of the behavioral-psychological and of the introspective language are (systemically) synonymous. If further factual discoveries should force upon us a radical revision of the conceptual system, then, conceivably, this claim of synonymity may have to be modified or even abandoned. In the meantime it is well to remember that the tentative indentifications which generally underlie synonymities of this type are among the most fruitful devices in the search for unifying explanations in the progress of science. The identification of light with a special kind of electromagnetic oscillating field; ferro-magnetism with the spin of electrons; of heat with molecular motion; of chemical valencies with certain dynamical features of the atoms; of the medium of inheritable traits with the gene-structure of the chromosomes; etc., these are only some of the more noteworthy cases in point.

One last critical question requires discussion. The entire preceding argument, it may be urged, depends upon the presupposition that the vocabulary of introspection is part of an intersubjective language and thus really interpreted behavioristically. Introspective terms are then introduced on the evidential basis of linguistic responses and are therefore in any case logically on a par with those terms that have their basis in non-linguistic responses of the organisms. Thus, it may be said, that the real difficulty of the mind-body problem has been avoided rather than resolved. This objection obviously implies that the language of introspection is to be taken as phenomenal, purely experiental and thus strictly subjective. My reply, very briefly, is this: The problem thus proposed is the epistemological question of the relation between the "private" (if not solipsistic) language of data (phenomena) to the language of "public", intersubjective "constructs" (thing-concepts). It is highly questionable as to whether the idea of a phenomenal language in this sense can even be consistently maintained, let alone fully elaborated. But to those who cling to this "Aufbau" phase of positivism I would offer the suggestion that there can be only a correspondence, but never a translation between the phenomenal language and the thing language. If introspective descriptions are not to be taken as referring to events which are at least in principle confirmable by the much more indirect route of behavioral (or physiological) evidence, then they are indeed severed from the language of intersubjective communication and doomed to solipsistic privacy. There is no bridge between such a private language and the language of science except one of isomorphic correspondence. Structurally the situation bears a certain resemblance to the one in the reconstruction of the rational numbers on the basis of the natural numbers. Certain ordered pairs of natural numbers are introduced, they define rational numbers. But the rational numbers (like 3/1, for example) which represent integers (3 in this example) merely correspond to them, but are not identical with them. This isomorphism here consists not only in the one-to-one correspondence of certain elements of one realm to all elements in another, but in the one-to-one correspondence of the results of all arithmetical operations with corresponding elements. The analogy with the (however much more complex) field of epistemology lies in the isomorphism between certain statements in the phenomenal language and those in the intersubjective scientific language. As Carnap pointed out long ago (116) epistemological reconstruction may be attempted in either of two ways. The protocol-propositions may be part of the system of the scientific language or they are outside of it. In the latter case we must have some statements in the scientific language that correspond to the protocol propositions. This correspondence, however, must not be confused with what is traditionally called psycho-physical or psycho-physiological parallelism. Parallelism has always been a doctrine according to which two different types of processes or two aspects of one and the same process are related by laws of coexistence or contemporaneity. The correspondence of the protocol propositions with propositions of the intersubjective system is a purely formal relation which arises exclusively out of the constructive definitions, involving differences in Russellian type-levels, by means of which the terms of the physical language are supposedly constituted out of terms belonging to the language of data. This is the position a consistent phenomenalist must take. But the many difficulties of that position have impelled Carnap and other physicalists to replace it by the reconstruction on an intersubjective basis. The analogy of this procedure in mathematics is of course the axiomatic method by means of which the total system of numbers (real numbers) is introduced and no problems of the "Aufbau"—type are then encountered. If the protocol propositions, i.e. the names and predicates occurring in them are part of the total symbolic system of the language of science, then we have here before us the sort of "realistic" reconstruction which underlies the systemic identity view of mind and body.

Résumé: Logical Empiricism in its present phase possesses the logical tools for a reformulation of the identity or double-language view of the mental and the physical. As in so many other issues of philosophy, this solution represents an equilibrium that has been reached only after several oscillations toward untenable extreme positions. The identity proposed is neither the reductive definitional one of phenomenalism or of behaviorism, nor is it an identity that presupposes a metaphysical realism. It is rather the hypothetical identity of the referents of terms whose evidential bases are respectively: introspective, behavioral or physiological. It is granted that the relations between the evidential indicators (linguistic responses, overt behavior and the data of neuro-physiology) must be interpreted as empirical laws. But this does not in the least preclude the identity of the factual reference of the concepts which characterize the causal processes and events in terms of which the facts in each sphere of evidence may become explainable and predictable to an ever increasing extent. It is this hypothetical, systemic, referential identity that has been overlooked by those who retreated to a timid parallelism. The alleged difficulties of the identity view are mainly due to a confusion of pictorial appeals with cognitive meanings. A more adequate discussion of the points touched upon as well as of the many related questions and difficulties would of course require much more space than is available here. Many of the books and articles in the attached bibliography may help the reader to round out what has been merely sketched in the present essay.

University of Minnesota.

Psychological Review
Vol. 62, No. 3, 1955

FUNCTIONALISM, PSYCHOLOGICAL THEORY, AND THE UNITING SCIENCES: SOME DISCUSSION REMARKS

HERBERT FEIGL

University of Minnesota and Minnesota Center for Philosophy of Science

Tolman and Brunswik (1, 24, 25), perhaps more deliberately than any other psychologists in the more recent development of the functionalistic approach in psychology, have emphasized the mutually substitutable nature of the mediating processes which occur between the environmental situation (stimuli, cues, means), and the responses (or achievements) in the behavior of organisms. Brunswik, with sovereign and perspicacious knowledge of the history and the systems of psychology, has illuminated in several publications (see especially 2, 3) the relationships of various ways in which scientific research may "focus" upon the subject matter in the biological-psychological area. The "thematic" and "tactical" differences of these various determinations of the type and level of research are not only compatible with the logical empiricist conception of the unity of science (for an exposition see Carnap, 5; Feigl, 6, 7), but actually support the conception of a unitary science precisely because of the obvious relationships between the assorted possibilities of focusing. It seems to me indeed only a tactical (if not a historical-terminological) question just what type or level of analysis one wishes to reserve for *psychology*, and which other focusings one wishes to delegate to biological ecology, physiology, neurology, or sociology and anthropology.

The mutual substitutibility ("vicarious functioning") of the mediating processes in organisms is indeed a striking feature. Biological thinking has borne its imprint for a long time. There is relatively little in the inorganic parts of nature (but of course a good deal in such inorganic artifacts as the servomechanisms) which invites the sort of concept formation which is so prevalent and fruitful in the bio-psycho-social sciences. Some imaginative scientists (not to mention several excessively imaginative philosophical cosmologists) have perceived principles of organization even in stars and atoms. But while Pauli's exclusion principle might indeed be viewed as basic to chemical organization, and Köhler's physical *Gestalten* (14) may describe a great deal of inorganic and organismic patterning, "vicarious functioning" or "means-ends" relationships seem primarily characteristic of the biological level (and everything above it in the pyramid of the sciences)—and of course in man-made machines.

Since scientific research zealously looks for regularities—be it for the sake of prediction, explanation, or control—it will fasten upon any sort of lawful relationships wherever it can find them. Regularities, however, are of many various forms and types (see Feigl, 10); there are those which we (rather sanguinely) consider *"basic,"* viz., the regularities formulated in the fundamental laws of theoretical physics; and there are other regularities which are obviously dependent on the structure of the systems (e.g., drops, bubbles, crystals, geysers, machines, organisms, social groups, etc.). Phenomena of homeostasis (e.g., regarding the relative stability of the blood temperature or the blood sugar level, etc., in mammals) provide a good example of regularities which depend on the structure (and many functions) of living organisms. The regularities of perception, learning, and motivation as studied in behavioristic psychology likewise depend on the "intact" structure and functioning of the organism. If the "life," i.e., the rise and decline, of fixed stars as studied in astrophysics had as much "teleological" and developmental regularities as does the life of organisms, we would probably have a scientific "ecology" and "theory of adaptation of the stars," as we have ecological, evolutionistic, and developmental biological and psychological disciplines.

Brunswik knows, perhaps better than anyone else, that the subject matter of a science can be "carved out" in a number of ways. Ultimately there is only one criterion by which scientists decide which ways of "focusing" are preferable: "By their fruits ye shall know them." But what sort of fruits are desirable depends on one's interests. Skinner's (22, 23) fruits do not satisfy Krech's (15) appetites—and vice versa. Skinner's interests are focused on prediction-for-the-sake-of-control; Krech's on understanding and explaining. Skinner, like many a positivist before him, has his eye especially on the practical applications of science. The systematic design of his experiments selects certain narrowly circumscribed segments of behavior and examines them under various conditions of drive, stimulation, etc. The laws which he establishes, e.g., regarding response frequencies, are thus highly reliable, though of course statistical, i.e., not strictly deterministic. Brunswik's own research on size constancy in visual perception, to which he referred in the present symposium (4), proceeds by representative sampling (thus achieving a "close-to-the-actual-life-situation character") and thus also results in statistical laws; but it is as little concerned with a detailed theory of the mediating processes as is Skinner's psychology of the "empty organism." The more ambitious attempts of Tolman, Hull, Spence, and others to derive behavior regularities from various systems of theoretical postulates strive for more comprehensive unifications than Skinner or Brunswik in their respective (and different) procedures could ever achieve. But the success of molar behavior theories is still quite problematic, even if for the present it is more distinct than that of neurophysiological theories.

In view of the controversial question of the fruitfulness of the more highly theoretical approaches, one can readily understand why Brunswik tends to limit the subject matter of psychology to the study of certain (statistically certifiable) teleological relations between environmental and achievement variables. There is much in the common life connotations of the term "psychology" which would justify this delimitation of subject matter. But Brunswik, in so restricting his enterprise, cannot plausibly claim to produce a psychological *theory*. As I have argued at some length elsewhere (9), it may well be that sooner or later we shall have to recognize that explanations of the mediating processes can be given fruitfully only in terms of neurophysiology. In order to prevent misunderstandings I wish to say, however, that I should be very happy (and perhaps not too surprised either) if typically *molar* behavior theories were destined to succeed better than they have thus far. After all, classical thermodynamics was, and still is, a very powerful theory—independently of its (even more powerful) molecular interpretation. Here, with some reservations, I share the hopes expressed by Postman (21) and Hilgard (13).

The teleological patterns of perception, learning, and action may lend themselves to explanations of more or less unitary form on the molar level. But an explanation of the very functioning of teleological mechanisms cannot be teleological itself (Nagel, 20); it will no doubt have to be of a relatively micro- or molecular character. This, I take it, is the lesson to be learned from cybernetics. The unification of the sciences which is being achieved along these lines is indeed much more exciting, though admittedly more problematic, than the merely methodological or empirical unity of science. With the abandonment of the earlier (logical positivist) thesis of the *translatability* of scientific theories into observation language, the more ambitious thesis of *unitary* science becomes again more vital and pertinent. The empiricist requirement, as we understand it today, demands only that the nomological net of scientific concepts and propositions be tied to observables in certain points. More accurately, the primitives of scientific theories are given their meaning by implicit definitions, i.e., by the postulates that combine them into a network with one another and with the observables.

Operationism, once helpful in eliminating metaphysical elements from the scientific enterprise, has tended to overstep the limits of its usefulness. There is, and here I agree with Krech (15, 16), in this day and age no longer any reason to be afraid of hypotheses. But (as I wish to remind Krech) it is of course wise to label as "promissory notes," "hopes for theoretical postulates," or "explanation sketches" what in present day neurophysiology can be regarded as no more than a program or blueprint for the derivation of behavior laws. Radical operationism has tended unduly to restrict psychological concept formation to purely "intervening variables" (see MacCorquodale and Meehl, 18), i.e., to dispositional concepts which are merely shorthand for empirical laws.

It is questionable whether there are any important concepts at all in psychological theory which can be adequately conceived as pure intervening variables in this sense. (For a suggestive discussion of "conventional constructs," i.e., concepts which are neither pure empirical dispositionals, nor physiological existentials, see Lindzey, 17.) It will not help much either if intervening variables are construed as "open concepts," i.e., as dispositionals introducible by sets of test-condition → test-result conditionals open to additional (partial) specifications of meaning of the same (conditional) form. It is the distinctive mark and advantage of any *theory* with genuine explanatory power that it enables us to *derive* the total set of those conditionals. What would otherwise remain a miscellaneous collection of empirical laws or correlations thus acquires the sort of unity which only a theory can engender. Existential hypotheses or conventional constructs thus play a vital and indispensable role in theory construction; the atomic hypotheses in physics and the neurophysiological hypotheses in the behavior sciences are in many ways analogous to one another as regards their unifying power (see Feigl, 8).

Unitary science, understood in the sense of nomological nets (tied to points of observation), has no sharply separable parts or provinces. Only practical divisions of labor, convenient lines of demarcation of research, can be recognized as separating psychology from biology or physiology on the one side and the social sciences on the other. Even if Brunswik's own predilections in defining the subject matter of his probabilistic-functionalistic psychology prevent him from attaining the theoretical level in his research, he has—in the capacity of a metatheorist (i.e., as a comparative methodologist of science)—provided extremely valuable suggestions regarding the place of psychology in the system of the united (or rather unit*ing*) sciences.

REFERENCES

1. BRUNSWIK, E. *Wahrnehmung und Gegenstandswelt.* Vienna: Deuticke, 1934.
2. BRUNSWIK, E. The conceptual focus of some psychological systems. *J. unified Sci.* (*Erkenntnis*), 1939, **8**, 36–49. (Also in Marx, 19.)
3. BRUNSWIK, E. *The conceptual framework of psychology.* Chicago: Univer. of Chicago Press, 1952. (*Int. Encycl. unified Sci.*, Vol. I, No. 10.)
4. BRUNSWIK, E. Representative design and probabilistic theory in a functional psychology. *Psychol. Rev.*, 1955, **62**, 193–217.
5. CARNAP, R. *Logical foundations of the unity of science.* Chicago: Univer. of Chicago Press, 1938. (*Int. Encycl. unified Sci.*, Vol. I, No. 1.) (Also in Feigl and Sellars, 12.)
6. FEIGL, H. Unity of science and unitary science. Paper read at Fifth International Congress for the Unity of Science, Cambridge, Mass., 1939.
7. FEIGL, H. The mind-body problem in the development of logical empiricism. *Rev. int. Phil.*, 1950, **4**. (Also in Feigl and Brodbeck, 11.)
8. FEIGL, H. Existential hypotheses. *Phil. Sci.*, 1950, **17**, 35–62.
9. FEIGL, H. Principles and problems of theory construction in psychology. In W. Dennis (Ed.), *Current trends in psychological theory.* Pittsburgh: Univer. of Pittsburgh Press, 1951. Pp. 179–213.
10. FEIGL, H. Notes on causality. In H. Feigl & M. Brodbeck (Eds.), *Readings in the philosophy of science.* New York: Appleton-Century-Crofts, 1953. Pp. 408–418.
11. FEIGL, H., & BRODBECK, M. (Eds.) *Readings in the philosophy of science.* New York: Appleton-Century-Crofts, 1953.
12. FEIGL, H., & SELLARS, W. (Eds.) *Readings in philosophical analysis.* New York: Appleton-Century-Crofts, 1949.
13. HILGARD, E. R. Discussion of probabilistic functionalism. *Psychol. Rev.*, 1955, **62**, 226–228.
14. KÖHLER, W. *Dynamics in psychology.* New York: Liveright, 1940.
15. KRECH, D. Discussion: theory and reductionism. *Psychol. Rev.*, 1955, **62**, 229–231.
16. KRECHEVSKY, I. Hypotheses in rats. *Psychol. Rev.*, 1932, **39**, 516–532.
17. LINDZEY, G. Hypothetical constructs. *Psychiatry*, 1953, **16**, 27–33.
18. MacCORQUODALE, K., & MEEHL, P. On a distinction between hypothetical constructs and intervening variables. *Psychol. Rev.*, 1948, **55**, 95–107. (Also in Feigl and Brodbeck, 11.)
19. MARX, M. H. (Ed.) *Psychological theory: contemporary readings.* New York: Macmillan, 1951.
20. NAGEL, E. Teleological explanation and teleological systems. In S. Ratner (Ed.), *Vision and action: essays in honor of Horace Kallen.* Rutgers Univer. Press, 1953. (Also in Feigl and Brodbeck, 11.)
21. POSTMAN, L. The probability approach and nomothetic theory. *Psychol. Rev.*, 1955, **62**, 218–225.
22. SKINNER, B. F. *The behavior of organisms.* New York: D. Appleton-Century, 1938.
23. SKINNER, B. F. *Science and human behavior.* New York: Macmillan, 1953.
24. TOLMAN, E. C. *Purposive behavior in animals and men.* New York: Century, 1932.
25. TOLMAN, E. C., & BRUNSWIK, E. The organism and the causal texture of the environment. *Psychol. Rev.*, 1935, **42**, 43–77.

14

Sociology and Physicalism

BY OTTO NEURATH

(TRANSLATED BY MORTON MAGNUS AND RALPH RAICO)

I. PHYSICALISM: A NON-METAPHYSICAL STANDPOINT

CONTINUING the work of Mach, Poincaré, Frege, Wittgenstein and others, the "Vienna Circle for the Dissemination of the Scientific World-Outlook (*Weltauffassung*) seeks to create a climate which will be free from metaphysics in order to promote scientific studies in all fields by means of logical analysis. It would be less misleading to speak of a "Vienna Circle for Physicalism," since "world" is a term which does not occur in the language of science, and since world-outlook (Weltauffassung) is often confused with world-view (Weltanschauung). All the representatives of the Circle are in agreement that "philosophy" does not exist as a discipline, alongside of science, with propositions of its own: the body of scientific propositions exhausts the sum of all meaningful statements.

When reduced to unified science, the various sciences are pursued in precisely the same manner as in their disassociation. Up to now their uniform logical character has not always been sufficiently emphasized. Unified science is the outgrowth of comprehensive *collective labor*—in the same way as the structure of chemistry, geology, biology or even mathematics and logic.

Unified science will be pursued in the same fashion as the individual sciences have been pursued hitherto. Thus, the "thinker without a school" will have no more significance than he had when the sciences were disunited. The individual can here achieve just as much or just as little with isolated notions as he could before. Every proposed innovation must be so formulated that it may be expected

This article, originally entitled "Soziologie im Physikalismus," first appeared in Volume II of *Erkenntnis* (1931/2). It is included in the present work with the kind permission of Mrs. Marie Neurath and Professor Rudolf Carnap.

to gain universal acceptance. Only through the cooperative effort of many thinkers do all its implications become clear. If it is false or meaningless, i.e., metaphysical, then, of course, it falls outside the range of unified science. Unified science, alongside of which there exists no "philosophy" or "metaphysics," is not the achievement of isolated individuals, but of a generation.

Some representatives of the "Vienna Circle" who, like all their colleagues in this group, explicitly declare that there are no peculiarly "philosophic truths," nevertheless still occasionally employ the word "philosophy." By this they mean to designate "philosophizing," the "operation whereby concepts are clarified." This concession to traditional linguistic usage, though understandable for a number of reasons, easily gives rise to misconceptions. In the present exposition the term is not employed. We are not here seeking to oppose a new "Weltanschauung" to an old one, or to improve on an old one by the clarification of concepts. The opposition, rather, is between all world-views and science which is "free of any world-view." In the opinion of the "Vienna Circle," the traditional edifice of metaphysics and other constructions of a similar nature, consist, insofar as they do not "accidentally" contain scientific statements, of meaningless sentences. But the objection to the expression, "philosophizing," is not merely a terminological one; the "clarification of the meaning of concepts" cannot be separated from the "scientific method," to which it belongs. The two are inextricably intertwined.

The contributions to unified science are closely interrelated, whether it be a question of thinking out the implications of new astronomical observation-statements, or of inquiring into the chemical laws which are applicable to certain digestive processes, or of re-examining the concepts of various branches of science in order to find out the degree to which they are already capable of being connected with one another, in the way that unified science demands. That is to say, every law in unified science must be capable of being connected, under given conditions, with every other law, in order to reach new formulations.

It is, of course, possible to delimit different kinds of laws from one another, as for instance, chemical, biological or sociological. *But one may not assert that the prediction of a concrete individual event depends solely on laws of one of these kinds.* Whether, for example, the burning down of a forest at a certain spot on the earth will proceed in a certain way depends just as much on the weather as on whether or not human beings will undertake certain measures. These measures, however, can only be predicted if the laws of human behavior are known. *That is to say, all types of laws must, under given conditions, be capable of being connected with one another.* All laws, whether chemical, climatological or sociological, must, therefore, be conceived of as constituents of a system, viz., of unified science.

For the construction of unified science a unified language (*"Einheitsprache"*)[1], with its unified syntax, is required. To the imperfections of syntax in the period preparatory to unified science one may trace the respective positions of particular schools and ages.

Wittgenstein and other proponents of the scientific world-outlook, who deserve great credit for their rejection of metaphysics, i.e., for the elimination of meaningless statements, are of the opinion that every individual, in order to arrive at scientific knowledge, has temporary need of meaningless word-sequences for "elucidation" (Wittgenstein, *Tractatus* 6. 54): "My propositions are elucidatory in this way: he who understands me finally recognizes them as senseless, when he has climbed out through them, on them, over them. (He must, so to speak, throw away the ladder after he has climbed up on it.)" This sentence seems to suggest that one must as it were undergo repeated purgations of meaningless, i.e., metaphysical, statements, that one must repeatedly make use of and then discard this ladder. Only with the help of elucidations, consisting of what are later recognized to be mere meaningless sequences of words, is one able to arrive at the unified language of science. These elucidations, which may, indeed, be pronounced metaphysical, do not, however, appear in isolation in Wittgenstein's writings: we find there further expressions which resemble less the rungs of a ladder than parts of an unobtrusively formulated subsidiary metaphysical doctrine. The conclusion of the *Tractatus*, "Whereof one cannot speak, thereof one must be silent," is, at least grammatically, misleading. It sounds as if there were a "something" of which one could not speak. We should rather say: if one really wishes to avoid the metaphysical attitude entirely, then one will "be silent," but not "about something."

We have no need of any metaphysical ladder of elucidation. We cannot follow Wittgenstein in this matter, although his great significance for logic is not, for that reason, to be less highly valued. We owe him, among other things, the distinction between "tautologies" and "statements about empirical events." Logic and mathematics show us what linguistic transformations are possible *without any*

---

1. Kurt Lewin has pointed out that the term has been employed, although in a somewhat different sense, by Franz Oppenheimer.

extension of meaning, independently of the way in which we choose to formulate the facts.

Logic and mathematics do not require any observation statements to complete their structures. Logical and mathematical errors can be eliminated without recourse to any outside field. This is not contradicted by the fact that empirical statements may be the occasion for such corrections. Let us suppose that a captain sails his ship on to a reef. All the rules of calculation have been correctly applied, and the reef is to be found on the maps. In this way an error in the logarithm tables, which was responsible for the misfortune, could be discovered, but it also could be discovered independently of such an experience.

In his "elucidations," which may also be characterized as "mythological introductory remarks," Wittgenstein seems to be attempting to investigate, as it were, a pre-linguistic state from the point of view of a pre-linguistic stage of development. These attempts must not only be rejected as meaningless; they are also not required as a preliminary step towards unified science. One part of language can, to be sure, be used to discuss other parts, but one cannot make pronouncements concerning language as a whole from a "not yet linguistic" standpoint, as Wittgenstein and certain representatives of the "Vienna Circle" seek to do. A part of these endeavors, although in a modified form, may be suitably incorporated into scientific work. The rest would have to be discarded.

Nor may language as a whole be set against "experience as a whole," "the world," or "the given." Thus, every statement of the kind, "The very possibility of science depends on the fact of order in the universe," is meaningless. Such statements cannot be salvaged by counting them as "elucidations," to which a somewhat less rigorous criterion applies. There is little difference between such an attempt and metaphysics in the conventional sense. The possibility of science is demonstrated by the existence of science. We extend its domain by augmenting the *body of scientific propositions,* by comparing new propositions with the legacy of past scientists, and thus creating a self-consistent system of unified science capable of being utilized for successful *prediction.* We cannot as deponents stand aside, as it were, from our depositions and serve simultaneously as plaintiff, defendant and judge.

That science keeps within the domain of propositions, that propositions are its starting point and terminus, is often conceded even by metaphysicians, of course with the rider that besides science there exists yet another domain, containing statements which are

to some degree figurative. In contrast to the dovetailing of science
and metaphysics which is so frequently proposed, this separation of
science and metaphysics (without, however, eliminating the latter)
is carried out by Reininger,[2] who also, when it comes to scientific
questions, adopts a position towards behaviorism which is similar
to that of the Vienna Circle.

Unified science formulates statements, corrects them, and makes
predictions. But it cannot anticipate its own future state. There is
no "true" system of statements as distinct from that which is accepted
at the present time. It would be meaningless to speak of such a thing
even as a limiting concept. We can only ascertain that we are operating today with the space-time system to which that of physics corresponds, and thus achieve successful predictions. This system of
statements is that of unified science. This is the point of view which
may be designated *physicalism*.[3] If this term should become established, then it would be advisable to speak of "physicalistic" when
one has in mind any spatio-temporal description framed in the spirit
of contemporary physics, e.g., a behavioristic description. The term
"physical" would then be reserved for "physical statement in the
narrower sense," e.g., for those of mechanics, electrodynamics, etc.
Ignoring all meaningless statements, the unified science proper to a
given historical period proceeds from proposition to proposition,
blending them into a self-consistent system which is an instrument
for successful prediction, and, consequently, for life.

II. The Unified Language of Physicalism

Unified science comprises all scientific *laws*. These are capable,
without exception, of being combined with one another. Laws are
not statements, but merely directions for proceeding from observation
statements to *predictions*. (Schlick)

Unified science expresses everything in its unified language,
common to the blind and the seeing, the deaf and the hearing. It is
"intersensual" and "intersubjective." It connects what the soliloquizer
asserts today with what he asserted yesterday, the statements he
makes when his ears are closed with those he makes when he opens
them. The only thing essential in language is *ordering*, something
present even in a Morse Code message. "Intersubjective" or "inter-
sensual" language depends above all on *ordering* ("next to," "between," etc.), that is, on that which is expressed in the symbol

2. *Metaphysik der Wirklichkeit*, 1931.
3. Cf. Otto Neurath, *Empirische Soziologie*, p. 2.

sequences of logic and mathematics. It is in this language that all predictions are formulated.

The unified language of unified science, which is derivable by and large from modifications of the language of everyday life, is the language of physics. In this connection, it is a matter of indifference for the uniformity of the language of physicalism what particular language the physics of a given period may use. It is of no significance whether it explicitly employs a four-dimensional continuum in its more highly refined formulations, whether it recognizes a spatio-temporal order of such a type that the locus of every event is precisely determined, or whether couplings of placer and velocity-dispersions, *whose precision is limited in principle*, figure as basic elements. It is essential only that the concepts of unified science, both where they are thought out in the most subtle detail and where the description remains imprecise, be made to share the current fate of fundamental physical concepts. It is precisely in this that the point of view of physicalism is expressed. But all predictions, in whose confirmation or rejection we see the measure of science, are reducible to observation-statements, to statements involving percipient individuals and objects emitting stimuli.

The belief that with the abandonment, as in modern physics, of the ideal of complete precision, the more or less complex relations which this yields provide a less intelligible picture than we should obtain by the introduction of hypothetical electron paths is probably due to our persistence in certain habitual ways of thinking.[4]

The unified language of physicalism confronts us wherever we make a scientific prediction on the basis of laws. When someone says that if he sees a certain color he will hear a certain sound, or *vice versa*, or when he speaks of the "red patch" next to the "blue patch," which will appear under certain conditions, he is already operating within the framework of physicalism. As a percipient he is a physical structure: he must localize perception, e.g., in the central nervous system or in some other place. Only in this way can he make predictions and reach agreement with others and with himself at different times. Every temporal designation is already a physical formulation.

Science endeavors to transform the statements of everyday life. They are presented to us as "agglomerations," consisting of physical-istic and pre-physicalistic components. We replace them by the "unification" of physicalistic language. If one says, for instance, "the

---

4. Cf. Concerning this, Philipp Frank, "Der Charakter der heutigen physikalischen Theorien," *Scientia*, March 1931.

screeching saw cuts through the blue wooden cube," "cube" is obviously an "intersensual" and "intersubjective" concept, equally available for the blind and the deaf. If a man soliloquizes and makes predictions, which he can himself control, he is able to compare what he said of the cube when he saw it with what he communicates in the dark when he touches it.

With the word "blue," on the other hand, there is, at first, a doubt as to how it is to be incorporated into the unified language. It can be used in the sense of the rate of vibration of electromagnetic waves. But it can also be used in the sense of a "field statement," meaning: when a seeing man (defined in a certain way) enters, as a test body, the range of this cube, he behaves in a certain manner, describable physicalistically; e.g., he says, "I see 'blue.'" While there may be doubt as to what people mean when they use "blue" in colloquial speech, "screeching" would be chiefly intended as a "field statement," i.e., as an expression in which the auditor is always included. Closer consideration, however, reveals that "cube," "blue" and "screeching" are all words of the same type.

Let us attempt to follow up our analysis by giving a more exact rendering of the above sentence, in accordance with physicalism, and reformulating it in a way that will make it more suitable for prediction.

"Here is a blue cube." (This formulation, like those which follow, may be restated as a physical formula, in which the locus is determined by means of coordinates.)

"Here is a screeching saw." (The screeching enters into the formulation, at first, only as vibrations of the saw and the air, which could be expressed in physical formulae.)

"Here is a percipient man." (Possibly a "field statement" could be added indicating that under certain conditions the percipient enters into a relationship with the physical blue and the physical screeching.)

This perceiving may perhaps be divided into:

"Neural changes are occurring here."

"Cerebral changes are occurring here in the perception area and, perhaps, in the speech area also." (It is immaterial for our purposes whether these areas can be defined locationally or whether they have to be defined structurally. Neither is it necessary to discuss whether changes in the speech area—the "speech-thought" of the behaviorists—are connected with the larynx or laryngeal innervation.)

Perhaps, in order to exhaust the physicalistic meaning of this simple sentence, something more has still to be added, e.g., particulars concerning time, or positional coordinates; but the essential thing is that, in every case, the additions should be statements involving physical concepts.

It would be a mistake to suppose, because physical formulae of a very complex nature, which are still not fully at our disposal, are required for the computation of certain correlations, that, therefore, the physicalistic expressions of everyday life must also be complex. Physicalistic everyday language will arise from existing common speech, only parts of which will have to be discarded; others will be integrated, while supplements will make up for certain deficiencies. The occurrence of a perception will be, from the outset, more closely connected than hitherto with the observation statement and with the identification of the object. The analysis of certain groups of statements, e.g., observation statements, will proceed in a different manner from before.

Children are capable of learning physicalistic every-day language. They are able to advance to the rigorous symbolic language of science and learn how to make successful predictions of all kinds, without having to resort to "elucidations" supposedly functioning as a meaningless introduction. It is a question of a more lucid mode of speech, so formulated as to omit such expressions, for instance as "illusion of the senses," which create so much confusion. But even though the physicalistic language has the capacity some day to become the universal language of social intercourse, we must continue to devote ourselves, for the present, to cutting away the metaphysical appendages from the "agglomerations" of our language and to defining physicalistically everything that remains. When the metaphysical cord is no longer present, much of what is left may present itself as disconnected heaps. The further use of such remnants would not be profitable, and a reconstruction would be indispensable.

We can often continue to make use of available "agglomerations" by reinterpreting them. But caution is required here: men who are ready enough to adjust their views, but at the same time comfort-loving, frequently console themselves with the belief that a great deal can be "systematically" reinterpreted. It is more than questionable whether it would be convenient to continue to employ terms like "instinct," "motive," "memory," "world," etc., attaching to them a wholly unusual sense, which one may easily forget when one goes on using these terms for the sake of peace. Certainly there are many cases in which a reconstruction of language is superfluous, or even dangerous. So long as one expresses oneself "approximately," one must guard against the desire to be, at the same time, excessively subtle.

Since the views presented here are most nearly similar to the ideas of Carnap, let it be emphasized that they exclude the special "phenomenal" language from which Carnap seeks to derive the physical language. The elimination of the "phenomenal" language, which does not even seem to be usable for "prediction"—the essence of science—in the form it has assumed up to now, will probably necessitate many modifications in his system of concept construction.* In the same way, we must exclude "methodological solipsism" (Carnap, Driesch), which seems to be an attenuated residue of idealistic metaphysics, a position from which Carnap himself constantly attempts to get away. The thesis of "methodological solipsism," as even Carnap would probably concede, cannot be scientifically formulated. Nor can it be used to indicate a particular standpoint, which would be an alternative to some other standpoint, because there exists only *one* physicalism, and everything susceptible of scientific formulation is contained in it.

There can be no contrasting of the "ego" or the "thinking personality," or anything else with "experience," "what is experienced," or "thought." The statements of physicalism are based on statements connected with seeing, hearing, feeling and other "sense perceptions" (as physical events), but also with "organic perceptions," which, for the most part, are only roughly noted. We can, of course, close our eyes, but we cannot stop the process of digestion, the circulation of the blood, or the occurrence of muscular innervations. What people are at pains to separate off as "the ego" are, in the language of physicalism, events of this sort also, of which we are not informed through our ordinary "external" senses. All "personality co-efficients" which separate one individual from another are of a physicalistic kind!

Although the "ego" cannot be set off against either the "world" or "thinking," one is able, without abandoning physicalism, to distinguish statements about the "physicalistically described person," besides those concerning the "physicalistically described cube," and can, under certain conditions, make "observation statements," thereby creating a substitute for the "phenomenal language." But careful investigation will show that the mass of *observation statements is contained in the mass of physical statements.*

---

* [This is a reference to the "Konstitutionssystem" elaborated by Carnap in his book *Der Logische Aufbau der Welt. Vide* the Introduction p. 24. Ed.]

The protocol statements of an astronomer or a chronicler (appearing as physical formulations) will, of course, be distinguished from statements having a precisely determined position in the context of a physical system, despite the fact that between the two there are fluid transitional stages. But there is no special "phenomenal" as opposed to physicalistic language. *Every one of our statements can, from the very outset, be a physicalistic one—and it is this that distinguishes what is said here from all the pronouncements of the "Vienna Circle,"* which otherwise constantly stresses the importance of predictions and their verification. Unified language is the language of predictions, which are the very heart of physicalism.

In a certain sense, the doctrine here proposed proceeds from a given condition of everyday language, which in the beginning is essentially physicalistic, and, in the usual course of events, is gradually developed in a metaphysical direction. This forms a point of contact with the "natural concept of the world" ("Natürlicher Weltbegriff") in Avenarius. The language of physicalism is, so to speak, in no way new; it is the language familiar to certain "naive" children and peoples.

It is always science as a system of statements which is at issue. *Statements are compared with statements,* not with "experiences," "the world," or anything else. All these meaningless *duplications* belong to a more or less refined metaphysics and are, for that reason, to be rejected. Each new statement is compared with the totality of existing statements previously coordinated. To say that a statement is correct, therefore, means that it can be incorporated in this totality. What cannot be incorporated is rejected as incorrect. The alternative to rejection of the new statement is, in general, one accepted only with great reluctance: the whole previous system of statements can be modified up to the point where it becomes possible to incorporate the new statement. Within unified science there is important work to be done in making transformations. The definition of "correct" and "incorrect" proposed here departs from that customary among the "Vienna Circle," which appeals to "meaning" and "verification." In our presentation we confine ourselves always to the sphere of linguistic thought. Systems of statements are subjected to transformation. Generalizing statements, however, as well as statements elaborated by means of determinate relations, can be compared with the totality of protocol statements.

Unified science thus comprehends a variety of types of statements. So, for example, whether one is dealing with "statements about reality," "hallucination statements," or "untruths" depends on the degree to which one can employ the statements in drawing inferences about physical events other than oral movements. One is confronted by an "untruth" when one can infer a certain excitation of the speech center of the brain, but not corresponding events in the perception centers; the latter events are, on the other hand, essential for a hallucination. If, besides excitation in the perception centers, one can also infer, in a manner to be specified, events outside the body, then one is dealing with "statements about reality." In this case, we can continue to employ the statement—for example, "a cat is sitting in this room"—as a physicalistic statement. A statement is always compared with another statement or with the system of statements, never with a "reality." Such a procedure would be metaphysical; it would be meaningless. "The" reality is not, however, replaced by "the" physicalistic system, but by groups of such systems, one of which is employed in practice.

From all this it becomes clear that within a consistent physicalism there can be no "theory of knowledge," at least not in the traditional form. It could consist only of defensive operations against metaphysics, that is to say, of the unmasking of meaningless phrases. Many of the problems of the theory of knowledge will, perhaps, be transformed into empirical questions in such a way that they can be accommodated in unified science.

This problem can no more be discussed here than the question of how all "statements" can be incorporated in physicalism as physicalistic constructions. "Two statements are equivalent" could, perhaps, be expressed in this way. Let a man be acted on by a system of commands connected with all sorts of statements, e.g., "If A behaves in such and such a way, do this and that." Now one can fix certain conditions, and observe that the *addition of a certain statement* produces the same change in his reactions as the addition of another. Then one will say that the first statement is *equivalent* to the second. When tautologies are added, the stimulus offered by the system of commands remains *unchanged*.

All this could be developed experimentally with the aid of a "thinking machine" such as Jevons proposed. By means of this machine, syntax could be formulated and logical errors automatically avoided. The machine would not even be able to write the sentence, "two times red is hard."

The views suggested here are best combined with a *behavioristic* orientation. One will not then speak of "thought," but of "speech-thought," i.e., of *statements as physical events*. Whether a perception statement about the past (e.g., "I recently heard a melody") can be traced back to a past speech-thought, or whether previous stimuli are only now evoking a reaction in speech-thought, is, in this regard, of no essential importance. All too often the discussion is conducted as if the refutation of some minor assertions of the behaviorists had somehow shaken the fundamental principle that only *physicalistic statements* have a meaning, i.e., can become part of unified science.

We begin with statements, and we conclude with statements. There are no "elucidations" which are not physicalistic statements. If someone wished to conceive of "elucidations" as exclamations, then, like whistles and caresses, they would be subject to no logical analysis. The physicalistic language, *unified language,* is the Alpha and Omega of all science. There is no "phenomenal language" beside the "physical language," no "methodological solipsism" beside some other possible position, no "philosophy," no "theory of knowledge," no new "Weltanschauung" beside the others: there is only *Unified Science,* with its laws and predictions.

III. SOCIOLOGY NO "MORAL SCIENCE"

Unified science makes predictions about the behavior of machines just as it does about that of animals, about the behavior of stones as about that of plants. Some of its complex statements we could analyze even today, while the analysis of others temporarily eludes us. There are "laws" of the behavior of animals and of machines. The "laws" of machines can be reduced to physical laws. But even in this sphere, a law in terms of mass and metrical measurement often suffices, without recourse to atoms or other elements. In the same way, the laws of the animal body are often so formulated that there is no need to fall back on micro-structural laws. Admittedly, where much has been hoped for from the investigation of macro-structural laws, they have often turned out to be inadequate: certain irregularities remain incalculable.

There is a constant search for *correlations* between magnitudes appearing in the physicalistic description of events. It makes no fundamental difference whether *statistical* or *nonstatistical* descriptions are involved. No matter whether one is investigating the statistical behavior of atoms, plants or animals, the methods employed in establishing correlations are always the same. As we saw above, all the laws of unified science must be capable of being connected with one another if they are to be equal to the task of serving, as often as possible, to *predict* individual events or certain groups of events.

This does away, at the outset, with any fundamental division of unified science, for instance, into the "natural sciences" and the "moral sciences," the latter being often referred to also in other ways, e.g., "Kulturwissenschaften" ("sciences of culture"). The theses by which it is intended to establish this division vary, but are always of a metaphysical character, that is, meaningless. It is senseless to speak of different "essences" reposing "behind" events. What cannot be expressed in terms of relations among elements cannot be expressed at all. It is consequently meaningless *to go beyond correlations and speak of the "essence of things."* Once it is understood what the unified language of science really means, there will be no more talk of "different kinds of causality." One can only compare the organization of one field and its laws with the organization of another, and ascertain, perhaps, that the laws in one field are more complex than those in another, or that certain modes of organization lacking in one are found in another; that, for example, certain mathematical formulae are required in one case but not in the other.

If the "natural sciences" cannot be delimited from the "moral sciences," it is even less possible to make the distinction between the "philosophy of nature" and the "philosophy of the moral sciences." Even leaving aside the fact that the former term is unsuitable because, as mentioned above, it still contains the word "philosophy," by "philosophy of nature" one can only understand a sort of introduction to the whole work of unified science. For how should "nature" be distinguished from "non-nature"?

One cannot even adduce the practical exigencies of everyday life or of the conduct of scientific investigations as justification for this dichotomy. Is the theory of human behavior seriously to be opposed to that of the behavior of all other objects? Is it seriously intended that the theory of human societies should be fitted into one discipline and the theory of animal societies into another? Are the natural sciences to deal with "cattle-breeding," "slavery" and "warfare" among ants, and the moral sciences with these same institutions among men? If this is not meant, then the distinction is no sharper than that between different "scientific fields" in the older sense.

Or is there something to be said, perhaps, for the linguistic usage according to which one simply speaks of "moral sciences" whenever "social sciences" are meant? But, to be consistent, one would have to count the theory of animal societies together with the theory of human societies as social sciences, and therefore as "moral sciences," an implication from which most people would recoil. And quite understandably so, for then where would be the great cleavage concealed behind all this, the cleavage depending on the maintenance of the centuries-old theological habit of thought which divides up all existence into at least two departments, e.g., a "noble"' and an "ignoble"? The dualism of "natural sciences" and "moral sciences," and the dualism of "philosophy of nature" and "philosophy of culture" are, in the last analysis, *residues of theology*.

The ancient languages are, on the whole, more physicalistic than the modern. They are full of magical elements, to be sure, but above all they treat "body" and "soul" as simply two forms of matter: the soul is a diminutive, shadowy body which issues from the mouth of the individual at death. It is theology which first replaces the contrast of "matter-soul" and "matter-body" with that of "non-matter-soul" and "matter-body," as well as "non-matter-God" and "matter-world," adding a whole hierarchy of subordinate and superior entities, natural and supernatural. The opposition of "natural" and "supernatural" can be formulated only by means of meaningless phrases. These phrases, because they are meaningless, do not contradict the statements of unified science; neither are they in accord with them. But they are certainly the cause of great confusion. It is when it is asserted that these expressions are just as meaningful as those of science that the trouble starts.[5]

What part the mental habit of theological dualism plays in the creation of such dichotomies can perhaps be gathered from the fact that as soon as one such division is discarded another easily establishes itself. The opposition of the "Is" and the "Ought," which is encountered especially among philosophers of law, may be mentioned here. In part, of course, this may be traced to the theological opposition of "Ideal" to "Reality." But the capacity of language for forming nouns facilitates all these meaningless schemes. One can, without violating syntactical rules, as serenely say "the Ought" as "the sword." And then people go on to make statements about this "Ought" just as they would about a "sword," or at least as they would about the "Is."

The "moral sciences," the "psychical world," the world of the "categorical imperative," the realm of *Einfühlung* (empathy), the realm of *Verstehen* ("the 'understanding' characteristic of the historian")—these are more or less interpenetrating, often mutually substitutable, expressions. Some authors prefer one group of meaningless phrases, some another, some combine and accumulate them.

---

5. Cf. Hans Hahn, *Überflüssige Wesenheiten* (Publications of the Verein Ernst Mach, Vol. II).

While such phrases provide only the marginal decorations of science in the case of many writers, with others they influence the entire body of their pronouncements. Even if the practical effect of the doctrines on which the school of "moral sciences" is based are not over-rated, even if the confusion in empirical investigation wrought by it is not exaggerated, still, in the systematic establishment of physicalism and sociology, clarity requires that a clean sweep be made here. It is the duty of the practitioners of unified science to take a determined position against such distinctions; this is not a matter for their arbitrary choice.

If there is uncertainty over these questions even among anti-metaphysically minded thinkers, it is partly connected with the fact that there does not exist sufficient clarity about the subject matter and method of "psychology." The detachment of the "moral "sciences" from other disciplines is concurrent with the separation of "psychical" objects from others in other fields. This detachment has only been systematically eliminated by *behaviorism*, which, in this essay, we always understand in the widest sense. Only physicalistic statements about human behavior are incorporated into its system. When the sociologist makes predictions about human groups in the same way as the behaviorist does about individual men or animals, sociology may appropriately be called *social behaviorism.*

Our conclusion is as follows: sociology is not a "moral science" or "the study of man's spiritual life" (Sombart's "Geisteswissenschaft") standing in fundamental opposition to some other sciences, called "natural sciences"; no, *as social behaviorism, sociology is a part of unified science.*

IV. SOCIOLOGY AS SOCIAL BEHAVIORISM

It is possible to speak in the same terms of men's painting, housebuilding, religion, agriculture, poetry. And yet, it is maintained again and again that "understanding" human beings is fundamentally different from "merely" observing them and determining regularities expressible as laws. The area of "understanding," of "empathy" with other personalities is closely connected with that traditionally claimed by the "moral sciences." We find here a resurrection of the division on grounds of principle—already eliminated on a previous level—between "internal" and "external" perceptions (experience, mind, etc.), which possess the same empirical character.

Philosophical literature, especially the literature of the philosophy of history, frequently insists that without "empathy" and "understanding" it would be impossible to pursue historical studies or comprehensively to arrange and describe human actions at all.

How can we attempt to dispose in general of these obstacles from the point of view of physicalism? It must be assumed from the outset that the persistent asseverations of many sociologists and philosophers of history concerning the unavoidability of recourse to "understanding" are also aimed at preserving the results of some very worthy scientific researches. Here, as so often elsewhere, it may be a case of a not easily disentangled combination of the dualistic habits of mind, originating in theology, and the actual procedure of science. It will be apparent to anyone familiar with the *monism of unified science* that even statements which are fully capable of formulation in physicalistic terms have been presented in an unphysicalistic form.

Sentences such as "I see a blue table in this room," and "I feel angry" do not lie far apart. The "I" is appropriately replaced by some personal name, since all such statements may be applied to anyone, and an "I-statement," therefore, must be capable of being asserted by someone else. Now we have: "There is a blue table in this room," and "There is anger in this man." The discussions concerning "primary" and "secondary" qualities are at an end when it is realized that, in the last analysis, all statements about qualities are of one type, only tautologies being excluded from the class of such statements. Then all statements about qualities become physicalistic statements. Besides these, there are tautologies, rules for the combination and connection of statements. The propositions of geometry can be interpreted as physicalistic statements or as tautologies, thereby removing many difficulties.

What, among other things, is characteristic of the sentence, "There is anger in this man"? Its peculiarity is that it is open only to inadequate analysis. It is as if someone were able to tell us, "Here is a severe storm" without being in a position to state in what manner it was composed of lightning, thunder, rain, etc., nor yet whether he arrived at his discoveries by means of his eyes, ears or nose.

When one speaks of anger, *organic perceptions* are made use of. Changes in the intestinal tract, internal secretions, blood pressure and muscle contraction are essentially equivalent to changes in the eye, ear or nose. In the systematic construction of behaviorism, a man's statement, "I am angry" is incorporated into physicalism *not only as the reaction* of the speaker, but also as the formulation of his "organic perceptions." Just as, from the enunciation of "color perceptions," one can infer physicalistic statements about retinal changes and other events, so from assertions about anger, i.e., about "organic perceptions," one can derive physicalistic statements about "intestinal changes," "changes in blood pressure," etc., phenomena which become known to others often only by means of such statements. This may be appended as a supplement to Carnap's discussions on this subject, where the full value of statements about "organic perceptions" (in the older sense) has not been taken into consideration.

If someone says that he requires this experience of "organic perceptions" in order to have empathy with another person, his statement is unobjectionable. That is to say, the employment of physicalistic statements concerning one's own body in making physicalistic statements about another's is completely in line with our scientific work, which throughout makes this sort of "extrapolation." Our commitment to induction leads us constantly to such extensions. The same principle is involved in making statements about the other side of the moon on the basis of our experience concerning the side which faces us. That is to say, one may speak of "empathy" in the physicalistic language if one means no more by it than that one draws inferences about physical events in other persons on the basis of formulations concerning organic changes in one's own body. What is involved here, as in so many other cases, is a physicalistic induction, the usual attempt at establishing certain correlations. The linguistic clarity achieved so far in regard to many of these events leaves, to be sure, much to be desired. One would come very close to the actual state of affairs if one were to say that the moral sciences are, above all, the sciences in which correlations are asserted between events which are very inadequately described and for which only complex names are available.

When we analyze the concepts of "understanding" and "empathy" more closely, everything in them that is usable in a physicalistic way proves to be a statement about order, exactly as in all sciences. The alleged distinction between "natural sciences" and "moral sciences," to the effect that the former concern themselves "only" with arrangement, the latter with understanding as well, is non-existent.

If, wherever non-metaphysical formulations are encountered, they are subjected to systematic formulation, physicalistic statements will be achieved throughout. There will no longer be a special sphere of the "psychical." It is a matter of indifference for the position here maintained whether certain individual theses of Watson's, Pavlov's or others are upheld or rejected. What is essential is that only

216

physicalistically formulated correlations be employed in the description of living things, whatever may be observed in these things.

It would be misleading to express this by saying that the distinction of "psychical" and "corporeal" no longer existed, but had been replaced by "something neutral." It is not at all a question of a "something," but simply of correlations of a physicalistic character. Only insufficient analysis can lead anyone to say something like: "It cannot yet be ascertained whether the whole sphere of the 'psychical' really admits of physicalistic expression. It is, after all, possible that here and there another type of formulation is required, i.e., concepts not physicalistically definable." This is the last remnant of belief in a "soul" as a separate form of being. When people have observed a running clock and then see it stop, they can easily make use of the capacity of language for creating nouns, and pose the problem, "Where has the 'movement' gone to?" And after it is explained to them that all that can be known about the clock is to be discovered through analysis of the relations between its parts and the surroundings, a sceptic may still object that, although he understands that speculation about the "movement" is pure metaphysics, he is still doubtful whether, for the solution of certain complicated problems relating to the operations of clocks, physicalism entirely suffices.

Without meaning to say that every sociologist must be trained in behaviorism, we can still demand of him that, if he wishes to avoid errors, he must be careful to formulate all his descriptions of human behavior in a wholly straightforward physicalistic fashion. Let him not speak of the "spirit of the age" if it is not completely clear that he means by it certain verbal combinations, forms of worship, modes of architecture, fashions, styles of painting, etc. That he undertakes to predict the behavior of men of other ages on the basis of his knowledge of his own behavior is wholly legitimate, even if sometimes misleading. But "empathy" may not be credited with any peculiar magical power transcending ordinary induction.

With inductions in this or that field, it is always a question of a *decision*. This decision may be characteristic of certain human groups or of whole ages, but is not itself logically deducible. Yet induction always leads, within the physicalistic sphere, to meaningful statements. It must not, for this reason, be confused with the *interpolation of metaphysical constructions*. There are many who concede that they formulate metaphysical constructions, i.e., that they insert meaningless verbal combinations, but nevertheless will not fully appreciate the damage caused by such a procedure. The elimination of such constructions in sociology and psychology, as well as in other fields, must be undertaken not only for the sake of freeing them of superfluities and of avoiding meaningless verbal combinations, which perhaps afford satisfaction to some. The elimination of metaphysics will become *scientifically fruitful* through *obviating the occasion for certain false correlations in the empirical sphere.* It will be seen that one is most likely to overestimate the significance of certain elements in the historical process, which are capable of a physicalistic formulation, when they are believed to be linked with certain metaphysical essences. People often expect from the priest of the transcendent God certain empirically controllable super-achievements which would not be deducible from empirical experience.

There are many who allege in favor of metaphysical constructions that, with their help, better predictions can be made. According to this view, one proceeds from physicalistically formulated observation-statements to the realm of metaphysical word-sequences. By the employment of certain rules which, in the metaphysical sphere, are applied to meaningless word-sequences, this process is supposed to result in predictions consonant with a system of protocol statements. *Even if* results are actually achieved in this way, metaphysics is not essential for prediction in this case, although it may perhaps act as a stimulus, like some narcotic. For if predictions can be made in this roundabout way, "then they can also be *directly* deduced *from the given data.* This is *clear from a purely logical* consideration: if Y follows from X, and Z follows from Y, then Z follows immediately from X."[6] Even if Kepler made use of the world of theological conceptions in arriving at the planetary orbits, this world of conceptions nevertheless does not enter into his scientific statements. Much the same is true of the highly productive fields of psychoanalysis and individual psychology, whose behavioristic transformation will *certainly be no easy task.*

When the metaphysical deviations from the main line of behaviorism have been distinguished in this way, the path will be cleared for a sociology *free of metaphysics.* Just as the behavior of animals can be studied no less than that of machines, stars and stones, so can the behavior of animal groups be investigated. It is possible to take into account both changes in individuals produced by "external" stimuli and those caused by "autonomous" changes "within" living things (e.g., the rhythmical course of a process), just as one can investigate the disintegration of radium, which is not

---

6. Otto Neurath, *Empirische Soziologie,* p. 57.

influenced by anything external, as well as the decomposition of a chemical compound through the addition of oxygen. Whether analogies to the disintegration of radium play a role within the human body need not be discussed here.

Sociology does not investigate purely statistical variations in animal or, above all, human groups; it is concerned with the *connections among stimuli* occurring between particular individuals. Sometimes, without analyzing these connections in detail, it can determine under certain conditions the total behavior of groups united by common stimuli, and make predictions by means of the laws obtained. How is "social behaviorism," unimpeded by metaphysics, to be pursued? *Just as every other actual science* is pursued. Naturally, in investigating human beings certain correlations result which are not encountered in the study of stars or machines. Social behaviorism attains to laws of a definite type peculiar to itself.

To pursue physicalistic sociology is not to transfer the laws of physics to living things and the groups they form, as some have considered feasible. It is possible to discover comprehensive sociological laws, as well as laws for narrower social areas, without having recourse to micro-structure, and thus being able to base these sociological laws on physical ones. *Whatever sociological laws are discovered without the aid of physical laws in the narrower sense are not necessarily altered by the addition of a subsequently discovered physical substructure.* The sociologist is completely unimpeded in his search for laws. The only stipulation is that he must always speak, in his predictions, of structures which are given in space and time.

V. Sociological Correlations

It is as little possible in sociology as in other sciences to state at the outset on the basis of purely theoretical considerations what correlations can be employed with a prospect of success. But it is demonstrable that certain traditional endeavors meet with consistent failure, while other methods, adapted to discovering correlations, are not at present sufficiently cultivated.

Of what type, then, are sociological correlations? How does one arrive, with a certain degree of reliability, at sociological predictions? In order to be able to predict the behavior of a group in a certain respect, it is often necessary to be acquainted with the total life of the group. Variations in the particular modes of behavior distinguishable in the totality of events, the construction of machines, the erection of temples, the forms of marriage, are not "autonomously" calculable. They must be regarded as parts of the whole that is investigated at any given time. In order to know how the construction of temples will change in the future, one must be familiar with the methods of production, the form of social organization, and the modes of religious behavior in the period which is taken as the starting point; one must know the transformation to which *all of these together* are subject.

Not all events prove equally resistant to being employed in such predictions. Given certain conditions, from the mode of production of a historical period one can often roughly infer the next phases in the development of the mode of production and the form of social organization. Then one is in a position to attempt with some success to make further predictions about religious behavior and similar matters with the aid of such previous predictions. Experience shows that the reverse procedure, on the other hand, meets with failure, i.e., it does not seem possible to derive predictions about the mode of production from predictions about religious behavior alone.

But, whether we direct our attention to the methods of production, to religious behavior, to the construction of buildings, or to music, we are always confronted with events which can be physicalistically described.

Many of the social institutions of an age can be properly accounted for only if their distant past is known, while others might, so to speak, be devised at any time given the appropriate stimuli. There is a certain sense in which the presence of cannons acts as a stimulus, producing armed turrets by way of reaction. The dress coats of our day, on the other hand, do not represent a reaction to dancing, and it is only with difficulty that they would be newly devised. But it is comprehensible to us that at some time in the past, a man dressed in a long-skirted coat became the inventor of the dress coat when the skirts of his coat flapped up while he was riding. The coherence between established customs is different in the two cases.

Just as one must be informed about the type of coherence in order to be able to make predictions, so one must know whether the detachment of a certain institution or segment from a social complex is easy or difficult, and whether, in the case of loss, it can be replaced. The state, for example, is a highly stable complex whose operations are, to a considerable extent, independent of the change-over of personnel: even if many judges and soldiers were to die, there would be new ones to take their place. A machine, on the contrary, does not generally replace wheels which have been removed from it.

It is a wholly physicalistic question to what extent the existence of specially conditioned individuals, deviating from the norm, assures the stability of the state structure. The related question of the degree to which such significant individuals are replaceable must be treated separately. The queen bee assumes a special position in the hive, but when a queen bee gets lost, there is the possibility that a new one will emerge. There are always, so to speak, latent queens. How is this in the case of human society?

The extent to which predictions about social complexes can be made without taking into consideration the fate of certain particularly prominent individuals is entirely a concrete sociological question. It is possible to maintain, with good reason, that the creation of bourgeois Europe, once the machine system had imparted to the modern capitalistic transformation its characteristic hue, was predictable at the end of the eighteenth century. On the other hand, one could hardly have predicted Napoleon's Russian campaign and the burning of Moscow. But it would, perhaps, be valid to say that if Napoleon had defeated Russia, the transformation of the social order would have proceeded in the same way as it did in fact proceed. Even a victorious Napoleon would have had to countenance the old feudalism of Central Europe to a certain degree and for a certain time, just as, on another occasion, he re-established the Catholic Church.

The extent to which prediction is possible, or relates to particular individuals, in no way affects the essence of social behaviorism. The movements of a leaf of paper in the wind are equally unpredictable, and yet kinematics, climatology and meteorology are all highly developed sciences. It is no part of the essence of a developed science to be capable of predicting every individual event. That the fate of a single leaf of paper, say, a breeze-blown thousand dollar bill, may especially interest us, is of little concern to scientific investigation. We need not discuss here whether a chronicle of the "accidental" paths of leaves in the wind could eventually lead to a theory of the paths of leaves. Many of the views associated with Rickert and allied thinkers yield no scientific laws even where they can be physicalistically interpreted.

Sociology, like every science, tracks down correlations which can be utilized for predictions. It seeks to lay down its basic conceptions as unambiguously and clearly as possible. One may attempt, for instance, to define groups through *"commercium"* and *"connubium."* One ascertains who trades with whom, or who marries whom. There may emerge clearly distinguishable areas of concentration, together with poorly occupied border areas. And then one could investigate the conditions under which such concentrations vary or even disappear. To discover the correlation of such areas of concentration with the processes of production obtaining at their respective periods is obviously a legitimate sociological task, which might be of importance for the theory of "classes."

One can investigate, for instance, under what conditions matriarchy, ancestor worship, agriculture and similar institutions arise, at what point the founding of cities begins, or what correlations exist between systematic theology and other human activities. One can also ask how the administration of justice is determined by social conditions, although it is questionable whether such limited divisions will exhibit sufficient law-like regularities. It may well be, for instance, that certain events occurring outside the field of law must be added to those involved in the administration of justice, if relations statable as laws are to be found.

What one group recognizes as law, another may regard as outside the legal order. Thus, only correlations among men's statements concerning the "law," or between their behavior and their statements can be established. But it is not possible, without special preliminary work, to contrast "legal events" as such with other events.

It is doubtful whether simple sociological correlations can be determined between the allowed interest rate, on the one hand, and the standard of living of a period, on the other; whether simpler relations do not appear when the "allowed interest rate" and "prohibited usury" are taken together. Thus, the modes of behavior on which unfavorable "legal" and "ethical" judgments are passed could be incorporated into sociology, and the judgments themselves could be included. These disciplines are in every sense branches of sociology, but they are quite different from the "ethics" and "jurisprudence" which are commonly cultivated. The latter yield few or no sociological correlations. They are predominantly metaphysical, or, where free of metaphysics, their methodology and arrangement of statements can only be explained as residues of theology. In part, they yield purely logical deductions, the extraction of certain injunctions from others, or of certain conclusions from given legal assumptions. But all this lies outside the sphere of ordered correlations.

VI. ETHICS AND JURISPRUDENCE AS
REMNANTS OF METAPHYSICS

In its origin, ethics is the discipline which seeks to determine the totality of divine injunctions. It attempts to find out, by means of a logical combination of commandments and prohibitions of a universal kind, whether a given individual act is commanded, permitted or forbidden. The "casuistry" of Catholic moral theologians has extensively elaborated this type of deduction. It is quite obvious that the indeterminateness of divine injunctions and the ambiguity of their meaning preclude any genuine scientific method. The great expenditure on logical deductions was, so to speak, squandered on a worthless object, even though, historically, it prepared the way for the coming logicizing period of science. If the God who issues the commands, as well as events in heaven and hell (which was located by many theologians at the center of the earth) are physicalistically defined, then one is dealing with a non-metaphysical discipline, to be sure, but a highly uncritical one.

But how is a discipline of "ethics" to be defined once God is eliminated? Is it possible to pass meaningfully to a "command-in-itself," to the "categorical imperative"? One might just as well talk of a "neighbor-in-himself without any neighbors," or of a "son-in-himself, who never had a father or mother."

How is one to distinguish certain injunctions or modes of behavior in order to make possible "a new ethics within the context of physicalism"? It seems to be impossible. Men can form joint resolutions and conduct themselves in certain ways, and it is possible to study the consequences of such actions. But what modes of behavior, what directives is one to distinguish as "ethical," so that correlations may then be set up?

The retention of an old name is based on the view that there is something abiding to be discovered, which is common to the old theological or metaphysical and the new empiricist disciplines. When all metaphysical elements, as well as whatever physicalistic theological elements it may contain, have been eliminated from ethics, there remain only statements about certain modes of human behavior or the injunctions directed by some men to others.

One could, however, also conceive of a discipline pursuing its investigations in a wholly behavioristic fashion as part of unified science. Such a discipline would seek to determine the reactions produced by the stimulus of a certain way of living, and whether such ways of living make men more or less happy. It is easy to imagine a thoroughly empirical "felicitology" (*Felicitologie*), on a behavioristic foundation, which could take the place of traditional ethics.

But a non-metaphysical ethics usually seeks to analyze, in one way or another, men's "motivations," as if this provided a suitable groundwork for relations statable as laws. What men assert about the "reasons" for their conduct, however, is essentially more dependent on contingencies than the general run of their behavior. When the general social conditions of a given period are known, the behavior of whole groups can be far more readily predicted than the rationale which individuals will adduce for their conduct. The modes of conduct will be defended in very different fashions, and very few, moreover, will note the correlation between the social situation and average conduct.

These "conflicts of motivations," for the most part metaphysically formulated, are avoided by an empirical sociology, which is intent upon fruitful work. This is the case with Marxism, the most productive sociology of the present day. It endeavors to establish correlations between the social situation and the behavior of entire classes, so that it can then account for the frequently changing verbal sequences which are supposed to "explain the motivation" of the scientifically law-abiding actions which are conditioned in this way. Since Marxism, in its descriptions of relations expressible as laws, makes as little use as possible of what men assert about themselves, the "events in their consciousness," their "ideology," it is related to those schools of "psychology" which accord to the "unconscious" in one form or another a prominent role. Thus it is that *psychoanalysis* and *individual psychology,* by virtue of the fact that they confute and eliminate the motivational psychology of consciousness (today quite obsolete), prepare the way for modern empirical sociology, which seeks, in the spirit of unified science, to discover correlations between actions and the factors that condition them.

And even if psychoanalysis and individual psychology in their present form contain very many metaphysical expressions, nevertheless, through their emphasis on the relation between behavior and its unconscious preconditions, they are precursors of the behavioristic way of thinking and of sociological methodology.

Thus it is permissible to ask whether a certain manner of living yields more or less happiness, since "happiness" can be described wholly behavioristically; it is valid to ask on what depend the demands which masses of men make of one another, what new demands are set, what modes of behavior will emerge in such a situation.

(Claims and modes of behavior in this regard are often fundamentally divergent.) All these are legitimate sociological formulations of problems. Whether it is advisable to characterize them as "ethical" need not be decided here.

The case of "jurisprudence" is a similar one, when it is understood as something other than the sociology of certain social phenomena. But when it takes up the task of establishing whether a system of claims is logically consistent, whether certain conclusions of the statute books can be harmonized with certain observation-statements about legal practice, we are concerned with purely logical investigations. When we determine that the rules of a chemist are logically compatible, we have not yet entered the sphere of the science of chemistry. In order to pursue chemistry, we must establish correlations between certain chemical events and certain temperatures, and other such things. The fact that, despite their essentially metaphysical preliminary formulations, the representatives of certain schools of jurisprudence can produce something logically and scientifically significant does not prevent our rejecting these formulations, as, for example, the following:

The *thinking* of mathematical or logical laws is a psychical act, but the object of mathematics—that which is thought of—is not something *psychical*, neither a mathematical nor a logical "soul," but a specific intellectual reality. For mathematics and logic abstract from the psychological fact of the thinking of such an object. In the same way the state, as the object of a special mode of thought to be distinguished from psychology, is a distinctive reality, but is not the fact of the thinking and willing of such an object. It is an ideal order, a specific system of norms. It resides not in the realm of *nature*—the realm of physical-psychical relations—but in the realm of spirit. The state as obligating authority is a value or—so far as the propositional expression of value is established—a norm, or system of norms. As such, it is essentially different from the specifically real fact of the conceiving or willing of the norm, which is characterized by indifference to value.[7]

Formulations of this type are connected with similar ones on "ethics" and related disciplines, without any attempt having been made to discover how the term "objective goals" is to be fitted into unified science, and without indicating any observation-statements through which "objective goals" as such might be determined. Again:

If the "general theory of the state" asks *what* the state is and *how* it is, i. e., what its possible basic forms and chief components are, politics

7. Kelsen, *Allgemeine Staatslehre*, pp. 14 ff.

225

asks *whether* the state is to be at all, and, if it is, which of its possible forms might be the best. Through this formulation of questions, politics exhibits itself as part of ethics, as the judgment of morality which sets objective goals for human conduct, i. e., which posits as obligatory the content of some actions. But politics, so far as it seeks means appropriate to the realization of these objective goals which are somehow established and assumed by virtue of their establishment; so far, that is, as it fixes those contents of conduct which are shown by experience to cause the effects corresponding substantially to the presupposed goals; to this extent, politics is not ethics, it is not addressed to the normative, prescriptive law. Rather, it is a *technology*, if the term may be used, social technology, and as such directed towards causal-type laws of the connection of means and end.[8]

Even after extensive alteration, most of these views cannot be employed within the body of an empirical sociology, i. e., of a social behaviorism. For what correlation is supposed to be asserted? One may object, however, that it is again a question of showing that combinations of certain rules and legal definitions are logically equivalent to other. definitions (although this is something not necessarily noticed at first sight). But then, while such demonstrations are certainly important for practical life, no special metaphysical discussions are required.

It is clear that these tautologies of the legal system will be less prominent when the basic spirit of unified science prevails. People will then be more interested in what effects certain measures produce, and less in whether the ordinances formulated in statute books are logically consistent. No special discipline, certainly, is required to test the logical compatibility of the rules for the administration of a hospital. What one wishes to know is how the joint operation of certain measures affects the standard of health, so that one may act accordingly.

VII. The Empirical Sociology of Marxism

The unified language of physicalism safeguards the scientific method. Statement is linked to statement, law to law. It has been shown how sociology can be incorporated in unified science no less than biology, chemistry, technology or astronomy. The fundamental separation of special "moral sciences" from the "natural sciences" has proved itself theoretically meaningless. But even a purely practical division, sharper than any of the many others that exist has been shown to be inappropriate and wholly uncalled for.

---

8. Kelsen, *op. cit.*, p. 27.

In this connection we have given a sketch of the concept of sociological correlations as applicable within a developed social behaviorism. We have seen that by virtue of this conception, disciplines such as "ethics" and jurisprudence" lose their traditional foundations. Without metaphysics, without distinctions explicable only through reference to metaphysical habits, these disciplines cannot maintain their independence. *Whatever elements of genuine science are contained in them become incorporated into the structure of sociology.*

In this science there gradually converge whatever useful protocol statements and laws economics, ethnology, history and other disciplines have to offer. Sometimes the fact that men alter their modes of reaction plays an important role in sociological thought, and sometimes the starting point is the fact that men do not change in their reaction behavior, but enter into modified relations with one another. Economics, for instance, reckons with a constant human type, and then investigates the consequences of the operation of the given economic order, e. g. the market mechanism. It seeks to determine how crises and unemployment arise, how net profits accrue, etc.

But when it is observed that the given economic order is altered by men, the need arises for sociological laws which describe this change. Investigation of the economic order and its operation is then not sufficient. It is necessary to investigate, in addition, the laws which determine the change in the economic order itself. How certain changes in the mode of production alter stimuli in such a way that men transform their traditional ways of living, often by means of revolutions, is a question investigated by sociologists of the most divergent schools. Marxism is, to a higher degree than any other present-day sociological theory, a system of empirical sociology. The most important Marxist theses employed for prediction are either already formulated in a fairly physicalistic fashion (so far as traditional language made this possible), or they can be so formulated, without the loss of anything essential.

We can see in the case of Marxism how sociological laws are sought for and how relations conformable to law are established. When one attempts to establish the correlation between the modes of production of successive periods and the contemporaneous forms of religious worship, the books, discourses, etc., then one is investigating the *correlation between physicalistic structures.* Marxism lays down, over and above the doctrine of physicalism (materialism) certain special doctrines. When it opposes *one* group of forms as "substructure" to another group as "superstructure" ("historical materialism" as a special physicalistic theory), it proceeds throughout its operations within the confines of social behaviorism. What is involved here is no opposition of the "material" to the "spiritual," i. e., of "essences" with "different types of causality."

The coming decades may be concerned in growing measure with the discovery of such correlations. Max Weber's prodigious attempt to demonstrate the emergence of capitalism from Calvinism clearly shows to how great an extent concrete investigation is obstructed by metaphysical formulations. To a proponent of social behaviorism it seems at once quite natural that certain verbal sequences—the formulation of certain divine commands—should be recognized as dependent on certain modes of production and power situations. But it does not seem very plausible that the way of life of vast numbers of human beings occupied with trade, industry and other matters, should be determined by verbal sequences of individual theologians, or by the deity's injunctions, always very vaguely worded, which the theologians transmit. And yet Max Weber was committed to this point of view. He sought to show that from the "spirit of Calvinism" was born the "spirit of capitalism" and with it the capitalist order.

A Catholic theologian, Kraus, has pointed out that such an overestimation of the influence of theological formulations can only be explained by the fact that he ascribed to spirit a sort of "magical" effect. In the work of Weber and other thinkers, "spirit" is regarded as very closely bound up with words and formulae. Thus we understand Weber's assiduous quest for crucial theological formulations of individual Calvinists, in which the origins of crucial capitalist formulations might be sought. The "rationalism" of one sphere is to spring from that of the other. That theological discourses and writings possess such enormous powers is a supposition which would be formally possible within a physicalistic system. *But experience proves otherwise.* In company with the Marxists, the Catholic theologian mentioned above points out the fact that theological subtleties exercise little influence on human behavior, indeed, that they are scarcely known to the average merchant or professional man. It would be much more plausible to suppose that in England, for example, merchants opposed to the royal monopoly, and usurers desiring to take interest at a higher rate than the Church of England permitted, readily gave their support to a doctrine and a party which turned against the Church and the crown allied with the Church. First the behavior of these men was, to a considerable extent, capitalistically oriented—then they became Calvinists. We should expect to find, in accordance with all our experience of theological doctrines at other times, that these doctrines were subsequently revised and adapted to the system of production and commerce. And, Kraus further shows, in opposition to Weber, that those theological formulations which are "compatible" with capitalism did not appear until later, while Calvinism in its original form was related rather to the dogmas of the anti-capitalist Middle Ages. *Weber's metaphysical starting-point impeded his scientific work, and determined unfavorably his selection of observation-statements.* But without a suitable selection of observation-statements there can be no fruitful scientific work.

Let us analyze a concrete case in somewhat greater detail. With what is the decline of slavery in the ancient world connected?

Many have been inclined to the view that Christian doctrine and the Christian way of life effected the disappearance of slavery, after the Stoic philosophers had impaired the conception of slavery as an eternal institution.

If such an assertion is meant to express a correlation, it is natural to consider, in the first place, whether or not Christianity and slavery are found together. It is then seen that the most oppressive forms of slavery appear at the beginning of the modern era, at a time when Christian states are everywhere expanding their power, when the Christian Churches are vigorous above all in the colonies. Because of the intervention of Catholic theologians motivated by humanitarian considerations, the preservation of the perishing Indian slaves of America was undertaken through the importation of sturdier Negro slaves brought to that continent in shiploads.

It would really be necessary to define in advance with a greater degree of precision what is meant, on the one hand, by "Christian," and, on the other, by "slavery." If the attempt is made to formulate the correlation between them with greater clarity, it must be said that statements of a certain type, religious behavior, etc., never appear in conjunction with the large-scale ownership of slaves. But in this connection, it would be necessary to lay down a definite mode of application. For a man can be a "slave" from the "juristic" standpoint, and, simultaneously, a "master" from the "sociological" point of view. Sociological concepts, however, may be linked only with other sociological concepts.

"Christian dogma" is an extraordinarily indeterminate concept. Many theologians have believed it possible to demonstrate, from the Bible, that God has condemned the Negroes to slavery: when

Ham treated his drunken father Noah irreverently, Noah cursed him and declared that he and his descendants were to be subject to his brothers Shem and Japheth and their descendants. Still other theologians have sought to discover in Christian doctrine arguments against slavery.

It is evident that the sociologist advances much further when he delimits a certain system of men, religious acts, dogmas, etc., and then notes whether it comes into being in conjunction with certain modes of social behavior. This is, of course, a very rough procedure. The attempt must be made to discover not only such simple correlations, but also correlations of greater complexity. Laws must be combined with one another, in order for it to be possible to produce certain predictions.

Some sociological "laws" are valid only for limited periods, just as, in biology, there are laws about ants and about lions in addition to more general laws. That is to say, we are not yet in a position to state precisely on what certain correlations depend: the phrase "historical period" refers to a complicated set of conditions which has not been analyzed. Much confusion is due to the opinion of some analytical sociologists that the laws which they had discovered had to possess the same character as chemical laws, i.e., that they had to hold true under all conceivable earthly conditions. But sociology is concerned for the most part with correlations valid for *limited periods of time*. Marx was justified in asserting that it is senseless to speak of a universal law of population, as Malthus did. But it is possible to state which law of population holds for any given sociological period.

When, for the purpose of clarifying the question, "How does the decline of slavery come about?" one analyzes the conflict between the Northern and Southern States over the freeing of the slaves, one is confronted by a conflict between industrial and plantation states. The emancipation inflicts serious injury on the plantation states. Shouldn't we expect a connection between the freeing of the slaves and the processes of production? How is such a notion to be made plausible?

One attempts to determine the conditions under which slavery offers the slave-owner advantages, and the conditions under which the contrary is the case. If those masters who free their slaves are asked why they do so, only a few will say that they oppose slavery because it does not yield sufficient advantages. Many will inform us, without hypocrisy, that they have been deeply impressed by reading a philosopher who championed the slaves. Others will describe in detail their conflicting motives, will perhaps explain that slavery would really be more advantageous to them, but that the desire to sacrifice, to renounce property, has led them, after a long inner conflict, to the difficult step of freeing their slaves. Anyone accustomed to operate in the spirit of social behaviorism will, above all, keep in mind the very complicated "stimulus" of the way of life based on slave-owning, and then proceed to investigate the "reaction"—retention or freeing of the slaves. He will employ the results of this inquiry to determine how far theological doctrines concerning the emancipation of slaves are to be recognized as "stimulus," how far as "reaction."

If it is shown that relatively simple correlations can be established between the effects of slavery on the masters' tenor of life and the behavior of the master toward the liberation of slaves, and that, as against this, no *simple correlations* can be laid down between the doctrines of the time and the behavior of the slave-owners, then preference will be given to the former mode of investigation.

Thus there will be examined under various conditions the relationship between hunting and slavery, agriculture and slavery, manufacture and slavery. It will be found, for example, that the possession of slaves generally offers no advantage where there are sufficiently numerous free workers who eagerly seek employment in order to avoid starvation. Columella, a Roman agrarian writer of the later period, bluntly says, for example, that the employment of slaves is disadvantageous to anyone who drains fever swamps in the Campagna: the sickness of a slave means loss of interest, while his death results in loss of capital. He goes on to say that it is possible, on the other hand, to obtain free workers on the market at any time, and that the employer is in no way burdened by their sickness or death.

When serious fluctuations in the business situation occur, entrepreneurs find it desirable to be in a position to drop free workers; slaves, like horses, must continue to be fed. When one reads in Strabo, therefore, that in antiquity papyrus shrubs in Egypt were already being cut down in order to maintain the monopoly price, one understands that the universal employment of free labor could not be far away.

The conditions which led to the fluctuating tendencies of early capitalist economic institutions can likewise be investigated. Correlation is added to correlation. It is seen that "free labor" and "the destruction of commodities" seem to be correlated under certain conditions. This is equally true of "plantation slavery" and "a con-

stant market." One can view the Civil War as a conflict between the
industrial North, which was not interested in slavery, and the cotton-
producing agrarian South, and thereby be able to make extensive
predictions.

This does not mean that the religious and ethical opponents
of slavery were lying when they said that they directly rejoiced in the
emancipation of the slaves, but not in the increase of industrial
profits which ensued in the North. That such a desire for the freeing of the slaves could develop at this time and find so rich a satisfaction is something which the empirical sociologist could deduce,
in broad outline, from the total economic situation.

The methods used in the elaboration of a theory of agricultural
economics have also been applied by several writers (theologians
among them) in the construction of a wholly empirical theory of
the "employment of natives," yielding all types of correlations.[9] In
combination with other relations expressible as laws one can make
all sorts of predictions concerning the fate of slavery in particular
countries and territories.

The distribution of grain to freemen but not to slaves, during
the later history of Rome, offered slave-owners an additional motive
for freeing their slaves. The former master could then re-employ the
freedman at a lower cost, and also use his support at elections. It
is likewise easy to understand how, during the decline of Rome, the
system of the "coloni" and serfdom emerge through the regression
of early capitalistic institutions. In order to undertake an enterprise
with slave labor, one had to have at one's disposal extensive financial
resources, since both workers and implements of production had
to be purchased. Under a regime of free labor, the purchase of
tools was sufficient. The system of the "coloni" required no invest-
ment at all from the owner, who was assured of dues of all kinds.
The "free" workers were forced by the whole social order to labor
—the death penalty was imposed for idleness—while each slave had
to be disciplined by his own master. The master had to protect the
health and life of his slave, to care for him just as he had to care
for a horse or a bullock, even when it was unruly.

We see how, by means of such analyses, correlations are established between general social conditions and certain modes of behavior of limited human groups. The "statements" which these
groups make about their own behavior are not essential to these
correlations; they can often be added with the help of additional

---

9. Cf. Otto Neurath, "Probleme der Kriegswirtschaftslehre," *Zeitschrift für die
gesamten Staatswissenschaften*, 1914, p. 474.

correlations. It is above all in Marxism that empirical sociology is pursued in this way.[10]

A system of empirical sociology in the spirit of social behaviorism, as it has been developed above all in the United States and the U.S.S.R., would have to direct its inquiries primarily to the typical "reactions" of whole groups. But significant historical movements are also often measured or evaluated without such analysis. And it may further be shown that through the development of certain institutions, through the increase in a certain magnitude, a reversal is produced which causes further changes to take a wholly different direction. The primitive "idea of progress," that every magnitude increases indefinitely, is untenable. One must consider the whole system of sociological magnitudes in all its complexity, and then note what changes are predictable. One cannot infer from the growth of large cities up to the present time that the process will continue approximately the same. Rapid growth is especially apt to release stimuli leading to a sudden cessation of growth and perhaps to the reconstitution of many small centers. The expansion of capitalist large-scale industry and the emergence of the proletarian masses dependent on those industries can lead to a situation where the whole capitalist mechanism moves through a series of economic crises towards its ultimate dissolution.

VIII. POSSIBILITIES OF PREDICTION

It is possible to state the extent to which predictions can be successfully made within the sphere of social behaviorism. It is evident that its various *"predictions," i.e., its scientific theories, are sociological events essentially dependent on the social and economic order.* It is only after this is understood that it becomes clear, for example, that under certain conditions certain predictions either do not emerge at all, or cannot be elaborated. Even when an individual believes that he divines the direction of further successful investigation, he can be prevented from finding the collaboration required for sociological research by the indifference, or even the opposition, of other men.

The approach of social changes is difficult to notice. In order to be able to make predictions about events of a new type, one must usually possess a certain amount of new experiences. It is often the changes in the historical process that first give the scientist the

10. Cf. for example, Ettore Ciccotti, *Der Untergang der Sklaverei in Altertum* (German translation by Oda Olberg), Berlin, 1910.

necessary data for further investigations. But since sociological investigations also play a certain role as stimuli and instruments in
the organization of living, the development of sociology is very
closely bound up with social conflicts. Only established schools of
sociology, requiring social support, can master, by means of collective
labor, the masses of material which must be adapted to a stricter
formulation of correlations. This presupposes that the powers which
finance such work are favorably inclined towards social behaviorism.

This is in general not the case today. Indeed, there exists in the
ruling classes an aversion to social, as well as to individual, behaviorism which is much more than a matter of a scientific doubt, which
would be comprehensible in view of the imperfections of this doctrine. The opposition of the ruling circles, which usually find support
in the universities of the capitalist countries, is explained sociologi-
cally, above all, by the fact that empirical sociology, through its
non-metaphysical attitude, reveals the meaninglessness of such expressions as "categorical imperative," "divine injunction," "moral
idea," "superpersonal state," etc. In doing this it undermines important doctrines which are useful in the maintenance of the prevailing
order. The proponents of "unified science" do not defend *one* world-
view among other world-views. Hence the question of tolerance
cannot be raised. They declare transcendental theology to be not
false, but meaningless. Without disputing the fact that powerful inspiration, and cheering and depressing effects, can be associated with
meaningless doctrines, they can in practice "let seven be a holy
number," since they do not harass the supporters of these doctrines.
But they cannot allow that these claims have any meaning at all,
however "hidden," i.e., that they can confirm or confute scientific
statements. Even if such reasoning by the pure scientist leaves metaphysics and theology undisturbed, it doubtlessly shakes the reverence
for them which is frequently demanded.

All the metaphysical entities whose injunctions men endeavored
to obey, and whose "holy" powers they venerated, disappear. In
their place there stands as an empirical substitute, confined within
the bounds of purely scientific formulations, the actual behavior of
groups, whose commands operate as empirical forces on individual
men. That groups of men lend strength to individual men pursuing
certain modes of action and obstruct others pursuing different modes,
is a statement which is wholly meaningful in the context of social
behaviorism.

The social behaviorist, too, makes commands, requests and
reproaches; *but he does not suppose that these utterances, when*

connected with propositions, can yield a system. Words can be employed like whistles, caresses and whiplashes; but when used in this way, they can neither agree with nor contradict propositions. *An injunction can never be deduced* from a system of propositions! This is no "limitation" of the scientific method: it is simply the result of logical analysis. That injunctions and predictions are so frequently linked follows from the fact that both are directed to the future. An injunction is an event which it is assumed will evoke certain changes in the future. A prediction is a statement which it is assumed will agree with a future statement.

The proponents of "unified science" seek, with the help of laws, to formulate predictions in the "unified language of physicalism." This takes place in the sphere of empirical sociology through the development of "social behaviorism." In order to attain to more useful predictions, one can immediately eliminate meaningless verbal sequences by the use of logic. But this is not sufficient. There must follow the elimination of all false formulations. The representatives of modern science, even after they have effected the elimination of metaphysical formulations, must still dispose of false doctrines, for example, astrology, magic, etc. In order to liberate someone from such ideas, the universal acknowledgment accorded to the rules of logic does not, as with the elimination of *meaningless* statements, suffice. One must, if one wishes to see one's own theory prevail, create the groundwork which will lead people to recognize the inadequacy of these theories, which, while *"also physicalistic,"* are *uncritical.*

The fruitfulness of social behaviorism is demonstrated by the establishment of new correlations and by the successful predictions made on the basis of them. Young people educated in the spirit of physicalism and its unified language will be spared many of the hindrances to scientific work to which we are still at present subjected. A single individual cannot create and employ this successful language, for it is the product of the labor of a generation. Thus, even in the form of social behaviorism, sociology will be able to formulate valid predictions on a large scale only when a generation trained in physicalism sets to work in all departments of science. Despite the fact that we can observe metaphysics on the increase, there is much to show that non-metaphysical doctrines are also spreading and constantly gaining ground as the new "superstructure" erected on the changing economic "substructure" of our age.[11]

---

11. Cf. Otto Neurath, "Physicalism, the Philosophy of the Vienna Circle," *The Monist*, October 1931.

THE FUNCTION OF GENERAL LAWS IN HISTORY

1. It is a rather widely held opinion that history, in contradistinction to the so-called physical sciences, is concerned with the description of particular events of the past rather than with the search for general laws which might govern those events. As a characterization of the type of problem in which some historians are mainly interested, this view probably can not be denied; as a statement of the theoretical function of general laws in scientific historical research, it is certainly unacceptable. The following considerations are an attempt to substantiate this point by showing in some detail that general laws have quite analogous functions in history and in the natural sciences, that they form an indispensable instrument of historical research, and that they even constitute the common basis of various procedures which are often considered as characteristic of the social in contradistinction to the natural sciences.

By a general law, we shall here understand a statement of universal conditional form which is capable of being confirmed or disconfirmed by suitable empirical findings. The term "law" suggests the idea that the statement in question is actually well confirmed by the relevant evidence available; as this qualification is, in many cases, irrelevant for our purpose, we shall frequently use the term "hypothesis of universal form" or briefly "universal hypothesis" instead of "general law," and state the condition of satisfactory confirmation separately, if necessary. In the context of this paper, a universal hypothesis may be assumed to assert a regularity of the following type: In every case where an event of a specified kind C occurs at a certain place and time, an event of a specified kind E will occur at a place and time which is related in a specified manner to the place and time of the occurrence of the first event. (The symbols "C" and "E" have been chosen to suggest the terms "cause" and "effect," which are often, though by no means always, applied to events related by a law of the above kind.)

2.1 The main function of general laws in the natural sciences is to connect events in patterns which are usually referred to as *explanation* and *prediction*.

The explanation of the occurrence of an event of some specific kind E at a certain place and time consists, as it is usually expressed, in indicating the causes or determining factors of E. Now the assertion that a set of events—say, of the kinds C_1, C_2, . . . , C_n— have caused the event to be explained, amounts to the statement that, according to certain general laws, a set of events of the kinds mentioned is regularly accompanied by an event of kind E. Thus, the scientific explanation of the event in question consists of

(1) a set of statements asserting the occurrence of certain events C_1, . . . C_n at certain times and places,

(2) a set of universal hypotheses, such that

(a) the statements of both groups are reasonably well confirmed by empirical evidence,

(b) from the two groups of statements the sentence asserting the occurrence of event E can be logically deduced.

In a physical explanation, group (1) would describe the initial and boundary conditions for the occurrence of the final event; generally, we shall say that group (1) states the *determining conditions* for the event to be explained, while group (2) contains the general laws on which the explanation is based; they imply the statement that whenever events of the kind described in the first group occur, an event of the kind to be explained will take place.

Illustration: Let the event to be explained consist in the cracking of an automobile radiator during a cold night. The sentences of group (1) may state the following initial and boundary conditions: The car was left in the street all night. Its radiator, which consists of iron, was completely filled with water, and the lid was screwed on tightly. The temperature during the night dropped from 39° F. in the evening to 25° F. in the morning; the air pressure was normal. The bursting pressure of the radiator material is so and so much.—Group (2) would contain empirical laws such as the following: Below 32° F., under normal atmospheric pressure, water freezes. Below 39.2° F., the pressure of a mass of water increases with decreasing temperature, if the volume remains constant or decreases; when the water freezes, the pressure again increases. Finally, this group would have to include a quantitative law concerning the change of pressure of water as a function of its temperature and volume.

From statements of these two kinds, the conclusion that the radiator cracked during the night can be deduced by logical reasoning; an explanation of the considered event has been established.

2.2 It is important to bear in mind that the symbols *"E," "C,"* *"C_1," "C_2,"* etc., which were used above, stand for kinds or properties of events, not for what is sometimes called individual events. For the object of description and explanation in every branch of empirical science is always the occurrence of an event of a certain *kind* (such as a drop in temperature by 14° F., an eclipse of the moon, a cell-division, an earthquake, an increase in employment, a political assassination) at a given place and time, or in a given empirical object (such as the radiator of a certain car, the planetary system, a specified historical personality, etc.) at a certain time.

What is sometimes called the complete description of an individual event (such as the earthquake of San Francisco in 1906 or the assassination of Julius Caesar) would require a statement of all the properties exhibited by the spatial region or the individual object involved, for the period of time occupied by the event in question. Such a task can never be completely accomplished.

A fortiori, it is impossible to explain an individual event in the sense of accounting for *all* its characteristics by means of universal hypotheses, although the explanation of what happened at a specified place and time may gradually be made more and more specific and comprehensive.

But there is no difference, in this respect, between history and the natural sciences: both can give an account of their subject-matter only in terms of general concepts, and history can "grasp the unique individuality" of its objects of study no more and no less than can physics or chemistry.

3. The following points result more or less directly from the above study of scientific explanation and are of special importance for the questions here to be discussed.

3.1 A set of events can be said to have caused the event to be explained only if general laws can be indicated which connect "causes" and "effect" in the manner characterized above.

3.2 No matter whether the cause-effect terminology is used or not, a scientific explanation has been achieved only if empirical laws of the kind mentioned under (2) in 2.1 have been applied.[1]

3.3 The use of universal empirical hypotheses as explanatory principles distinguishes genuine from pseudo-explanation, such as, say, the attempt to account for certain features of organic behavior by reference to an entelechy, for whose functioning no laws are offered, or the explanation of the achievements of a given person in terms of his "mission in history," his "predestined fate," or simi-

---

[1] Maurice Mandelbaum, in his generally very clarifying analysis of relevance and causation in history (*The Problem of Historical Knowledge*, New York, 1938, Chs. 7, 8) seems to hold that there is a difference between the "causal analysis" or "causal explanation" of an event and the establishment of scientific laws governing it in the sense stated above. He argues that "scientific laws can only be formulated on the basis of causal analysis," but that "they are not substitutes for full causal explanations" (*l.c.*, p. 238). For the reasons outlined above, this distinction does not appear to be justified: every "causal explanation" is an "explanation by scientific laws"; for in no other way than by reference to empirical laws can the assertion of a causal connection between certain events be scientifically substantiated.

lar notions. Accounts of this type are based on metaphors rather than laws; they convey pictorial and emotional appeals instead of insight into factual connections; they substitute vague analogies and intuitive "plausibility" for deduction from testable statements and are therefore unacceptable as scientific explanations.

Any explanation of scientific character is amenable to objective checks; these include

(a) an empirical test of the sentences which state the determining conditions;

(b) an empirical test of the universal hypotheses on which the explanation rests;

(c) an investigation of whether the explanation is logically conclusive in the sense that the sentence describing the event to be explained follows from the statements of groups (1) and (2).

4. The function of general laws in *scientific prediction* can now be stated very briefly. Quite generally, prediction in empirical science consists in deriving a statement about a certain future event (for example, the relative position of the planets to the sun, at a future date) from (1) statements describing certain known (past or present) conditions (for example, the positions and momenta of the planets at a past or present moment), and (2) suitable general laws (for example, the laws of celestial mechanics). Thus, the logical structure of a scientific prediction is the same as that of a scientific explanation, which has been described in 2.1. In particular, prediction no less than explanation throughout empirical science involves reference to universal empirical hypotheses.

The customary distinction between explanation and prediction rests mainly on a pragmatical difference between the two: While in the case of an explanation, the final event is known to have happened, and its determining conditions have to be sought, the situation is reversed in the case of a prediction: here, the initial conditions are given, and their "effect"—which, in the typical case, has not yet taken place—is to be determined.

In view of the structural equality of explanation and prediction, it may be said that an explanation as characterized in 2.1 is not complete unless it might as well have functioned as a prediction: If the final event can be derived from the initial conditions and universal hypotheses stated in the explanation, then it might as well have been predicted, before it actually happened, on the basis of a knowledge of the initial conditions and the general laws. Thus, e.g., those initial conditions and general laws which the astronomer would adduce in explanation of a certain eclipse of the sun are such that they might also have served as a sufficient basis for a forecast of the eclipse before it took place.

However, only rarely, if ever, are explanations stated so completely as to exhibit this predictive character (which the test referred to under (c) in 3.3 would serve to reveal). Quite commonly, the explanation offered for the occurrence of an event is incomplete. Thus, we may hear the explanation that a barn burnt down "because" a burning cigarette was dropped in the hay, or that a certain political movement has spectacular success "because" it takes advantage of widespread racial prejudices. Similarly, in the case of the broken radiator, the customary way of formulating an explanation would be restricted to pointing out that the car was left in the cold, and the radiator was filled with water.—In explanatory statements like these, the general laws which confer upon the stated conditions the character of "causes" or "determining factors" are completely omitted (sometimes, perhaps, as a "matter of course"), and, furthermore, the enumeration of the determining conditions of group (1) is incomplete; this is illustrated by the preceding examples, but even by the earlier analysis of the broken-radiator case: as a closer examination would reveal, even that much more detailed statement of determining conditions and universal hypotheses would require amplification in order to serve as a sufficient basis for the deduction of the conclusion that the radiator broke during the night.

In some instances, the incompleteness of a given explanation may be considered as inessential. Thus, e.g., we may feel that the explanation referred to in the last example could be made complete if we so desired; for we have reasons to assume that we know the kind of determining conditions and of general laws which are relevant in this context.

Very frequently, however, we encounter "explanations" whose incompleteness can not simply be dismissed as inessential. The methodological consequences of this situation will be discussed later (especially in 5.3 and 5.4).

5.1 The preceding considerations apply to *explanation in history* as well as in any other branch of empirical science. Historical explanation, too, aims at showing that the event in question was not "a matter of chance," but was to be expected in view of certain antecedent or simultaneous conditions. The expectation referred to is not prophecy or divination, but rational scientific anticipation which rests on the assumption of general laws.

If this view is correct, it would seem strange that while most historians do suggest explanations of historical events, many of them deny the possibility of resorting to any general laws in history.

It is, however, possible to account for this situation by a closer study of explanation in history, as may become clear in the course of the following analysis.

5.2 In some cases, the universal hypotheses underlying a historical explanation are rather explicitly stated, as is illustrated by the italicized passages in the following attempt to explain the tendency of government agencies to perpetuate themselves and to expand (italics the author's) :

As the activities of the government are enlarged, more people develop a vested interest in the continuation and expansion of governmental functions. *People who have jobs do not like to lose them; those who are habituated to certain skills do not welcome change; those who have become accustomed to the exercise of a certain kind of power do not like to relinquish their control—* if anything, *they want to develop greater power and correspondingly greater prestige.* . . . Thus, government offices and bureaus, once created, in turn institute drives, not only to fortify themselves against assault, but to enlarge the scope of their operations.[2]

Most explanations offered in history or sociology, however, fail to include an explicit statement of the general regularities they presuppose; and there seem to be at least two reasons which account for this:

First, the universal hypotheses in question frequently relate to individual or social psychology, which somehow is supposed to be familiar to everybody through his everyday experience; thus, they are tacitly taken for granted. This is a situation quite similar to that characterized in 4.

Second, it would often be very difficult to formulate the underlying assumptions explicitly with sufficient precision and at the same time in such a way that they are in agreement with all the relevant empirical evidence available. It is highly instructive, in examining the adequacy of a suggested explanation, to attempt a reconstruction of the universal hypotheses on which it rests. Particularly, such terms as "hence," "therefore," "consequently," "because," "naturally," "obviously," etc., are often indicative of the tacit presupposition of some general law: they are used to tie up the initial conditions with the event to be explained; but that the latter was "naturally" to be expected as "a consequence" of the stated conditions follows only if suitable general laws are presupposed. Consider, for example, the statement that the Dust Bowl farmers migrate to California "because" continual drought and sandstorms render their existence increasingly precarious,

[2] Donald W. McConnell, *Economic Behavior;* New York, 1939; pp. 894–895.

and because California seems to them to offer so much better living conditions. This explanation rests on some such universal hypothesis as that populations will tend to migrate to regions which offer better living conditions. But it would obviously be difficult accurately to state this hypothesis in the form of a general law which is reasonably well confirmed by all the relevant evidence available. Similarly, if a particular revolution is explained by reference to the growing discontent, on the part of a large part of the population, with certain prevailing conditions, it is clear that a general regularity is assumed in this explanation, but we are hardly in a position to state just what extent and what specific form the discontent has to assume, and what the environmental conditions have to be, to bring about a revolution. Analogous remarks apply to all historical explanations in terms of class struggle, economic or geographic conditions, vested interests of certain groups, tendency to conspicuous consumption, etc.: All of them rest on the assumption of universal hypotheses [3] which connect certain characteristics of individual or group life with others; but in many cases, the content of the hypotheses which are tacitly assumed in a given explanation can be reconstructed only quite approximately.

5.3 It might be argued that the phenomena covered by the type of explanation just mentioned are of a statistical character, and that therefore only probability hypotheses need to be assumed in their explanation, so that the question as to the "underlying general laws" would be based on a false premise. And indeed, it seems possible and justifiable to construe certain explanations offered in history as based on the assumption of probability hypotheses rather than of general "deterministic" laws, i.e., laws in the form of universal conditionals. This claim may be extended to many of the explanations offered in other fields of empirical science as well. Thus, e.g., if Tommy comes down with the measles two weeks after his brother, and if he has not been in the company of other persons having the measles, we accept the explanation that he caught the disease from his brother. Now, there is a general hypothesis underlying this explanation; but it can hardly be said to be a general law to the effect that any person who has not had the measles before will get them without fail if he stays in the company of somebody else who has the measles; that a contagion will occur can be asserted only with a high probability.

---

[3] What is sometimes, misleadingly, called an explanation by means of a certain *concept* is, in empirical science, actually an explanation in terms of *universal hypotheses* containing that concept. "Explanations" involving concepts which do not function in empirically testable hypotheses—such as "entelechy" in biology, "historic destination of a race" or "self-unfolding of absolute reason" in history—are mere metaphors without cognitive content.

Many an explanation offered in history seems to admit of an analysis of this kind: if fully and explicitly formulated, it would state certain initial conditions, and certain probability hypotheses,[4] such that the occurrence of the event to be explained is made highly probable by the initial conditions in view of the probability hypotheses. But no matter whether explanations in history be construed as "causal" or as "probabilistic" in character, it remains true that in general the initial conditions and especially the universal hypotheses involved are not clearly indicated, and can not unambiguously be supplemented. (In the case of probability hypotheses, for example, the probability values involved will at best be known quite roughly.)

5.4 What the explanatory analyses of historical events offer is, then, in most cases not an explanation in one of the meanings developed above, but something that might be called an *explanation sketch*. Such a sketch consists of a more or less vague indication of the laws and initial conditions considered as relevant, and it needs "filling out" in order to turn into a full-fledged explanation. This filling-out requires further empirical research, for which the sketch suggests the direction. (Explanation sketches are common also outside of history; many explanations in psychoanalysis, for instance, illustrate this point.)

Obviously, an explanation sketch does not admit of an empirical test to the same extent as does a complete explanation; and yet, there is a difference between a scientifically acceptable explanation sketch and a pseudo-explanation (or a pseudo-explanation sketch). A scientifically acceptable explanation sketch needs to be filled out by more specific statements; but it points into the direction where these statements are to be found; and concrete research may tend to confirm or to infirm those indications; i.e., it may show that the kind of initial conditions suggested are actually relevant; or it may reveal that factors of a quite different nature have to be taken into account in order to arrive at a satisfactory explanation.—The filling-out process required by an explanation sketch will, in general, assume the form of a gradually increasing precision of the formulations involved; but at any stage of this process, those formulations will have some empirical import: it will be possible to indicate, at least roughly, what kind of evidence would be relevant in testing

---

[4] E. Zilsel, in a very stimulating paper on "Physics and the Problem of Historico-Sociological Laws" (*Philosophy of Science*, Vol. 8, 1941, pp. 567–579), suggests that all specifically historical laws are of a statistical character similar to that of the "macro-laws" in physics. The above remarks, however, are not restricted to specifically historical laws since explanation in history rests to a large extent on non-historical laws (cf. section 8 of this paper).

them, and what findings would tend to confirm them. In the case of non-empirical explanations or explanation sketches, on the other hand—say, by reference to the historical destination of a certain race, or to a principle of historical justice—the use of empirically meaningless terms makes it impossible even roughly to indicate the type of investigation that would have a bearing upon those formulations, and that might lead to evidence either confirming or infirming the suggested explanation.

5.5 In trying to appraise the soundness of a given explanation, one will first have to attempt to reconstruct as completely as possible the argument constituting the explanation or the explanation sketch. In particular, it is important to realize what the underlying explaining hypotheses are, and to judge of their scope and empirical foundation. A resuscitation of the assumptions buried under the gravestones "hence," "therefore," "because," and the like will often reveal that the explanation offered is poorly founded or downright unacceptable. In many cases, this procedure will bring to light the fallacy of claiming that a large number of details of an event have been explained when, even on a very liberal interpretation, only some broad characteristics of it have been accounted for. Thus, for example, the geographic or economic conditions under which a group lives may account for certain general features of, say, its art or its moral codes; but to grant this does not mean that the artistic achievements of the group or its system of morals has thus been explained in detail; for this would imply that from a description of the prevalent geographic or economic conditions alone, a detailed account of certain aspects of the cultural life of the group can be deduced by means of specifiable general laws.

A related error consists in singling out one of several important groups of factors which would have to be stated in the initial conditions, and then claiming that the phenomenon in question is "determined" by and thus can be explained in terms of that one group of factors.

Occasionally, the adherents of some particular school of explanation or interpretation in history will adduce, as evidence in favor of their approach, a successful historical prediction which was made by a representative of their school. But though the predictive success of a theory is certainly relevant evidence of its soundness, it is important to make sure that the successful prediction is in fact obtainable by means of the theory in question. It happens sometimes that the prediction is actually an ingenious guess which may have been influenced by the theoretical outlook of its author, but which can not be arrived at by means of his theory alone. Thus, an adherent of a quite metaphysical "theory" of history may have a sound feeling for historical developments and may be able to make correct predictions, which he will even couch in the terminology of his theory, though they could not have been attained by means of it. To guard against such pseudo-confirming cases would be one of the functions of test (c) in 3.3.

6. We have tried to show that in history no less than in any other branch of empirical inquiry, scientific explanation can be achieved only by means of suitable general hypotheses, or by theories, which are bodies of systematically related hypotheses. This thesis is clearly in contrast with the familiar view that genuine explanation in history is obtained by a method which characteristically distinguishes the social from the natural sciences, namely, *the method of empathetic understanding:* The historian, we are told, imagines himself in the place of the persons involved in the events which he wants to explain; he tries to realize as completely as possible the circumstances under which they acted, and the motives which influenced their actions; and by this imaginary self-identification with his heroes, he arrives at an understanding and thus at an adequate explanation of the events with which he is concerned.

This method of empathy is, no doubt, frequently applied by laymen and by experts in history. But it does not in itself constitute an explanation; it rather is essentially a heuristic device; its function is to suggest certain psychological hypotheses which might serve as explanatory principles in the case under consideration. Stated in crude terms, the idea underlying this function is the following: The historian tries to realize how he himself would act under the given conditions, and under the particular motivations of his heroes; he tentatively generalizes his findings into a general rule and uses the latter as an explanatory principle in accounting for the actions of the persons involved. Now, this procedure may sometimes prove heuristically helpful; but its use does not guarantee the soundness of the historical explanation to which it leads. The latter rather depends upon the factual correctness of the empirical generalizations which the method of understanding may have suggested.

Nor is the use of this method indispensable for historical explanation. A historian may, for example, be incapable of feeling himself into the rôle of a paranoiac historic personality, and yet he may well be able to explain certain of his actions; notably by reference to the principles of abnormal psychology. Thus, whether the historian is or is not in a position to identify himself with his historical hero, is irrelevant for the correctness of his explanation; what counts, is the soundness of the general hypotheses involved, no matter whether they were suggested by empathy or by a strictly behavioristic procedure. Much of the appeal of the "method of understanding" seems to be due to the fact that it tends to present the phenomena in question as somehow "plausible" or "natural" to us; [5] this is often done by means of attractively worded metaphors. But the kind of "understanding" thus conveyed must clearly be separated from scientific understanding. In history as anywhere else in empirical science, the explanation of a phenomenon consists in subsuming it under general empirical laws; and the criterion of its soundness is not whether it appeals to our imagination, whether it is presented in suggestive analogies, or is otherwise made to appear plausible—all this may occur in pseudo-explanations as well—but exclusively whether it rests on empirically well confirmed assumptions concerning initial conditions and general laws.

7.1 So far, we have discussed the importance of general laws for explanation and prediction, and for so-called understanding in history. Let us now survey more briefly some other procedures of historical research which involve the assumption of universal hypotheses.

Closely related to explanation and understanding is the so-called *interpretation of historical phenomena* in terms of some particular approach or theory. The interpretations which are actually offered in history consist either in subsuming the phenomena in question under a scientific explanation or explanation sketch; or in an attempt to subsume them under some general idea which is not amenable to any empirical test. In the former case, interpretation clearly is explanation by means of universal hypotheses; in the latter, it amounts to a pseudo-explanation which may have emotive appeal and evoke vivid pictorial associations, but which does not further our theoretical understanding of the phenomena under consideration.

7.2 Analogous remarks apply to the procedure of ascertaining the *"meaning"* of given historical events; its scientific import consists in determining what other events are relevantly connected with the event in question, be it as "causes," or as "effects"; and the statement of the relevant connections assumes, again, the form of explanations or explanation sketches which involve universal hypotheses; this will be seen more clearly in the subsequent section.

7.3 In the historical explanation of some social institutions great emphasis is laid upon an analysis of the *development* of the institution up to the stage under consideration. Critics of this approach

---

[5] For a criticism of this kind of plausibility, cf. Zilsel, *l.c.*, pp. 577–578, and sections 7 and 8 in the same author's "Problems of Empiricism," in *International Encyclopedia of Unified Science*, Vol. II, 8.

have objected that a mere description of this kind is not a genuine explanation. This argument may be given a slightly different aspect in terms of the preceding reflections: A description of the development of an institution is obviously not simply a statement of *all* the events which temporally preceded it; only those events are meant to be included which are *"relevant"* to the formation of that institution. And whether an event is relevant to that development is not a question of the value attitude of the historian, but an objective question depending upon what is sometimes called a causal analysis of the rise of that institution.[6] Now, the causal analysis of an event consists in establishing an explanation for it, and since this requires reference to general hypotheses, so do assumptions about relevance, and, consequently, so does the adequate analysis of the historical development of an institution.

7.4 Similarly, the use of the notions of *determination* and of *dependence* in the empirical sciences, including history, involves reference to general laws.[7] Thus, e.g., we may say that the pressure of a gas depends upon its temperature and volume, or that temperature and volume determine the pressure, in virtue of Boyle's law. But unless the underlying laws are stated explicitly, the assertion of a relation of dependence or of determination between certain magnitudes or characteristics amounts at best to claiming that they are connected by some unspecified empirical law; and that is a very meager assertion indeed: If, for example, we know only that there is some empirical law connecting two metrical magnitudes (such as length and temperature of a metal bar), we can not even be sure that a change of one of the two will be accompanied by a change of the other (for the law may connect the same value of the "dependent" or "determined" magnitude with different values of the other), but only that with any specific value of one

---

[6] See the detailed and clear exposition of this point in M. Mandelbaum's book; l.c., Chs. 6–8.

[7] According to Mandelbaum, history, in contradistinction to the physical sciences, consists ''not in the formulation of laws of which the particular case is an instance, but in the description of the events in their actual determining relationships to each other; in seeing events as the products and producers of change'' (l.c., pp. 13–14). This is essentially a view whose untenability has been pointed out already by Hume; it is the belief that a careful examination of two specific events alone, without any reference to similar cases and to general regularities, can reveal that one of the events produces or determines the other. This thesis does not only run counter to the scientific meaning of the concept of determination which clearly rests on that of general law, but it even fails to provide any objective criteria which would be indicative of the intended relationship of determination or production. Thus, to speak of empirical determination independently of any reference to general laws means to use a metaphor without cognitive content.

of the variables, there will always be associated one and the same value of the other; and this is obviously much less than most authors mean to assert when they speak of determination or dependence in historical analysis.

Therefore, the sweeping assertion that economic (or geographic, or any other kind of) conditions "determine" the development and change of all other aspects of human society, has explanatory value only in so far as it can be substantiated by explicit laws which state just what kind of change in human culture will regularly follow upon specific changes in the economic (geographic, etc.) conditions. Only the establishment of concrete laws can fill the general thesis with scientific content, make it amenable to empirical tests, and confer upon it an explanatory function. The elaboration of such laws with as much precision as possible seems clearly to be the direction in which progress in scientific explanation and understanding has to be sought.

8. The considerations developed in this paper are entirely neutral with respect to the problem of *"specifically historical laws"*: neither do they presuppose a particular way of distinguishing historical from sociological and other laws, nor do they imply or deny the assumption that empirical laws can be found which are historical in some specific sense, and which are well confirmed by empirical evidence.

But it may be worth mentioning here that those universal hypotheses to which historians explicitly or tacitly refer in offering explanations, predictions, interpretations, judgments of relevance, etc., are taken from *various* fields of scientific research, in so far as they are not pre-scientific generalizations of everyday experiences. Many of the universal hypotheses underlying historical explanation, for instance, would commonly be classified as psychological, economical, sociological, and partly perhaps as historical laws; in addition, historical research has frequently to resort to general laws established in physics, chemistry, and biology. Thus, e.g., the explanation of the defeat of an army by reference to lack of food, adverse weather conditions, disease, and the like, is based on a—usually tacit—assumption of such laws. The use of tree rings in dating events in history rests on the application of certain biological regularities. Various methods of testing the authenticity of documents, paintings, coins, etc., make use of physical and chemical theories.

The last two examples illustrate another point which is relevant in this context: Even if a historian should propose to restrict his research to a *"pure description"* of the past, without any attempt at offering explanations, statements about relevance and determination, etc., he would continually have to make use of general laws. For the object of his studies would be the past—forever inaccessible to his direct examination. He would have to establish his knowledge by indirect methods: by the use of universal hypotheses which connect his present data with those past events. This fact has been obscured partly because some of the regularities involved are so familiar that they are not considered worth mentioning at all; and partly because of the habit of relegating the various hypotheses and theories which are used to ascertain knowledge about past events, to the "auxiliary sciences" of history. Quite probably, some of the historians who tend to minimize, if not to deny, the importance of general laws for history, are actuated by the feeling that only "genuinely historical laws" would be of interest for history. But once it is realized that the discovery of historical laws (in some specified sense of this very vague notion) would not make history methodologically autonomous and independent of the other branches of scientific research, it would seem that the problem of the existence of historical laws ought to lose some of its weight.

The remarks made in this section are but special illustrations of two broader principles of the theory of science: first, the separation of "pure description" and "hypothetical generalization and theory-construction" in empirical science is unwarranted; in the building of scientific knowledge the two are inseparably linked. And, second, it is similarly unwarranted and futile to attempt the demarcation of sharp boundary lines between the different fields of scientific research, and an autonomous development of each of the fields. The necessity, in historical inquiry, to make extensive use of universal hypotheses of which at least the overwhelming majority come from fields of research traditionally distinguished from history is just one of the aspects of what may be called the methodological unity of empirical science.

Carl G. Hempel.

Queens College,
New York.

ADOLF GRÜNBAUM

Historical Determinism, Social Activism, and Predictions in the Social Sciences

IN a recent issue of this *Journal*, M. Roshwald maintained that it was *self-contradictory* for Marx to combine a belief in the need for exhorting men to establish socialism with the thesis of the historical inevitability of its establishment.[1] Presumably, Roshwald's criticism of this feature of Marx's theory does not depend on the merits of Marx's particular prognosis of the career of industrialism, but derives from Roshwald's view that it is inconsistent for *any* deterministic socio-political theory to *advocate* a social activism with the aim of thereby bringing about a future state whose eventuation the theory in question regards as assured by historical causation. Since

[1] This *Journal*, November 1955, **6**, 191, n. 3. A similar contention is found in George L. Kline's 'A Philosophical Critique of Soviet Marxism', *The Review of Metaphysics*, 1955, **9**, 100.

236

Roshwald's thesis does not depend on whether the explanatory variables of the historical process are held to be economic, climatic, demographic, geopolitical, or the inscrutable will of God, it applies not only to the Marxian view but also, *mutatis mutandis*, to each of the following : (i) Augustine's (and Calvin's) belief in divine fore-ordination, when coupled with the advocacy of Christian virtue, (ii) the advocacy on the part of a British economist, who pays taxes in England, of the passage of a certain tax-law by Parliament for the purpose of assuring a specified income distribution, when combined with a deterministic *prediction* by him of the passage of that tax-law and of the ensuing desired income distribution, (iii) Justice Holmes' dictum that the inevitable comes to pass through effort.

In the present note, I wish to challenge (1) Roshwald's contention that historical determinism is logically incompatible with the advocacy of social activism ; in addition, I shall deal critically with the following claims : (2) Roshwald's assertions that ' An ethics, . . . , combined with a strictly deterministic philosophy, would have no practical significance as a motivating force ' and that determinism ' implies the practical futility of discussion about the best way to be chosen by men in forming their social relations ',[1] (3) Robert Merton's contention that the ' self-stultifying ' and ' self-fulfilling ' predictions encountered in the social sciences are ' peculiar to human affairs ' and are ' not found among predictions about the world of nature '.[2] The inclusion of a rebuttal of Merton's statement by means of counter-examples is prompted by the fact that his statement might otherwise be adduced in support of Roshwald's cardinal tenet that ' social science . . . seems as radically different from natural science as man is from inanimate objects '.[3]

(1) Although the predictions made by a contemporary (Marxian or anti-Marxian) historical determinist concerning the social organisation of industrial society and those made by our British economist pertain to a society of which these forecasters are members and are thus *self-referential*, they are made by social prophets who, *qua* deterministic forecasters, consider their own society *ab extra* rather than as active contributors to its destiny. But the predictions made from that theoretically external perspective are predicated on the prior fulfilment of certain initial conditions which include the presence in that society of men who are dissatisfied with the existing state of affairs and are therefore actively seeking the future realisation of the predicted social state. To ignore that the determinist rests his social prediction in part on the existence of the latter initial condition, just as much as a physicist makes a prediction of a thermal expansion conditional upon the presence of heat, is to commit the fallacy of equating determinism with

---

[1] Roshwald, op. cit., p. 192

[2] Robert K. Merton, *Social Theory and Social Structure*, Glencoe, 1949, p. 122

[3] Roshwald, op. cit. pp. 202-203

fatalism.[1] Now, in actual fact, the social forecasters whom we are considering are not only spectatorial theoretical analysts of the society whose future they are predicting but, *qua* members of that society, also participate in the fashioning of its destiny. On what grounds then does Roshwald feel entitled to maintain that it is logically inconsistent for them, *qua* participating citizens, to advocate that action be taken by their fellow-citizens to create the social system whose advent they are predicting on the basis of their theory ? Is it not plain now that Roshwald's charge derives its semblance of plausibility from his confusion of determinism with fatalism in the context of self-referential predictions ?

(2) Roshwald's denial of the practical significance of an ethics which is coupled with determinism is tantamount to asserting that a belief in determinism is *causally* incompatible with the determinist's possession of the psychological incentives necessary for taking the action required to implement the moral directives of his ethics. This contention is both empirically false and inconsistent with Roshwald's own premises. For Abraham Lincoln's view that his own beliefs (ethical and other) were causally determined did not weaken in the least his desire to abolish Negro slavery, as demanded by his ethical theory, and similarly for Augustine, Calvin, Spinoza and a host of lesser men. Moreover, Roshwald's mention of ' practical significance as a motivating force ' shows that he is actually affirming a *causal* connection between two kinds of psychic states : a belief in determinism and indolent futilitarianism. But this thesis not only contradicts his correct observation that ' *psychologically* the necessity of the " final victory " moved the adherents of the [Marxist] creed to participate in its realisation and served [as] a powerful source of revolutionary activity ' [2] but constitutes a covert and unwitting invocation of psychological determinism to which Roshwald himself is hardly entitled.

Equally untenable is his claim that it is practically futile for determinists to weigh alternative modes of social organisation with a view to optimising their own social arrangements. For the determinist does *not* maintain, in fatalist fashion, that the future state of society is independent of the decisions which men make in response to (i) facts (both physical and social), (ii) their own *interpretation* of these facts (which, of course, is often false), and (iii) their value-objectives. It is precisely because, on the deterministic theory, human decisions *are* causally dependent upon these factors that deliberation concerning optimal courses of action and social arrangements can be reasonably expected to issue in successful action rather than lose its

---

[1] For a discussion of this fallacy and of related issues, see A. Grünbaum, ' Causality and the Science of Human Behavior ', *American Scientist*, 1952, 40, 671 [reprinted in Feigl and Brodbeck (eds.), *Readings in the Philosophy of Science*, New York, 1953, pp. 766-778] and ' Time and Entropy ', *American Scientist*, 1955, 43, 568-570.

[2] Roshwald, op. cit., pp. 191-192, n. 3

238

significance by adventitiousness. Roshwald's objection here springs from the false supposition that if our beliefs and decisions have causes, these causes force the beliefs in question upon us, against our better judgment, as it were, thus rendering the attempt to exercise that judgment futile.

(3) It remains to consider Robert Merton's interpretation of what John Venn has called ' suicidal prophecies ' in the social sciences. As an example of such a prophecy, Merton cites a government economist's distant forecast of an oversupply of wheat which, upon becoming publicly known, induces individual wheat growers so to curtail their initially planned production as to invalidate the economist's forecast.[1] Since failure to take cognisance of the possible ' perturbational ' influence of the dissemination of a social forecast may issue in that prediction's *spurious disconfirmation*, Merton rightly points out that the possibility that a social prophecy stultify itself by becoming known creates a problem for the reliable testing of social predictions. It would, of course, be an error to infer that the phenomenon of suicidal prophecies encountered in the social studies constitutes evidence against determinism, since the dissemination of these prophecies alters the initial conditions on which the forecasts were predicated and would not prevent another forecaster, at least in principle, from correctly predicting the outcome on the basis of the actual, modified initial conditions.[2] Merton does not commit *this* error. But he does make the equally vulnerable claim that self-stultifying and self-fulfilling prophecies are endemic to the domain of human affairs and are ' not found among predictions about the world of nature '. As evidence, he cites the fact that a ' meteorologist's prediction of continued rainfall has until now not perversely led to the occurrence of a drought ' and that ' predictions of the return of Halley's comet do not influence its orbit '.[3] To be sure, these particular predictions of purely physical phenomena are not self-stultifying any more than those social predictions whose success is essentially independent of whether they are made public or not. But instead of confining ourselves to commonplace meteorological and astronomical phenomena, consider the goal-directed behaviour of a servo-mechanism like a homing device which employs feedback and is subject to automatic fire control. Clearly every phase of the

[1] Merton, op. cit., p. 122

[2] The dissemination of the *adjusted* prediction may then, in turn, require that the *latter* forecast be revised to allow for the effects of *its* publication, and so on *ad infinitum*. Fortunately, successful prediction *is* possible nonetheless in such cases under very general conditions, as has recently been shown by E. Grunberg and F. Modigliani in an article ' The Predictability of Social Events ' [*The Journal of Political Economy* (U.S.A.), 1954, **62**, 465] to which Professor H. Feigl has kindly called my attention. These authors elaborate their thesis by reference to the public prediction at time t of the price of some commodity that will prevail at time $t + 1$ and then generalise this analysis of the one-variable case to cover the correct public prediction of the values of n variables. [3] ibid., p. 181

239

operation of such a device constitutes an exemplification of one or more purely *physical* principles. Yet the following situation is *allowed* by these very principles : a computer predicts that, in its present course, the missile will miss its target, and the communication of this information to the missile in the form of a new set of instructions induces it to alter its course and thereby to reach its target, contrary to the computer's original prediction. How does this differ, in principle, from the case in which the government economist's forecast of an oversupply of wheat has the effect of instructing the wheat growers to alter their original planting intentions ?

Corresponding remarks can be made concerning self-fulfilling prophecies, which Merton likewise believes to be confined to the realm of human affairs. Such prophecies are characterised by the fact that an *initially false* rumour is believed and thereby creates the conditions of its own fulfilment and spurious confirmation : when the depositors of a bank whose assets are initially entirely adequate give credence to a false rumour of insolvency, this belief can inspire a run on the bank with resulting actual insolvency. Merton shows illuminatingly how the pattern characteristic of self-fulfilling predictions serves to buttress prejudices against minority groups by the mechanism of spurious confirmation.[1]

Is there any analogue to such predictions in the domain of physical phenomena ? It is at least physically possible, though not very likely, that the reception of a false message (e.g. a false warning of an impending collision) will actuate a servo-mechanism so as to realise the very conditions which the originally false message predicted.

It would be unavailing to object that the counter-examples which I have adduced against Merton's contention involve physical artifacts which depend for their existence upon having been constructed by human beings. For the phenomena in question fall entirely within the purview of physical laws and thus invalidate the widely-held belief, espoused by Merton, that, in principle, self-stultifying and self-fulfilling prophecies are 'not found among predictions about the world of nature '.

ADOLF GRÜNBAUM[+]

[+] Currently, he is the Andrew Mellon Professor of Philosophy, Research Professor of Psychiatry, and Chairman of the Center for Philosophy of Science at the University of Pittsburgh.

HISTORICAL EXPLANATION IN THE SOCIAL SCIENCES *

J. W. N. WATKINS

1 *Introduction*

THE hope which originally inspired methodology was the hope of
finding a method of enquiry which would be both necessary and
sufficient to guide the scientist unerringly to truth. This hope has
died a natural death. Today, methodology has the more modest task
of establishing certain rules and requirements which are necessary to
prohibit some wrong-headed moves but insufficient to guarantee
success. These rules and requirements, which circumscribe scientific
enquiries without steering them in any specific direction, are of the
two main kinds, formal and material. So far as I can see, the formal
rules of scientific method (which comprise both logical rules and
certain realistic and fruitful stipulations) are equally applicable to all
the empirical sciences. You cannot, for example, deduce a universal
law from a finite number of observations whether you are a physicist,
a biologist, or an anthropologist. Again, a single comprehensive
explanation of a whole range of phenomena is preferable to isolated
explanations of each of those phenomena, whatever your field of
enquiry. I shall therefore confine myself to the more disputable (I had
nearly said ' more disreputable ') and metaphysically impregnated part
of methodology which tries to establish the appropriate *material*
requirements which the *contents* of the premisses of an explanatory
theory in a particular field ought to satisfy. These requirements may
be called regulative principles. Fundamental differences in the subject-
matters of different sciences—differences to which formal methodological rules are impervious—ought, presumably, to be reflected in the
regulative principles appropriate to each science. It is here that the
student of the methods of the social sciences may be expected to have
something distinctive to say.

---

* A revised version of a paper read at the First Annual Conference of the
Philosophy of Science Group, Manchester, on 23rd September 1956. Footnotes
have been added subsequently.

104

An example of a regulative principle is mechanism, a metaphysical theory which governed thinking in the physical sciences from the seventeenth century until it was largely superseded by a wave or field world-view. According to mechanism, the ultimate constituents of the physical world are impenetrable particles which obey simple mechanical laws. The existence of these particles cannot be explained —at any rate by science. On the other hand, every complex physical thing or event is the result of a particular configuration of particles and can be explained in terms of the laws governing their behaviour in conjunction with a description of their relative positions, masses, momenta, etc. There may be what might be described as unfinished or half-way explanations of large-scale phenomena (say, the pressure inside a gas-container) in terms of other large-scale factors (the volume and temperature of the gas) ; but we shall not have arrived at rock-bottom explanations of such large-scale phenomena until we have deduced their behaviour from statements about the properties and relations of particles.

This is a typically metaphysical idea (by which I intend nothing derogatory). True, it is confirmed, even massively confirmed, by the huge success of mechanical theories which conform to its require-ments. On the other hand, it is untestable. No experiment could overthrow it. If certain phenomena—say, electromagnetic pheno-mena—seem refractory to this mechanistic sort of explanation, this refractoriness can always (and perhaps rightly) be attributed to our inability to find a successful mechanical model rather than to an error in our metaphysical intuition about the ultimate constitution of the physical world. But while mechanism is weak enough to be compatible with any *observation* whatever, while it is an untestable and unempirical principle, it is strong enough to be incompatible with various conceiv-able physical *theories*. It is this which makes it a *regulative*, non-vacuous metaphysical principle. If it were compatible with everything it would regulate nothing. Some people complain that regulative principles discourage research in certain directions, but that is a part of their purpose. You cannot encourage research in one direction without discouraging research in rival directions.

I am not an advocate of mechanism but I have mentioned it because I am an advocate of an analogous principle in social science, the principle of methodological individualism.[1] According to this

[1] Both of these analogous principles go back at least to Epicurus. In recent times methodological individualism has been powerfully defended by Professor

principle, the ultimate constituents of the social world are individual people who act more or less appropriately in the light of their dispositions and understanding of their situation. Every complex social situation, institution, or event is the result of a particular configuration of individuals, their dispositions, situations, beliefs, and physical resources and environment. There may be unfinished or half-way explanations of large-scale social phenomena (say, inflation) in terms of other large-scale phenomena (say, full employment); but we shall not have arived at rock-bottom explanations of such large-scale phenomena until we have deduced an account of them from statements about the dispositions, beliefs, resources, and inter-relations of individuals. (The individuals may remain anonymous and only typical dispositions, etc., may be attributed to them.) And just as mechanism is contrasted with the organicist idea of physical fields, so methodological individualism is contrasted with sociological holism or organicism. On this latter view, social systems constitute ' wholes ' at least in the sense that some of their large-scale behaviour is governed by macro-laws which are essentially *sociological* in the sense that they are *sui generis* and not to be explained as mere regularities or tendencies resulting from the behaviour of interacting individuals. On the contrary, the behaviour of individuals should (according to sociological holism) be explained at least partly in terms of such laws (perhaps in conjunction with an account, first of individuals' rôles within institutions and secondly of the functions of institutions within the whole social system). If methodological individualism means that human beings are supposed to be the only moving agents in history, and if sociological holism means that some superhuman agents or factors are supposed to be at work in history, then these two alternatives are exhaustive. An example of such a superhuman, sociological factor is the alleged long-term cyclical wave in economic life which is supposed to be self-propelling, uncontrollable, and inexplicable in terms of human activity, but in terms of the fluctuations of which such large-scale phenomena as wars, revolutions, and mass emigration, and such

---

F. A. Hayek in his *Individualism and Economic Order* and *The Counter-Revolution of Science*, and by Professor K. R. Popper in his *The Open Society and its Enemies* and ' The Poverty of Historicism ' *Economica*, 1944-45, 11-12. Following in their footsteps I have also attempted to defend methodological individualism in ' Ideal Types and Historical Explanation ' this *Journal*, 1952, 3, 22, reprinted in *Readings in the Philosophy of Science*, ed. Feigl and Brodbeck, New York, 1953. This article has come in for a good deal of criticism, the chief items of which I shall try to rebut in what follows.

106

psychological factors as scientific and technological inventiveness can, it is claimed, be explained and predicted.

I say 'and predicted' because the irreducible sociological laws postulated by holists are usually regarded by them as laws of social development, as laws governing the dynamics of a society. This makes holism well-nigh equivalent to historicism, to the idea that a society is impelled along a pre-determined route by historical laws which cannot be resisted but which can be discerned by the sociologist. The holist-historicist position has, in my view, been irretrievably damaged by Popper's attacks on it. I shall criticise this position only in so far as this will help me to elucidate and defend the individualistic alternative to it. The central assumption of the individualistic position —an assumption which is admittedly counter-factual and meta-physical—is that no social tendency exists which could not be altered *if* the individuals concerned both wanted to alter it and possessed the appropriate information. (They might want to alter the tendency but, through ignorance of the facts and/or failure to work out some of the implications of their action, fail to alter it, or perhaps even intensify it.) This assumption could also be expressed by saying that no social tendency is somehow imposed on human beings 'from above' (or 'from below')—social tendencies are the product (usually undesigned) of human characteristics and activities and situations, of people's ignorance and laziness as well as of their knowledge and ambition. (An example of a social tendency is the tendency of industrial units to grow larger. I do not call 'social' those tendencies which are deter-mined by uncontrollable physical factors, such as the alleged tendency for more male babies to be born in times of disease or war.) [1]

---

[1] The issue of holism *versus* individualism in social science has recently been presented as though it were a question of the existence or non-existence of irre-ducibly social *facts* rather than of irreducibly sociological *laws*. (See M. Mandelbaum, 'Societal Facts', *The British Journal of Sociology*, 1955, 6, and E. A. Gellner, 'Explanations in History', *Aristotelian Society*, Supplementary Volume 30, 1956.) This way of presenting the issue seems to me to empty it of most of its interest. If a new kind of beast is discovered, what we want to know is not so much whether it falls outside existing zoological categories, but how it behaves. People who insist on the existence of social facts but who do not say whether they are governed by sociological laws, are like people who claim to have discovered an unclassified kind of animal but who do not tell us whether it is tame or dangerous, whether it can be domesticated or is unmanageable. If an answer to the question of social facts could throw light on the serious and interesting question of sociological laws, then the question of social facts would also be serious and interesting. But this is not so. On the one hand, a holist may readily admit (as I pointed out in my ' Ideal

107

My procedure will be : first, to de-limit the sphere in which methodological indivisualism works in two directions ; secondly, to clear methodological individualism of certain misunderstandings ; thirdly, to indicate how fruitful and surprising individualistic explanations can be and how individualistic social theories can lead to sociological discoveries ; and fourthly, to consider in somewhat more detail how, according to methodological individualism, we should frame explanations, first for social regularities or repeatable processes, and secondly for unique historical constellations of events.

2 Where Methodological Individualism Does not Work

There are two areas in which methodological individualism does not work.

The first is a probability situation where accidental and unpredictable irregularities in human behaviour have a fairly regular and predictable overall result.[1] Suppose I successively place 1,000 individuals facing north in the centre of a symmetrical room with two exits, one east, the other west. If about 500 leave by one exit and about 500 by the other I would not try to explain this in terms of tiny undetectable west-inclining and east-inclining differences in the individuals, for the same reason that Popper would not try to explain

Types ' paper, which Gellner criticises) that all observable social facts *are* reducible to individual facts and yet hold that the latter are invisibly governed by irreducibly sociological laws. On the other hand, an individualist may readily admit (as Gellner himself says) that some large social facts are simply too complex for a full reduction of them to be feasible, and yet hold that individualistic explanations of them are in principle possible, just as a physicist may readily admit that some physical facts (for instance, the precise blast-effects of a bomb-explosion in a built-up area) are just too complex for accurate prediction or explanation of them to be feasible and yet hold that precise explanations and predictions of them in terms of existing scientific laws are in principle possible.

This revised way of presenting the holism *versus* individualism issue does not only divert attention from the important question. It also tends to turn the dispute into a purely verbal issue. Thus Mandelbaum is able to prove the existence of what he calls ' societal facts ' because he defines psychological facts very narrowly as ' facts concerning the thoughts and actions of specific human beings ' (op. cit. p. 307). Consequently, the *dispositions* of *anonymous* individuals which play such an important rôle in individualistic explanations in social science are ' societal facts ' merely by definition.

[1] Failure to exclude probability-situations from the ambit of methodological individualism was an important defect of my ' Ideal Types ' paper. Here, Gellner's criticism (op. cit. p. 163) does hit the nail on the head.

108

the fact that about 500 balls will topple over to the west and about 500 to the east, if 1,000 balls are dropped from immediately above a north-south blade, in terms of tiny undetectable west-inclining and east-inclining differences in the balls. For in both cases such an ' explanation ' would merely raise the further problem : why should these west-inclining and east-inclining differences be distributed approximately *equally* among the individuals and among the balls ?

Those statistical regularities in social life which are inexplicable in individualistic terms for the sort of reason I have quoted here are, in a sense, inhuman, the outcome of a large number of sheer *accidents*. The outcome of a large number of decisions is usually much less regular and predictable because variable human factors (changes of taste, new ideas, swings from optimism to pessimism) which have little or no influence on accident-rates are influential here. Thus Stock Exchange prices fluctuate widely from year to year, whereas the number of road-accidents does not fluctuate widely. But the existence of these actuarial regularities does not, as has often been alleged, support the historicist idea that defenceless individuals like you and me are at the chance mercy of the inhuman and uncontrollable tendencies of our society. It does not support a secularised version of the Calvinist idea of an Almighty Providence who picks people at random to fill His fixed damnation-quota. For we can control these statistical regularities in so far as we can alter the conditions on which they depend. For example, we could obviously abolish road-accidents if we were prepared to prohibit motor-traffic.

The second kind of social phenomenon to which methodological individualism is inapplicable is where some kind of physical connection between people's nervous systems short-circuits their intelligent control and causes automatic, and perhaps in some sense appropriate, bodily responses. I think that a man may more or less literally smell danger and instinctively back away from unseen ambushers ; and individuality seems to be temporarily submerged beneath a collective physical *rapport* at jive-sessions and revivalist meetings and among panicking crowds. But I do not think that these spasmodic mob-organisms lend much support to holism or constitute a very serious exception to methodological individualism. They have a fleeting existence which ends when their members put on their mufflers and catch the bus or otherwise disperse, whereas holists have conceived of a social whole as something which endures through generations of men ; and whatever holds together typical long-lived institutions, like a bank or a legal

109

system or a church, it certainly is not the physical proximity of their members.

3 Misunderstandings of Methodological Individualism

I will now clear methodological individualism of two rather widespread misunderstandings.

It has been objected that in making individual dispositions and beliefs and situations the terminus of an explanation in social science, methodological individualism implies that a person's psychological make-up is, so to speak, God-given, whereas it is in fact conditioned by, and ought to be explained in terms of, his social inheritance and environment.[1] Now methodological individualism certainly does not prohibit attempts to explain the formation of psychological characteristics; it only requires that such explanations should in turn be *individualistic*, explaining the formation as the result of a series of conscious or unconscious responses by an individual to his changing situation. For example, I have heard Professor Paul Sweezey, the Harvard economist, explain that he became a Marxist because his father, a Wall Street broker, sent him in the 1930's to the London School of Economics to study under those staunch liberal economists, Professors Hayek and Robbins. This explanation is perfectly compatible with methodological individualism (though hardly compatible, I should have thought, with the Marxist idea that ideologies reflect class-positions) because it interprets his ideological development as a human response to his situation. It is, I suppose, psycho-analysts who have most systematically worked the idea of a thorough individualist and historical explanation of the formation of dispositions, unconscious fears and beliefs, and subsequent defence-mechanisms, in terms of responses to emotionally charged, and especially childhood, situations.

My point could be put by saying that methodological individualism encourages *innocent* explanations but forbids *sinister* explanations of the widespread existence of a disposition among the members of a social group. Let me illustrate this by quoting from a reply I made to Goldstein's criticisms.

[1] Thus Gellner writes : ' The real oddity of the reductionist [i.e. the methodological individualist's] case is that it seems to preclude *a priori* the possibility of human dispositions being the dependent variable in an historical explanation—when in fact they often or always are ' (op. cit. p. 165). And Mr Leon J. Goldstein says that in making human dispositions methodologically primary I ignore their cultural conditioning. (*The Journal of Philosophy*, 1956, 53, 807.)

Suppose that it is established that Huguenot traders were relatively prosperous in 17th-century France and that this is explained in terms of a wide-spread disposition among them (a disposition for which there is independent evidence) to plough back into their businesses a larger proportion of their profits than was customary among their Catholic competitors. Now this explanatory disposition might very well be explained in its turn—perhaps in terms of the general thriftiness which Calvinism is said to encourage, and/or in terms of the fewer alternative outlets for the cash resources of people whose religious disabilities prevented them from buying landed estates or political offices. (I cannot vouch for the historical accuracy of this example.)

I agree that methodological individualism allows the formation, or ' cultural conditioning ', of a widespread disposition to be explained only in terms of other human factors and not in terms of something *inhuman*, such as an alleged historicist law which impels people willynilly along some pre-determined course. But this is just the antihistoricist point of methodological individualism.

Unfortunately, it is typically a part of the programme of Marxist and other historicist sociologies to try to account for the formation of ideologies and other psychological characteristics in strictly sociological and non-psychological terms. Marx, for instance, professed to believe that feudal ideas and *bourgeois* ideas are more or less literally generated by the water-mill and the steam-engine. But no description, however complete, of the productive apparatus of a society, or of any other non-psychological factors, will enable you to deduce a single psychological conclusion from it, because psychological statements logically cannot be deduced from wholly non-psychological statements. Thus whereas the mechanistic idea that explanations in physics cannot go behind the impenetrable particles is a prejudice (though a very understandable prejudice), the analogous idea that an explanation which begins by imputing some social phenomenon to human factors cannot go on to explain those factors in terms of some inhuman determinant of them is a necessary truth. That the human mind develops under various influences the methodological individualist does not, of course, deny. He only insists that such development must be explained ' innocently ' as a series of responses by the individual to situations and not ' sinisterly ' and illogically as a direct causal outcome of non-psychological factors, whether these are neurological factors, or impersonal sociological factors alleged to be at work in history.

Another cause of complaint against methodological individualism

III

265

is that it has been confused with a narrow species of itself (Popper calls it 'psychologism') and even, on occasion, with a still narrower sub-species of this (Popper calls it the 'Conspiracy Theory of Society').[1] Psychologism says that all large-scale social characteristics are not merely the intended or unintended result of, but a *reflection* of, individual characteristics.[2] Thus Plato said that the character and make-up of a *polis* is a reflection of the character and make-up of the kind of soul predominant in it. The conspiracy theory says that all large-scale social phenomena (do not merely reflect individual characteristics but) are deliberately brought about by individuals or groups of individuals.

Now there are social phenomena, like mass unemployment, which it would not have been in anyone's interest deliberately to bring about and which do not appear to be large-scale social reflections or magnified duplicates of some individual characteristic. The practical or technological or therapeutic importance of social science largely consists in explaining, and thereby perhaps rendering politically manageable, the unintended and unfortunate consequences of the behaviour of inter-

---

[1] See K. R. Popper, *The Open Society and its Enemies*, 2nd edn., 1952, ch. 14

[2] I am at a loss to understand how Gellner came to make the following strange assertion : '. . . Popper refers to both " psychologism " which he condemns, and " methodological individualism ", which he commends. When in the articles discussed [i.e., my " Ideal Types " paper] " methodological individualism" is worked out more fully than is the case in Popper's book, it seems to me to be indistinguishable from " Psychologism ".' Finding no difference between methodological individualism and a caricature of methodological individualism, Gellner has no difficulty in poking fun at the whole idea : ' Certain tribes I know have what anthropologists call a segmentary patrilineal structure, which moreover maintains itself very well over time. I could " explain " this by saying that the tribesmen have, all or most of them, dispositions whose effect is to maintain the system. But, of course, not only have they never given the matter much thought, but it also might very well be impossible to isolate anything in the characters and conduct of the individual tribesmen which *explains* how they come to maintain the system' (op. cit. p. 176). Yet this example actually suggests the lines along which an individualistic explanation might be found. The very fact that the tribesmen *have never given the matter much thought*, the fact that they accept their inherited system uncritically, may constitute an important part of an explanation of its stability. The explanation might go on to pin-point certain rules—that is firm and widespread dispositions—about marriage, inheritance, etc., which help to regularise the tribesmen's behaviour towards their kinsmen. How they come to share these common dispositions could also be explained individualistically in the same sort of way that I can explain why my young children are already developing a typically English attitude towards policemen.

112

acting individuals.   From this pragmatic point of view, psychologism and the conspiracy theory are unrewarding doctrines.   Psychologism says that only a change of heart can put a stop to, for example, war (I think that this is Bertrand Russell's view).   The conspiracy theory, faced with a big bad social event, leads to a hunt for scape-goats.   But methodological individualism, by imputing unwanted social phenomena to individuals' responses to their situations, in the light of their dispositions and beliefs, suggests that we may be able to make the phenomena disappear, not by recruiting good men to fill the posts hitherto occupied by bad men, nor by trying to destroy men's socially unfortunate dispositions while fostering their socially beneficial dispositions, but simply by altering the situations they confront.   To give a current example, by confronting individuals with dearer money and reduced credit the Government may (I do not say will) succeed in halting inflation without requiring a new self-denying attitude on the part of consumers and without sending anyone to prison.

4 Factual Discoveries in Social Science

To explain the unintended but *beneficial* consequences of individual activities—by 'beneficial consequences' I mean social consequences which the individuals affected *would* endorse *if* they were called on to choose between their continuation or discontinuation—is usually a task of less practical urgency than the explanation of undesirable consequences.   On the other hand, this task may be of greater theoretical interest.   I say this because people who are painfully aware of the existence of unwanted social phenomena may be oblivious of the unintended but beneficial consequences of men's actions, rather as a man may be oblivious of the good health to which the smooth functioning of his digestion, nervous system, circulation, etc., give rise. Here, an explanatory social theory may surprise and enlighten us not only with regard to the connections between causes and effect but with regard to the existence of the effect itself.   By showing that a certain economic system contains positive feed-back leading to increasingly violent oscillations and crises an economist may explain a range of well-advertised phenomena which have long been the subject of strenuous political agitation.   But the economists who first showed that a certain kind of economic system contains negative feed-back which tends to iron out disturbances and restore equilibrium, not only explained, but also revealed the existence of, phenomena which had hardly been remarked upon before.[1]

I will speak of organic-like social behaviour where members of some social system (that is, a collection of people whose activities disturb and influence each other) mutually adjust themselves to the situations created by the others in a way which, without direction from above, conduces to the equilibrium or preservation or development of the system. (These are again evaluative notions, but they can also be given a 'would-be-endorsed-if' definition.) Now such far-flung organic-like behaviour, involving people widely separated in space and largely ignorant of each other, cannot be simply observed. It can only be theoretically reconstructed—by deducing the distant social consequences of the typical responses of a large number of interacting people to certain repetitive situations. This explains why individualistic-minded economists and anthropologists, who deny that societies really are organisms, have succeeded in piecing together a good deal of unsuspected organic-like social behaviour, from an examination of individual dispositions and situations, whereas sociological holists, who insist that societies really are organisms, have been noticeably unsuccessful in convincingly displaying any organic-like social behaviour—they cannot observe it and they do not try to reconstruct it individualistically.

There is a parallel between holism and psychologism which explains their common failure to make surprising discoveries. A large-scale social characteristic should be explained, according to psychologism, as the manifestation of analogous small-scale psychological tendencies in individuals, and according to holism as the manifestation of a large-scale tendency in the social whole. In both cases, the *explicans* does little more than duplicate the *explicandum*. The methodological individualist, on the other hand, will try to explain the large-

---

[1] This sentence, as I have since learnt from Dr A. W. Phillips, is unduly complacent, for it is very doubtful whether an economist can ever *show* that an economic system containing negative feed-back will be stable. For negative feed-back may produce either a tendency towards equilibrium, or increasing oscillations, according to the numerical values of the parameters of the system. But numerical values are just what economic measurements, which are usually ordinal rather than cardinal, seldom yield. The belief that a system which contains negative feed-back, but whose variables cannot be described quantitatively, is stable may be based on faith or experience, but it cannot be shown mathematically. See A. W. Phillips, 'Stabilisation Policy and the Time-Forms of Lagged Responses', *The Economic Journal*, 1957, 67.

114

scale effect as the *indirect*, unexpected, complex product of individual factors none of which, singly, may bear any resemblance to it at all. To use hackneyed examples, he may show that a longing for peace led, in a certain international situation, to war, or that a government's desire to improve a bad economic situation by balancing its budget only worsened the situation. Since Mandeville's *Fable of the Bees* was published in 1714, individualistic social science, with its emphasis on unintended consequences, has largely been a sophisticated elaboration on the simple theme that, in certain situations, selfish private motives may have good social consequences and good political intentions bad social consequences.[1]

Holists draw comfort from the example of biology, but I think that the parallel is really between the biologist and the methodological individualist. The biologist does not, I take it, explain the large changes which occur during, say, pregnancy, in terms of corresponding large teleological tendencies in the organism, but physically, in terms of small chemical, cellular, neurological, etc., changes, none of which bears any resemblance to their joint and seemingly planful outcome.

5 How Social Explanations Should be Framed

I will now consider how regularities in social life, such as the trade cycle, should be explained according to methodological individualism. The explanation should be in terms of individuals and their situations; and since the process to be explained is repeatable, liable to recur at various times and in various parts of the world, it follows that only very general assumptions about human dispositions can be employed in its explanation. It is no use looking to abnormal psychology for an explanation of the structure of interest-rates—everyday experience

[1] A good deal of unmerited opposition to methodological individualism seems to spring from the recognition of the undoubted fact that individuals often run into social obstacles. Thus the conclusion at which Mandelbaum arrives is ' that there are societal facts which exercise external constraints over individuals ' (op. cit. p. 317). This conclusion is perfectly harmonious with the methodological individualist's insistence that plans often miscarry (and that even when they do succeed, they almost invariably have other important and unanticipated effects). The methodological individualist only insists that the social environment by which any particular individual is confronted and frustrated and sometimes manipulated and occasionally destroyed is, if we ignore its physical ingredients, made up of other *people*, their habits, inertia, loyalties, rivalries, and so on. What the methodological individualist denies is that an individual is ever frustrated, manipulated or destroyed or borne along by irreducible sociological or historical *laws*.

115

must contain the raw material for the dispositional (as opposed to the situational) assumptions required by such an explanation. It may require a stroke of genius to detect, isolate, and formulate precisely the dispositional premisses of an explanation of a social regularity. These premisses may state what no one had noticed before, or give a sharp articulation to what had hitherto been loosely described. But once stated they will seem obvious enough. It took years of groping by brilliant minds before a precise formulation was found for the principle of diminishing marginal utility. But once stated, the principle—that the less, relatively, a man has of one divisible commodity the more compensation he will be disposed to require for foregoing a small fixed amount of it—is a principle to which pretty well everyone will give his consent. Yet this simple and almost platitudinous principle is the magic key to the economics of distribution and exchange.

The social scientist is, here, in a position analogous to that of the Cartesian mechanist.[1] The latter never set out to discover new and unheard-of physical principles because he believed that his own principle of action-by-contact was self-evidently ultimate. His problem was to discover the typical physical configurations, the mechanisms, which, operating according to this principle, produce the observed regularities of nature. His theories took the form of models which exhibited such regularities as the outcome of ' self-evident' physical principles operating in some hypothetical physical situation. Similarly, the social scientist does not make daring innovations in psychology but relies on familiar, almost ' self-evident ' psychological material. His skill consists, first in spotting the relevant dispositions, and secondly in inventing a simple but realistic model which shows how, in a precise type of situation, those dispositions generate some typical regularity or process. (His model, by the way, will also show that in this situation certain things cannot happen. His negative predictions of the form, ' If you've got this you can't have that as well ' may be of great practical importance.) The social scientist can now explain in principle historical examples of this regular process, provided his model does in fact fit the historical situation.

This view of the explanation of social regularities incidentally clears up the old question on which so much ink has been spilt about whether the so-called ' laws ' of economics apply universally or only to a particular ' stage ' of economic development. The simple answer is that the economic principles displayed by economists' models apply

[1] I owe this analogy to Professor Popper

116

only to those situations which correspond with their models ; but a single model may very well correspond with a very large number of historical situations widely separated in space and time.

In the explanation of regularities the same situational scheme or model is used to reconstruct a number of historical situations with a similar structure in a way which reveals how typical dispositions and beliefs of anonymous individuals generated, on each occasion, the same regularity.[1]   In the explanation of a unique constellation of events the individualistic method is again to reconstruct the historical situation, or connected sequence of situations, in a way which reveals how (usually both named and anonymous) individuals, with their beliefs and dispositions (which may include peculiar personal dispositions as well as typical human dispositions), generated, in this particular situation, the joint product to be explained.   I emphasise *dispositions*, which are open and law-like, as opposed to *decisions*, which are occurrences, for this reason.   A person's set of dispositions ought, under varying conditions, to give rise to appropriately varying descisions. The subsequent occurrence of an appropriate decision will both confirm, and be explained by, the existence of the dispositions.   Suppose that a historical explanation (of, say, the growth of the early Catholic Church) largely relies on a particular decision (say, the decision of Emperor Constantine to give Pope Silvester extensive temporal rights in Italy).   The explanation is, so far, rather *ad hoc :* an apparently arbitrary *fiat* plays a key rôle in it.   But if this decision can in turn be explained as the offspring of a marriage of a set of dispositions (for instance, the Emperor's disposition to subordinate all rival power to himself) to a set of circumstances (for instance, the Emperor's recognition that Christianity could not be crushed but could be tamed if it became the official religion of the Empire), and if the existence of these dispositions and circumstances is convincingly supported by independent evidence, then the area of the arbitrarily given, of sheer brute fact in history, although it can never be made to vanish, will have been significantly reduced.

The London School of Economics and Political Science
The University of London

---

[1] This should rebut Gellner's conclusion that methodological individualism would transform social scientists into ' biographers *en grande série* ' (op. cit. p. 176).

117

11. THE LOGIC OF

FUNCTIONAL ANALYSIS

1. INTRODUCTION

EMPIRICAL SCIENCE, in all its major branches, seeks not only to *describe* the phenomena in the world of our experience, but also to *explain* or *understand* them. While this is widely recognized, it is often held, however, that there exist fundamental differences between the explanatory *methods* appropriate to the different fields of empirical science. In the physical sciences, according to this view, all explanation is achieved ultimately by reference to causal or correlational antecedents; whereas in psychology and the social and historical disciplines—and, according to some, even in biology—the establishment of causal or correlational connections, while desirable and important, is not sufficient. Proper understanding of the phenomena studied in these fields is held to require other types of explanation.

One of the explanatory methods that have been developed for this purpose is that of functional analysis, which has found extensive use in biology, psychology sociology, and anthropology. This procedure raises problems of considerable interest for the comparative methodology of empirical science. The present essay is an attempt to clarify some of these problems; its object is to examine the logical structure of functional analysis and its explanatory and predictive significance by means of a confrontation with the principal characteristics of the explanatory procedures used in the physical sciences. We begin therefore with a brief examination of the latter.

This article is reprinted with some changes by permission from Llewellyn Gross, Editor, *Symposium on Sociological Theory.* New York: Harper & Row, 1959.

2. NOMOLOGICAL EXPLANATION: DEDUCTIVE AND INDUCTIVE

In a beaker filled to the brim with water at room temperature, there floats a chunk of ice which partly extends above the surface. As the ice gradually melts, one might expect the water in the beaker to overflow. Actually the water level remains unchanged. How is this to be explained? The key to an answer is provided by Archimedes' principle, according to which a solid body floating in a liquid displaces a volume of liquid which has the same weight as the body itself. Hence the chunk of ice has the same weight as the volume of water its submerged portion displaces. Since melting does not affect the weights involved, the water into which the ice turns has the same weight as the ice itself, and hence, the same weight as the water initially displaced by the submerged portion of the ice. Having the same weight, it also has the same volume as the displaced water; hence the melting ice yields a volume of water that suffices exactly to fill the space initially occupied by the submerged part of the ice. Therefore, the water level remains unchanged.

This account (which deliberately disregards certain effects of small magnitude) is an example of an argument intended to explain a given event. Like any explanatory argument, it falls into two parts, which will be called the *explanans* and the *explanandum*.[1] The latter is the statement, or set of statements, describing the phenomenon to be explained; the former is the statement, or set of statements, adduced to provide an explanation. In our illustration, the explanandum states that at the end of the process, the beaker contains only water, with its surface at the same level as at the beginning. To explain this, the explanans adduces, first of all, certain laws of physics; among them, Archimedes' principle; laws to the effect that at temperatures above 0°C. and atmospheric pressure, a body of ice turns into a body of water having the same weight; and the law that, at any fixed temperature and pressure, amounts of water that are equal in weight are also equal in volume.

---

1. These terms are given preference over the more familiar words 'explicans' and 'explicandum,' in order to reserve the latter for use in the context of philosophical explication in the technical sense proposed by R. Carnap; see, for example, his *Logical Foundations of Probability* (Chicago: University of Chicago Press, 1950), secs. 1-3. The terms 'explanans' and 'explanandum' were introduced, for this reason, in an earlier article: Carl G. Hempel and P. Oppenheim, "Studies in the Logic of Explanation," *Philosophy of Science*, 15 (1948), pp. 135-75 (reprinted in the present volume). While that article does not deal explicitly with inductive explanation, its first four sections contain various further considerations on deductive explanation that are relevant to the present study. For a careful critical examination of some points of detail discussed in the earlier article, such as especially the relation between explanation and prediction, see the essay by I. Scheffler, "Explanation, Prediction, and Abstraction," *The British Journal for the Philosophy of Science*, 7 (1957), pp. 293-309, which also contains some interesting comments bearing on functional analysis.

In addition to these laws, the explanans contains a second group of statements; these describe certain particular circumstances which, in the experiment, precede the outcome to be explained; such as the facts that at the beginning, there is a chunk of ice floating in a beaker filled with water; that the water is at room temperature; and that the beaker is surrounded by air at the same temperature and remains undisturbed until the end of the experiment.

The explanatory import of the whole argument lies in showing that the outcome described in the explanandum was to be expected in view of the antecedent circumstances and the general laws listed in the explanans. More precisely, the explanation may be construed as an argument in which the explanandum is deduced from the explanans. Our example then illustrates what we will call explanation by deductive subsumption under general laws, or briefly, *deductive-nomological explanation*. The general form of such an explanation is given by the following schema:

$$(2.1) \qquad \left. \begin{array}{c} L_1, L_2, \ldots, L_m \\ \\ C_1, C_2, \ldots, C_n \end{array} \right\} \text{Explanans}$$
$$\overline{\qquad\qquad E \qquad\qquad \text{Explanandum}}$$

Here, $L_1, L_2, \ldots, L_m$ are general laws and $C_1, C_2, \ldots, C_n$ are statements of particular fact; the horizontal line separating the conclusion E from the premises indicates that the former follows logically from the latter.

In our example, the phenomenon to be explained is a particular event that takes place at a certain place and time. But the method of deductive subsumption under general laws lends itself also to the explanation of what might be called "general facts" or uniformities, such as those expressed by laws of nature. For example, the question why Galileo's law holds for physical bodies falling freely near the earth's surface can be answered by showing that the law refers to a special case of accelerated motion under gravitational attraction, and that it can be deduced from the general laws for such motion (namely, Newton's laws of motion and of gravitation) by applying these to the special case where two bodies are involved, one of them the earth and the other the falling object, and where the distance between their centers of gravity equals the length of the earth's radius. Thus, an explanation of the regularities expressed by Galileo's law can be achieved by deducing the latter from the Newtonian laws and from statements specifying the mass and the radius of the earth; the latter two yield the value of the constant acceleration of free fall near the earth.

It might be helpful to mention one further illustration of the role of deductive-nomological explanation in accounting for particular facts as well as for general uniformities or laws. The occurrence of a rainbow on a given occasion can be deductively explained by reference to (1) certain particular determining conditions, such as the presence of raindrops in the air, sunlight falling on these drops, the observer facing away from the sun, etc., and (2) certain general laws, especially those of optical reflection, refraction, and dispersion. The fact that these laws hold can be explained in turn by deduction from the more comprehensive principles of, say, the electromagnetic theory of light.

Thus, the method of deductive-nomological explanation accounts for a particular event by subsuming it under general laws in the manner represented by the schema (2.1); and it can similarly serve to explain the fact that a given law holds by showing that the latter is subsumable, in the same fashion, under more comprehensive laws or theoretical principles. In fact, one of the main objectives of a theory (such as, say, the electromagnetic theory of light) is precisely to provide a set of principles—often expressed in terms of "hypothetical," not directly observable, entities (such as electric and magnetic field vectors) —which will deductively account for a group of antecedently established "empirical generalizations" (such as the laws of rectilinear propagation, reflection, and refraction of light). Frequently, a theoretical explanation will show that the empirical generalizations hold only approximately. For example, the application of Newtonian theory to free fall near the earth yields a law that is like Galileo's except that the acceleration of the fall is seen not to be strictly constant, but to vary slightly with geographical location, altitude above sea level, and certain other factors.

The general laws or theoretical principles that serve to account for empirical generalizations may in turn be deductively subsumable under even more comprehensive principles; for example, Newton's theory of gravitation can be subsumed, as an approximation, under that of the general theory of relativity. Obviously, this explanatory hierarchy has to end at some point. Thus, at any time in the development of empirical science, there will be certain facts which, at that time, are not explainable; these include the most comprehensive general laws and theoretical principles then known and, of course, many empirical generalizations and particular facts for which no explanatory principles are available at the time. But this does not imply that certain facts are intrinsically unexplainable and thus must remain unexplained forever: any particular fact as yet unexplainable, and any general principle, however comprehensive, may subsequently be found to be explainable by subsumption under even more inclusive principles.

Causal explanation is a special type of deductive nomological explanation; for a certain event or set of events can be said to have caused a specified "effect" only if there are general laws connecting the former with the latter in such a way that, given a description of the antecedent events, the occurrence of the effect can be deduced with the help of the laws. For example, the explanation of the lengthening of a given iron bar as having been caused by an increase in its temperature amounts to an argument of the form (2.1) whose explanans includes (*a*) statements specifying the initial length of the bar and indicating that the bar is made of iron and that its temperature was raised, (*b*) a law pertaining to the increase in the length of any iron bar with rising temperature.[2]

Not every deductive-nomological explanation is a causal explanation, however. For example, the regularities expressed by Newton's laws of motion and of gravitation cannot properly be said to *cause* the free fall of bodies near the earth's surface to satisfy Galileo's laws.

Now we must consider another type of explanation, which again accounts for a given phenomenon by reference to general laws, but in a manner which does not fit the deductive pattern (2.1). When little Henry catches the mumps, this might be explained by pointing out that he contracted the diesase from a friend with whom he played for several hours just a day before the latter was confined with a severe case of mumps. The particular antecedent factors here invoked are Henry's exposure and, let us assume, the fact that Henry had not had the mumps before. But to connect these with the event to be explained, we cannot adduce a general law to the effect that under the conditions just mentioned, the exposed person invariably contracts the mumps: what can be asserted is only that the disease will be transmitted with high statistical probability. Again, when a neurotic trait in an adult is psychoanalytically explained by reference to critical childhood experiences, the argument explicitly or implicitly claims that the case at hand is but an exemplification of certain general laws governing the development of neuroses. But surely, whatever specific laws of this kind might be adduced at present can purport, at the very best, to express probabilistic trends rather than deterministic uniformities: they may be construed as *laws of statistical form*, or briefly as *statistical laws*, to the effect that, given the childhood experiences in question—plus, presumably, certain particular environmental conditions in later life—there is such and such a statistical probability that a specified kind of neurosis will develop. Such

---

2. An explanation by means of laws which are causal in the technical sense of theoretical physics also has the form (2.1) of a deductive-nomological explanation. In this case, the laws invoked must meet certain conditions as to mathematical form, and $C_1, C_2, \ldots, C_n$ express so-called boundary conditions. For a fuller account of the concepts of causal law and of causality as understood in theoretical physics, see, for example, H. Margenau, *The Nature of Physical Reality* (New York: McGraw-Hill Book Company, Inc., 1950), Chapter 19; or Ph. Frank, *Philosophy of Science* (Englewood Cliffs, N. J.: Prentice-Hall, Inc., 1957), Chapters 11, 12.

statistical laws differ in form from strictly universal laws of the kind mentioned in our earlier examples of explanatory arguments. In the simplest case, a *law of strictly universal form*, or briefly, a *universal law*, is a statement to the effect that in *all* cases satisfying certain antecedent conditions A (e.g., heating of a gas under constant pressure), an event of a specified kind B (e.g., an increase in the volume of the gas) will occur; whereas a law of statistical form asserts that the probability for conditions A to be accompanied by an event of kind B has some specific value p.

Explanatory arguments which, in the manner just illustrated, account for a phenomenon by reference to statistical laws are not of the strictly deductive type (2.1). For example, the explanans consisting of information about Henry's exposure to the mumps and of a statistical law about the transmission of this disease does not logically imply the conclusion that Henry catches the mumps; it does not make that conclusion necessary, but, as we might say, more or less probable, depending upon the probability specified by the statistical laws. An argument of this kind, then, accounts for a phenomenon by showing that its occurrence is highly probable in view of certain particular facts and statistical laws specified in the explanans. An account of this type will be called an *explanation by inductive subsumption under statistical laws*, or briefly, an *inductive explanation*.

Closer analysis shows that inductive explanation differs from its deductive counterpart in several important respects;[3] but for the purposes of the following discussion, our sketchy account of explanation by statistical laws will suffice.

The two types of explanation we have distinguished will both be said to be varieties of *nomological explanation;* for either of them accounts for a given phenomenon by "subsuming it under laws," i.e., by showing that its occurrence could have been inferred—either deductively or with a high probability—by applying certain laws of universal or of statistical form to specified antecedent circumstances. Thus, a nomological explanation shows that we might in fact have *predicted* the phenomenon at hand, either deductively or with a high probability, if, at an earlier time, we had taken cognizance of the facts stated in the explanans.

But the predictive power of a nomological explanation goes much farther than this: precisely because its explanans contains general laws, it permits

---

3. For details, see section 3 of the essay "Aspects of Scientific Explanation" in this volume. Some stimulating comments on explanation by means of statistical laws will be found in S. E. Gluck, "Do Statistical Laws Have Explanatory Efficacy?" *Philosophy of Science*, 22 (1955), 34–38. For a much fuller analysis of the logic of statistical inference, see R. B. Braithwaite, *Scientific Explanation* (Cambridge: Cambridge University Press, 1953), chapters V, VI, VII. For a study of the logic of inductive inference in general, Carnap's *Logical Foundations of Probability, op. cit.,* is of great importance.

predictions concerning occurrences other than that referred to in the explanandum. In fact, such predictions provide a means of testing the empirical soundness of the explanans. For example, the laws invoked in a deductive explanation of the form (2.1) imply that the kind of event described in E will recur whenever and wherever circumstances of the kind described by C_1, $C_2, \ldots, C_n$ are realized; e.g., when the experiment with ice floating in water is repeated, the outcome will be the same. In addition, the laws will yield predictions as to what is going to happen under certain specifiable conditions which differ from those mentioned in $C_1, C_2, \ldots, C_n$. For example, the laws invoked in our illustration also yield the prediction that if a chunk of ice were floating in a beaker filled to the brim with concentrated brine, which has a greater specific gravity than water, some of the liquid would overflow as the ice was melting. Again, the Newtonian laws of motion and of gravitation, which may be used to explain various aspects of planetary motion, have predictive consequences for a variety of totally different phenomena, such as free fall near the earth, the motion of a pendulum, the tides, and many others.

This kind of account of further phenomena which is made possible by a nomological explanation is not limited to future events; it may refer to the past as well. For example, given certain information about the present locations and velocities of the celestial bodies involved, the principles of Newtonian mechanics and of optics yield not only predictions about future solar and lunar eclipses, but also "postdictions," or "retrodictions," about past ones. Analogously, the statistical laws of radioactive decay, which can function in various kinds of predictions, also lend themselves to retrodictive use; for example, in the dating, by means of the radiocarbon method, of a bow or an ax handle found in an archaeological site.

A proposed explanation is scientifically acceptable only if its explanans is capable of empirical test, i.e., roughly speaking, if it is possible to infer from it certain statements whose truth can be checked by means of suitable observational or experimental procedures. The predictive and postdictive implications of the laws invoked in a nomological explanation clearly afford an opportunity for empirical tests; the more extensive and varied the set of implications that have been borne out by empirical investigation, the better established will be the explanatory principles in question.

3. THE BASIC PATTERN OF FUNCTIONAL ANALYSIS

Historically speaking, functional analysis is a modification of teleological explanation, i.e., of explanation not by reference to causes which "bring about" the event in question, but by reference to ends which determine its course. Intuitively, it seems quite plausible that a teleological approach might be required for an adequate understanding of purposive and other goal-directed behavior; and teleological explanation has always had its advocates in this context. The trouble with the idea is that in its more traditional forms, it fails to meet the minimum scientific requirement of empirical testability. The neovitalistic idea of entelechy or of vital force is a case in point. It is meant to provide an explanation for various characteristically biological phenomena, such as regeneration and regulation, which according to neovitalism cannot be explained by physical and chemical laws alone. Entelechies are conceived as goal-directed nonphysical agents which affect the course of physiological events in such a way as to restore an organism to a more or less normal state after a disturbance has occurred. However, this conception is stated in essentially metaphorical terms: no testable set of statements is provided (i) to specify the circumstances in which an entelechy will supervene as an agent directing the course of events otherwise governed by physical and chemical laws, and (ii) to indicate precisely what observable effects the action of an entelechy will have in such a case. And since neovitalism thus fails to state general laws as to when and how entelechies act, it cannot explain any biological phenomena; it can give us no grounds to expect a given phenomenon, no reasons to say: "Now we see that the phenomenon had to occur." It yields neither predictions nor retrodictions: the attribution of a biological phenomenon to the supervention of an entelechy has no testable implications at all. This theoretical defect can be thrown into relief by contrasting the idea of entelechy with that of a magnetic field generated by an electric current, which may be invoked to explain the deflection of a magnetic needle. A magnetic field is not directly observable any more than an entelechy; but the concept is governed by strictly specifiable laws concerning the strength and direction, at any point, of the magnetic field produced by a current flowing through a given wire, and by other laws determining the effect of such a field upon a magnetic needle in the magnetic field on the earth. And it is these laws which, by their predictive and retrodictive import, confer explanatory power upon the concept of magnetic field. Teleological accounts referring to entelechies are thus seen to be pseudo-explanations. Functional analysis, as will be seen, though often formulated in teleological terms, need not appeal to such problematic entities and has a definitely empirical core.

The kind of phenomenon that a functional analysis[4] is invoked to explain

---

4. For the account of functional analysis presented in this section, I have obtained much stimulation and information from the illuminating essay "Manifest and Latent Functions" in R. K. Merton's book, *Social Theory and Social Structure* (New York: The Free Press; revised and enlarged edition, 1957), 19–84. Each of the passages from this work which is referred to in the present essay may also be found in the first edition (1949), on a page with approximately the same number.

is typically some recurrent activity or some behavior pattern in an individual or a group, such as a physiological mechanism, a neurotic trait, a culture pattern or a social institution. And the principal objective of the analysis is to exhibit the contribution which the behavior pattern makes to the preservation or the development of the individual or the group in which it occurs. Thus, functional analysis seeks to understand a behavior pattern or a sociocultural institution by determining the role it plays in keeping the given system in proper working order or maintaining it as a going concern.

By way of a simple and schematized illustration, consider first the statement:

(3.1) The heartbeat in vertebrates has the function of circulating blood through the organism.

Before examining the possibilities of its explanatory use, we should ask ourselves: What does the statement *mean*? What is being asserted by this attribution of function? It might be held that all the information conveyed by a sentence such as (3.1) can be expressed just as well by substituting the word "effect" for the word "function." But this construal would oblige us to assent also to the statement:

(3.2) The heartbeat has the function of producing heart sounds; for the heartbeat has that effect.

Yet a proponent of functional analysis would refuse to assert (3.2), on the ground that heart sounds are an effect of the heartbeat which is of no importance to the functioning of the organism; whereas the circulation of the blood effects the transportation of nutriment to, and the removal of waste from, various parts of the organism—a process that is indispensable if the organism is to remain in proper working order, and indeed if it is to stay alive. Thus understood, the import of the functional statement (3.1) might be summarized as follows:

(3.3) The heartbeat has the effect of circulating the blood, and this ensures the satisfaction of certain conditions (supply of nutriment and removal of waste) which are necessary for the proper working of the organism.

We should notice next that the heart will perform the function here attributed to it only if certain conditions are met by the organism and by its environment. For example, circulation will fail if there is a rupture of the aorta; the blood can carry oxygen only if the environment affords an adequate supply of oxygen and the lungs are in proper condition; it will remove certain kinds of waste only if the kidneys are reasonably healthy; and so forth. Most of the conditions that would have to be specified here are usually left unmentioned, partly no doubt because they are assumed to be satisfied as a matter of course in situations in which the organism normally finds itself. But in part, the omission reflects lack of relevant knowledge, for an explicit specification of the relevant conditions would require a theory in which (*a*) the possible states of organisms and of their environments could be characterized by the values of certain physicochemical or perhaps biological "variables of state," and in which (b) the fundamental theoretical principles would permit the determination of that range of internal and external conditions within which the pulsations of the heart would perform the function referred to above.[5] At present, a general theory of this kind, or even one that could deal in this fashion with some particular class of organisms, is unavailable, of course.

Also, a full restatement of (3.1) in the manner of (3.3) calls for criteria of what constitutes "proper working," "normal functioning," and the like, of the organism at hand; for the function of a given trait is here construed in terms of its causal relevance to the satisfaction of certain necessary conditions of proper working or survival of the organism. Here again, the requisite criteria are often left unspecified—an aspect of functional analysis whose serious implications will be considered later (in section 5).

The considerations here outlined suggest the following schematic characterization of a functional analysis:

(3.4) *Basic pattern of a functional analysis:* The object of the analysis is some "item" i, which is a relatively persistent trait or disposition (e.g., the beating of the heart) occurring in a system s (e.g., the body of a living vertebrate); and the analysis aims to show that s is in a state, or internal condition, c_i and in an environment representing certain external conditions c_e such that under conditions c_i and c_e (jointly to be referred to as c) the trait i has effects which satisfy some "need" or "functional requirement" of s, i.e., a condition n which is necessary for the system's remaining in adequate, or effective, or proper, working order.

Let us briefly consider some examples of this type of analysis in psychology and in sociological and anthropological studies. In psychology, it is especially psychoanalysis which shows a strong functional orientation. One clear instance is Freud's functional characterization of the role of symptom formation. In *The Problem of Anxiety*, Freud expresses himself as favoring a conception according to which "all symptom formation would be brought about solely in order to avoid anxiety; the symptoms bind the psychic energy which otherwise would be discharged as anxiety."[6] In support of this view, Freud points out that if an agoraphobic who has usually been accompanied when going out is left alone in the street, he will suffer an attack of anxiety, as will the compulsion neurotic, who, having touched something, is prevented from washing

5. For a fuller statement and further development of this point, see the essay "A Formalization of Functionalism" in E. Nagel, *Logic Without Metaphysics* (New York; The Free Press, 1957), 247-83. Part I of that study offers a detailed analysis of Mertons' essay mentioned in Note 4.

6. S. Freud, *The Problem of Anxiety* (Transl. by H. A. Bunker. New York: Psychoanalytic Quarterly Press, and W. W. Norton & Company, Inc., 1936), p. 111.

his hands. "It is clear, therefore, that the stipulation of being accompanied and the compulsion to wash has as their purpose, and also their result, the averting of an outbreak of anxiety."[7] In this account, which is put in strongly teleological terms, the system *s* is the individual under consideration; *i* his agoraphobic or compulsive behavior pattern; *n* the binding of anxiety, which is necessary to avert a serious psychological crisis that would make it impossible for the individual to function adequately.

In anthropology and sociology the object of functional analysis is, in Merton's words, "a *standardized* (i.e., patterned and repetitive) item, such as social roles, institutional patterns, social processes, cultural pattern, culturally patterned emotions, social norms, group organization, social structure, devices for social control, *etc.*"[8] Here, as in psychology and biology, the function, i.e., the stabilizing or adjusting effect, of the item under study may be one not consciously sought (and indeed, it might not even be consciously recognized) by the agents; in this case, Merton speaks of *latent* functions—in contradistinction to *manifest* functions, i.e., those stabilizing objective effects which are intended by participants in the system.[9] Thus, e.g., the rain-making ceremonials of the Hopi fail to achieve their manifest meteorological objective, but they "may fulfill the latent function of reinforcing the group identity by providing a periodic occasion on which the scattered members of a group assemble to engage in a common activity."[10]

Radcliffe-Brown's functional analysis of the totemic rites of certain Australian tribes illustrates the same point:

> To discover the social function of the totemic rites we have to consider the whole body of cosmological ideas of which each rite is a partial expression. I believe that it is possible to show that the social structure of an Australian tribe is connected in a very special way with these cosmological ideas and that the maintenance of its continuity depends on keeping them alive, by their regular expression in myth and rite.
>
> Thus, any satisfactory study of the totemic rites of Australia must be based not simply on the consideration of their ostensible purpose . . . , but on the discovery of their meaning and of their social function.[11]

---

7. *Ibid.*, p. 112.

8. Merton, *op. cit.*, p. 50 (Author's italics).

9. *Ibid.*, p. 51. Merton defines manifest functions as those which are both intended and recognized, and latent functions as those which are neither intended nor recognized. But this characterization allows for functions which are neither manifest nor latent; e.g., those which are recognized though not intended. It would seem to be more in keeping with Merton's intentions, therefore, to base the distinction simply on whether or not the stabilizing effect of the given item was deliberately sought.

10. *Ibid.*, pp. 64–65.

11. A. R. Radcliffe-Brown, *Structure and Function in Primitive Society* (London: Cohen and West Ltd., 1952), 145.

Malinowski attributes important latent functions to religion and to magic: he argues that religious faith establishes and enhances mental attitudes such as reverence for tradition, harmony with environment, and confidence and courage in critical situations and at the prospect of death—attitudes which, embodied and maintained by cult and ceremonial, have "an immense biological value." He points out that magic, by providing man with certain ready-made rituals, techniques, and beliefs, enables him "to maintain his poise and his mental integrity in fits of anger, in the throes of hate, of unrequited love, of despair and anxiety. The function of magic is to ritualize man's optimism, to enhance his faith in the victory of hope over fear."[12]

There will soon be occasion to add to the preceding examples from psychoanalysis and anthropology some instances of functional analysis in sociology. To illustrate the general character of the procedure, however, the cases mentioned so far will suffice: they all exhibit the basic pattern outlined in (3.4). From our examination of the form of functional analysis we now turn to an appraisal of its significance as a mode of explanation.

4. THE EXPLANATORY IMPORT OF FUNCTIONAL ANALYSIS

Functional analysis is widely considered as achieving an *explanation* of the "items" whose functions it studies. Malinowski, for example, says of the functional analysis of culture that it "aims at the explanation of anthropological facts at all levels of development by their function . . ."[13] and he adds, in the same context: "To explain any item of culture, material or moral, means to indicate its functional place within an institution, . . ."[14] At another place, Malinowski speaks of the "functional explanation of art, recreation, and public ceremonials."[15]

Radcliffe-Brown, too, considers functional analysis as an explanatory

12. B. Malinowski, *Magic, Science and Religion, and Other Essays* (Garden City, N.Y.: Doubleday Anchor Books, 1954), p. 90. For an illuminating comparison of Malinowski's views on the functions of magic and religion with those advanced by Radcliffe-Brown, see G. C. Homans, *The Human Group* (New York: Harcourt, Brace & World, Inc., 1950), 321 ff. (Note also Homan's general comments on "the functional theory," *ibid.*, pp. 268–72.) This issue and other aspects of functional analysis in anthropology are critically examined in the following article, which confronts some specific applications of the method with programmatic declarations by its proponents: Leon J. Goldstein, "The Logic of Explanation in Malinowskian Anthropology," *Philosophy of Science*, 24 (1957), 156–66.

13. B. Malinowski, "Anthropology," *Encyclopaedia Britannica*, First Supplementary volume (London and New York: The Encyclopaedia Britannica, Inc., 1926), 132.

14. *Ibid.*, p. 139.

15. B. Malinowski, *A Scientific Theory of Culture, and Other Essays* (Chapel Hill: University of North Carolina Press, 1944), 174.

method, though not as the only one suited for the social sciences: "Similarly one 'explanation' of a social system will be its history, where we know it—the detailed account of how it came to be what it is and where it is. Another 'explanation' of the same system is obtained by showing (as the functionalists attempt to do) that is it a special exemplification of laws of social physiology or social functioning. The two kinds of explanation do not conflict, but supplement one another."[16]

Apart from illustrating the attribution of explanatory import to functional analysis, this passage is of interest because it stresses that a functional analysis has to rely on general laws. This is shown also in our schematic characterization (3.4): the statements that i, in the specified setting c, has effects that satisfy n, and that n is a necessary condition for the proper functioning of the system, both involve general laws. For a statement of causal connection this is well known; and the assertion that a condition n constitutes a functional prerequisite for a state of some specified kind (such as proper functioning) is tantamount to the statement of a law to the effect that whenever condition n fails to be satisfied, the state in question fails to occur. Thus, explanation by functional analysis requires reference to laws.[17]

What explanatory import may properly be claimed for functional analysis? Suppose, then, that we are interested in explaining the occurrence of a trait i in

---

16. Radcliffe-Brown, *op. cit.*, p. 186. For an analysis of the idea of historic-genetic explanation, referred to in this passage, see section 7 of the essay "Aspects of Scientific Explanation", in this volume.

17. Malinowski, at one place in his writings, endorses a pronouncement which might appear to be at variance with this conclusion: "Description cannot be separated from explanation, since in the words of a great physicist, 'explanation in nothing but condensed description'." (Malinowski, "Anthropology," *op. cit.*, p. 132.) He seems to be referring here to the views of Ernst Mach or of Pierre Duhem, who took a similar position on this point. Mach conceived the basic objective of science as the brief and economic description of recurrent phenomena and considered laws as a highly efficient way of compressing, as it were, the description of an infinitude of potential particular occurrences into a simple and compact formula. But, thus understood, the statement approvingly quoted by Malinowski is, of course, entirely compatible with our point about the relevance of laws for functional explanation.

Besides, a law can be called a description only in a Pickwickian sense. For even so simple a generalization as "All vertebrates have hearts" does not describe any particular individual, such as Rin-Tin-Tin, as being a vertebrate and having a heart; rather, it asserts of Rin-Tin-Tin and of any other object, whether vertebrate or not—that *if* it is a vertebrate *then* it has a heart. Thus, the generalization has the import of an indefinite set of conditional statements about particular objects. In addition, a law might be said to imply statements about "potential events" which never actually take place. The gas law, for example, implies that if a given body of gas were to be heated under constant pressure at time t, its volume would increase. But if in fact the gas is not heated at t this statement can hardly be said to be a description of any particular event.

a system s (at a certain time t), and that the following functional analysis is offered:

- (a) At t, s functions adequately in a setting of kind c (characterized by specific internal and external conditions)
- (b) s functions adequately in a setting of kind c only if a certain necessary condition, n, is satisfied

(4.1)

- (c) If trait i were present in s then, as an effect, condition n would be satisfied
- (d) (Hence), at t, trait i is present in s

For the moment, let us leave aside the question as to what precisely is meant by statements of the types (a) and (b), and especially by the phrase "s functions adequately"; these matters will be examined in section 5. Right now, we will concern ourselves only with the *logic* of the argument; i.e., we will ask whether (d) formally follows from (a), (b), (c), just as in a deductive-nomological explanation the explanandum follows from the explanans. The answer is obviously in the negative, for, to put it pedantically, the argument (4.1) involves the fallacy of affirming the consequent in regard to premise (c). More explicitly, the statement (d) could be validly inferred if (c) asserted that *only* the presence of trait i could effect satisfaction of condition n. As it is, we can infer merely that condition n must be satisfied in some way or other at time t; for otherwise by reason of (b), the system s could not be functioning adequately in its setting, in contradiction to what (a) asserts. But it might well be that the occurrence of any one of a number of alternative items would suffice no less than the occurrence of i to satisfy requirement n, in which case the account provided by the premises of (4.1) simply fails to explain why the trait i rather than one of its alternatives is present in s at t.

As has just been noted, this objection would not apply if premise (c) could be replaced by the statement that requirement n can be met *only* by the presence of trait i. And indeed, some instances of functional analysis seem to include the claim that the specific item under analysis is, in this sense, functionally indispensable for the satisfaction of n. For example, Malinowski makes this claim for magic when he asserts that "magic fulfills an indispensable function within culture. It satisfies a definite need which cannot be satisfied by any other factors of primitive civilization," and again when he says about magic that "without its power and guidance early man could not have mastered his practical difficulties as he has done, nor could man have advanced to the higher stages of culture. Hence the universal occurrence of magic in primitive societies and its enormous sway. Hence we do find magic an invariable adjunct of all important activities."[18]

18. Malinowski, "Anthropology," *op. cit.*, p. 136; and *Magic, Science and Religion, and Other Essays*, *op. cit.*, p. 90. (Note the explanatory claim implicit in the use of the word "hence.")

However, the assumption of functional indispensability for a given item is highly questionable on empirical grounds: in all concrete cases of application, there do seem to exist alternatives. For example, the binding of anxiety in a given subject might be effected by an alternative symptom, as the experience of psychiatrists seems to confirm. Similarly, the function of the rain dance might be subserved by some other group ceremonial. And interestingly, Malinowski himself, in another context, invokes "the principle of limited possibilities, first laid down by Goldenweiser. Given a definite cultural need, the means of its satisfaction are small in number, and therefore the cultural arrangement which comes into being in response to the need is determined within narrow limits."[19] This principle obviously involves at least a moderate liberalization of the conception that every cultural item is functionally indispensable. But even so, it may still be too restrictive. At any rate, sociologists such as Parsons and Merton have assumed the existence of "functional equivalents" for certain cultural items; and Merton, in his general analysis of functionalism, has insisted that the conception of the functional indispensability of cultural items be replaced explicitly by the assumption of "functional alternatives, or functional equivalents, or functional substitutes."[20] This idea, incidentally, has an interesting parallel in the "principle of multiple solutions" for adaptational problems in evolution. This principle, which has been emphasized by functionally oriented biologists, states that for a given functional problem (such as that of perception of light) there are usually a variety of possible solutions, and many of these are actually used by different—and often closely related—groups of organisms.[21]

It should be noted here that, in any case of functional analysis, the question whether there are functional equivalents to a given item i has a definite meaning only if the internal and external conditions c in (4.1) are clearly specified. Otherwise, any proposed alternative to i, say i', could be denied the status of a functional equivalent on the ground that, being different from i, the item i' would have certain effects on the internal state and the environment of s which would not be brought about by i; and that therefore, if i' rather than i were realized, s would not be functioning in the same internal and external situation. Suppose, for example, that the system of magic of a given primitive

19. B. Malinowski, "Culture," *Encyclopedia of the Social Sciences*, IV (New York: The Macmillan Company, 1931), 626.

20. Merton, *op. cit.*, p. 34. Cf. also T. Parsons, *Essays in Sociological Theory, Pure and Applied* (New York: The Free Press, 1949), 58. For an interesting attempt to establish the existence of functional alternatives in a specific case, see R. D. Schwartz, "Functional alternatives to inequality," *American Sociological Review*, 20 (1955), 424–30.

21. See G. G. Simpson, *The Meaning of Evolution* (New Haven: Yale University Press, 1949), 164 ff., 190, 342–43; and G. G. Simpson, C. S. Pittendrigh, L. H. Tiffany, *Life* (New York: Harcourt, Brace & World, Inc., 1957), 437.

group were replaced by an extension of its rational technology plus some modification of its religion, and that the group were to continue as a going concern. Would this establish the existence of a functional equivalent to the original system of magic? A negative answer might be defended on the ground that as a result of adopting the modified pattern, the group had changed so strongly in regard to some of its basic characteristics (i.e., its internal state, as characterized by c_i, had been so strongly modified) that it was not the original kind of primitive group any more; and that there simply was no functional equivalent to magic which would leave all the "essential" features of the group unimpaired. Consistent use of this type of argument would safeguard the postulate of the functional indispensability of every cultural item against any conceivable empirical disconfirmation—but at the cost of turning it from an empirical hypothesis into a covert definitional truth.

That unilluminating procedure certainly must be eschewed. But what can a functional analysis in the general manner of (4.1) establish if the possibility of functional equivalents of i is not thus ruled out by definitional fiat?[22] Let I be the class of all those items which are empirically sufficient for n under the circumstances indicated in (4.1), so that an item j will be included in I just in case its realization in system s under conditions of kind c would be empirically sufficient to ensure the satisfaction of requirement n. (The qualification 'empirically' is to indicate that the satisfaction of n by j must be a matter of empirical fact and not just of pure logic. This proviso excludes from I trivial items, such as n itself.) The class I will then be a class of functional equivalents in the sense mentioned above. Let us now replace premise (c) in (4.1) by the following statement:

(c′) I is the class of all empirically sufficient conditions for the fulfillment of requirement n in the context determined by system s in setting c.

What the premises (a), (b), and (c′) enable us to infer is then at best this:

(4.2) Some one of the items included in class I is present in system s at time t

But this conclusion offers no grounds for expecting the occurrence of any particular item from I rather than of one of its functional equivalents. And strictly, even the weak conclusion (4.2) is warranted only on the further premise that the class I is not empty, i.e., that there is at least one item whose occurrence would, by law, ensure satisfaction of n.

Thus, functional analysis surely does not account in the manner of a deductive argument for the presence of the particular item i that it is meant to explain. Perhaps, then, it could more adequately be construed as an inductive argument which exhibits the occurrence of i as highly probable under the circumstances

---

22. (Added in 1964.) The balance of this section has been revised to remedy a flaw in the original version, called to my attention by Professor John R. Gregg.

described in the premises? Might it not be possible, for example, to add to the premises of (4.1) a further statement to the effect that the functional prerequisite n can be met only by i and by a few specifiable functional alternatives? And might not these premises make the presence of i highly probable? This course is hardly promising, for in most, if not all, concrete cases it would be impossible to specify with any precision the range of alternative behavior patterns, institutions, customs, or the like that would suffice to meet a given functional prerequisite or need. And even if that range could be characterized, there is no satisfactory method in sight for dividing it into some finite number of cases and assigning a probability to each of these.

Suppose, for example, that Malinowski's general view of the function of magic is correct: how are we to determine, when trying to explain the system of magic of a given group, all the different systems of magic and alternative cultural patterns which would satisfy the same functional requirements for the group as does the actually existing system of magic? And how are we to ascribe probabilities of occurrence to each of these potential functional equivalents? Clearly, there is no satisfactory way of answering these questions, and practitioners of functional analysis do not claim to achieve their explanation in this extremely problematic fashion.

Nor is it any help to construe the general laws implicit in the statements (b) and (c) in (4.1) as statistical rather than strictly universal in form, i.e., as expressing connections that are very probable, but do not hold universally; for the premises thus obtained again would not preclude functional alternatives of i (each of which would make satisfaction of n highly probable), and thus the basic difficulty would remain: the premises taken jointly could still not be said to make the presence just of i highly probable.

In sum then, the information typically provided by a functional analysis of an item i affords neither deductively nor inductively adequate grounds for expecting i rather than one of its alternatives. The impression that a functional analysis does provide such grounds, and thus explains the occurrence of i, is no doubt at least partly due to the benefit of hindsight: when we seek to explain an item i, we presumably know already that i has occurred.

As was noted a moment ago, however, functional analysis might be construed as a deductive explanation with a very weak explanandum, thus:

(4.3)

    (a)   At time t, system s functions adequately in a setting of kind c

    (b) s functions adequately in a setting of kind c only if requirement n is satisfied

    (c') I is the class of empirically sufficient conditions for n, in the context determined by s and c; and I is not empty

    (d') Some one of the items included in I is present in s at t

This kind of inference is rather trivial, however, except when we have additional knowledge about the items contained in class I. Suppose for example that at time t, a certain dog (system s) is in good health in a "normal" kind of setting c which precludes the use of such devices as artificial hearts, lungs, and kidneys. Suppose further that in a setting of kind c, the dog can be in good health only if his blood circulates properly (condition n). Then schema (4.3) leads in effect only to the conclusion that in some way or other, the blood is being kept circulating properly in the dog at t—hardly a very illuminating result. If however, we have additional knowledge of the ways in which the blood may be kept circulating under the circumstances and if we know, for example, that the only feature that would ensure proper circulation (the only item in class I) is a properly working heart, then we may draw the much more specific conclusion that at t the dog has a properly working heart. But if we make explicit the further knowledge here used by expressing it as an additional premise, then our argument can be restated in the form (4.1), except that premise (c) has been replaced by the statement that i is the *only* trait by which n can be satisfied in setting c; and, as was pointed out above, the conclusion (d) of (4.1) does follow in this case.

In general, however, additional knowledge of the kind here referred to is not available, and the explanatory import of functional analysis is then limited to the precarious role schematized in (4.3).

5. THE PREDICTIVE IMPORT OF FUNCTIONAL ANALYSIS

We noted earlier the predictive significance of nomological explanation; now we will ask whether functional analysis can be put to predictive use.

First of all, the preceding discussion shows that the information which is typically provided by a functional analysis yields at best premises of the forms (a), (b), (c) in (4.1); and these afford no adequate basis for the deductive or inductive prediction of a sentence of the form (d) in (4.1). Thus, functional analysis no more enables us to predict than it enables us to explain the occurrence of a particular one of the items by which a given functional requirement can be met.

Second, even the much less ambitious explanatory schema (4.3) cannot readily be put to predictive use; for the derivation of the weak conclusion (e) relies on the premise (a); and if we wish to infer (e) with respect to some future time t, that premise is not available, for we do not know whether s will or will not be functioning adequately at that time. For example, consider a person developing increasingly severe anxieties, and suppose that a necessary condition for his adequate functioning is that his anxiety be bound by neurotic symptoms, or be overcome by other means. Can we predict that one or another of the modes of "adjustment" in the class *I* thus roughly characterized will actually come to pass? Clearly not, for we do not know whether the person in question will in fact continue to function adequately or will suffer some more or less serious breakdown, perhaps to the point of self-destruction.

It is of interest to note here that a somewhat similar limitation exists also for the predictive use of nomological explanations, even in the most advanced branches of science. For example, if we are to predict, by means of the laws of classical mechanics, the state in which a given mechanical system will be at a specified future time t, it does not suffice to know the state of the system at some earlier time t_0, say the present; we also need information about the boundary conditions during the time interval from t_0 to t, i.e., about the external influences affecting the system during that time. Similarly, the "prediction," in our first example, that the water level in the beaker will remain unchanged as the ice melts assumes that the temperature of the surrounding air will remain constant, let us say, and that there will be no disturbing influences such as an earthquake or a person upsetting the beaker. Again when we predict for an object dropped from the top of the Empire State Building that it will strike the ground about eight seconds later, we assume that during the period of its fall, the object is acted upon by no forces other than the gravitational attraction of the earth. In a full and explicit formulation then, nomological predictions such as these would have to include among their premises statements specifying the boundary conditions obtaining from t_0 up to the time t to which the prediction refers. This shows that even the laws and theories of the physical sciences do not actually enable us to predict certain aspects of the future exclusively on the basis of certain aspects of the present: the prediction also requires certain assumptions about the future. But in many cases of nomological prediction, there are good inductive grounds, available at t_0, for the assumption that during the time interval in question the system under study will be practically "closed," i.e., not subject to significant outside interference (this case is illustrated, for example, by the prediction of eclipses) or that the boundary conditions will be of a specified kind—a situation illustrated by predictions of events occurring under experimentally controlled conditions.

The predictive use of (4.3) likewise requires a premise concerning the future, namely (a); but there is often considerable uncertainty as to whether (a) will in fact prove to be true. Furthermore, if in a particular instance there should be good inductive grounds for considering (a) as true, the forecast yielded by (4.3) is still rather weak; for the argument then leads from the inductively warranted assumption that the system will be properly functioning at t to the "prediction" that a certain condition n, which is empirically necessary for such functioning, will be satisfied at t in some way or other.

The need to include assumptions about the future among the premises of predictive arguments can be avoided, in nomological predictions as well as in those based on functional analysis, if we are satisfied with predictive conclusions which are not categorical, but only conditional, or hypothetical, in character. For example, (4.3) may be replaced by the following argument, in which premise (*a*) is avoided at the price of conditionalizing the conclusion:

(*b*) System *s* functions adequately in a setting of kind *c* only if condition *n* is satisfied

(5·1) (*c'*) *I* is the class of empirically sufficient conditions for *n* in the context determined by *s* and *c*; and *I* is not empty

(*d"*) If *s* functions adequately in a setting of kind *c* at time *t*, then some one of the items in class *I* is present in *s* at *t*

This possibility deserves mention because it seems that at least some of the claims made by advocates of functional analysis may be construed as asserting no more than that functional analysis permits such conditional predictions. This may be the intent, for example, of Malinowski's claim: "If such [a functional] analysis discloses to us that, taking an individual culture as a coherent whole, we can state a number of general determinants to which it has to conform, we shall be able to produce a number of predictive statements as guides for field-research, as yardsticks for comparative treatment, and as common measures in the process of cultural adaptation and change."[23] The statements specifying the determinants in question would presumably take the form of premises of type (*b*); and the "predictive statements" would then be hypothetical.

Many of the predictions and generalizations made in the context of functional analysis, however, do not have this conditional form. They proceed from a statement of a functional prerequisite or need to the categorical assertion of the occurrence of some trait, institution, or other item presumably sufficient to meet the requirement in question. Consider, for example, Sait's functional explanation of the emergence of the political boss: "Leadership is necessary; and *since* it does not develop readily within the constitutional framework, the boss provides it in a crude and irresponsible form from the outside."[24] Or take Merton's characterization of one function of the political machine: referring to various specific ways in which the political machine can serve the interests of business, he concludes, "These 'needs' of business, as presently constituted, are not adequately provided for by conventional and culturally approved social structures; *consequently*, the extra-legal but more-or-less efficient organization

23. Malinowski, *A Scientific Theory of Culture, and Other Essays, op. cit.*, p. 38.

24. E. M. Sait, "Machine, Political," *Encyclopedia of the Social Sciences*, IX (New York: The Macmillan Company, 1933), p. 659. (Italics supplied.)

of the political machine comes to provide these services."[25] Each of these arguments, which are rather typical of the functionalist approach, is an inference from the existence of a certain functional prerequisite to the categorical assertion that the prerequisite will be satisfied in some way. What is the basis of the inferential claims suggested by the words, 'since' and 'consequently' in the passages just quoted? When we say that *since* the ice cube was put into warm water it melted; or that the current was turned on, and *consequently*, the ammeter in the circuit responded, these inferences can be explicated and justified by reference to certain general laws of which the particular cases at hand are simply special instances; and the logic of the inferences can be exhibited by putting them into the form of the schema (2.1). Similarly, each of the two functionalist arguments under consideration clearly seems to presuppose a general principle to the effect that, within certain limits of tolerance or adaptability, a system of the kind under analysis will—either invariably or with high probability—satisfy, by developing appropriate traits, the various functional requirements (necessary conditions for its continued adequate operation) that may arise from changes in its internal state or in its environment. Any assertion of this kind, no matter whether of strictly universal or of statistical form, will be called a (*general*) *hypothesis of self-regulation*.

Unless functional analyses of the kind just illustrated are construed as implicitly proposing or invoking suitable hypotheses of self-regulation, it remains quite unclear what connections the expressions 'since,' 'consequently,' and others of the same character are meant to indicate, and how the existence of those connections in a given case is to be objectively established.

Conversely, if a precise hypothesis of self-regulation for systems of a specified kind is set forth, then it becomes possible to explain, and to predict categorically, the satisfaction of certain functional requirements simply on the basis of information concerning antecedent needs; and the hypothesis can then be objectively tested by an empirical check of its predictions. Take, for example, the statement that if a hydra is cut into several pieces, most of these will grow into complete hydras again. This statement may be considered as a hypothesis concerning a specific kind of self-regulation in a particular kind of biological system. It can clearly be used for explanatory and predictive purposes, and indeed the success of the predictions it yields confirms it to a high degree.

We see, then, that whenever functional analysis is to serve as a basis for categorical prediction or for generalizations of the type quoted from Sait and from Merton, it is of crucial importance to establish appropriate hypotheses of self-regulation in an objectively testable form.

25. Merton, *op. cit.*, p. 76. (Italics supplied.)

The functionalist literature does contain some explicitly formulated general-izations of the kind here referred to. Merton, for example, after citing the passage from Sait quoted above, comments thus: "Put in more generalized terms, *the functional deficiencies of the official structure generate an alternative (un-official) structure to fulfill existing needs somewhat more effectively.*"[26] This statement seems clearly intended to make explicit a hypothesis of self-regulation that might be said to underlie Sait's specific analysis and to provide the rationale for his 'since'. Another hypothesis of this kind is suggested by Radcliffe-Brown: "it may be that we should say that ... a society that is thrown into a condition of functional disunity or inconsistency ... will not die, except in such com-paratively rare instances as an Australian tribe overwhelmed by the white man's destructive force, but will continue to struggle toward ... some kind of social health. . . ."[27]

But, as was briefly suggested above, a formulation proposed as a hypothesis of self-regulation can serve as a basis for explanation or prediction only if it is sufficiently definite to permit objective empirical test. And indeed many of the leading representatives of functional analysis have expressed their concern to develop hypotheses and theories which meet this requirement. Malinowski, for example, in his essay significantly entitled "A Scientific Theory of Culture," insists that "each scientific theory must start from and lead to observation. It must be inductive and it must be verifiable by experience. In other words, it must refer to human experiences which can be defined, which are public, that is, accessible to any and every observer, and which are recurrent, hence fraught with inductive generalizations, that is, predictive."[28] Similarly, Murray and Kluckhohn have this to say about the basic objective of their functionally oriented theory, and indeed about any scientific "formulation," of personality: "the general purposes of formulation are three: (1) to *explain* past and present events; (2) to *predict* future events (the conditions being specified); and (3) to serve, if required, as a basis for the selection of effective measures of *control.*"[29]

Unfortunately, however, the formulations offered in the context of con-crete functional analyses quite often fall short of these general standards. Among the various ways in which those conditions may be violated, two call for special consideration because of their pervasiveness and central importance in functional analysis. They will be referred to as (i) *inadequate specification of scope*, and

26. Merton, *op. cit.*, p. 73. (Author's italics.)
27. Radcliffe-Brown, *op. cit.*, p. 183.
28. Malinowski, *A Scientific Theory of Culture, and Other Essays, op. cit.*, p. 67.
29. Henry A. Murray and Clyde Kluckhohn, "Outline of a Conception of Personality," in Clyde Kluckhohn and Henry A. Murray, eds., *Personality in Nature, Society, and Culture* (New York: Knopf, 1950), pp. 3–32; quotation from p. 7; authors' italics.

(ii) *nonempirical use of functionalist key terms* (such as 'need,' 'functional require-ment,' 'adaptation,' and others). We will consider these two defects in turn: the former in the balance of the present section, the latter in the next.

Inadequate specification of scope consists in failure to indicate clearly the kind of system to which the hypothesis refers, or the range of situations (the limits of tolerance) within which those systems are claimed to develop traits that will satisfy their functional requirements. Merton's formulation, for exam-ple, does not specify the class of social systems and of situations to which the proposed generalization is meant to apply; as it stands, therefore, it cannot be put to an empirical test or to any predictive use.

The generalization tentatively set forth by Radcliffe-Brown has a similar shortcoming. Ostensibly, it refers to any society whatever, but the conditions under which social survival is claimed to occur are qualified by a highly indefinite "except" clause, which precludes the possibility of any reasonably clear-cut test. The clause might even be used to protect the proposed generalization against any conceivable disconfirmation: If a particular social group should "die," this very fact might be held to show that the disruptive forces were as overwhelming as in the case of the Australian tribe mentioned by Radcliffe-Brown. Systematic use of this methodological strategy would, of course, turn the hypothesis into a covert tautology. This would ensure its truth, but at the price of depriving it of empirical content: thus construed, the hypothesis can yield no explanation or prediction whatever.

A similar comment is applicable to the following pronouncement by Malinowski, in which we italicize the dubious qualifying clause: "When we consider any culture *which is not on the point of breaking down or completely disrupted, but which is a normal going concern*, we find that need and response are directly related and tuned up to each other."[30]

To be sure, Radcliffe-Brown's and Malinowski's formulations do not *have to* be construed as covert tautologies, and their authors no doubt intended them as empirical assertions; but, in this case, the vagueness of the qualifying clauses still deprives them of the status of definite empirical hypotheses that might be used for explanation or prediction.

6. THE EMPIRICAL IMPORT OF FUNCTIONALIST TERMS AND HYPOTHESES

A second flaw that may vitiate the scientific role of a proposed hypotheses of self-regulation consists in using key terms of functional analysis, such as

30. Malinowski, *A Scientific Theory of Culture, and Other Essays, op. cit.*, p. 94.

'need' and 'adequate (proper) functioning'[31] in a nonempirical manner, i.e., without giving them a clear "operational definition," or more generally, without specifying objective criteria of application for them.[32] If functionalist terms are used in this manner, then the sentences containing them have no clear empirical meaning; they lead to no specific predictions and thus cannot be put to an objective test; nor, of course, can they be used for explanatory purposes.

A consideration of this point is all the more important here because the functionalist key terms occur not only in hypotheses of self-regulation, but also in functionalist sentences of various other kinds, such as those of the types (a), (b), and (d'') in (4.1), (4.3), and (5.1). Nonempirical use of functionalist terms may, therefore, bar sentences of these various kinds from the status of scientific hypotheses. We turn now to some examples.

Consider first the terms 'functional prerequisite' and 'need,' which are used as more or less synonymous in the functionalist literature, and which serve to define the term 'function' itself. "Embedded in every functional analysis is some conception, tacit or expressed, of the functional requirements of the system under observation",[33] and indeed, "a definition [of function] is provided by showing that human institutions, as well as partial activities within these, are related to primary, that is, biological, or derived, that is, cultural needs. Function means, therefore, always the satisfaction of a need. . . ."[34]

How is this concept of need defined? Malinowski gives an explicit answer: "By need, then, I understand the system of conditions in the human organism, in the cultural setting, and in the relation of both to the natural environment, which are sufficient and necessary for the survival of group and organism."[35] This definition sounds clear and straightforward; yet it is not even quite in accord with Malinowski's own use of the concept of need. For he distinguishes,

31. In accordance with a practice followed widely in contemporary logic, we will understand by terms certain kinds of words or other linguistic expressions, and we will say that a term expresses or signifies a concept. For example, we will say that the term 'need' signifies the concept of need. As this illustration shows, we refer to, or mention, a linguistic expression by using a name for it which is formed by simply enclosing the expression in single quotes.

32. A general discussion of the nature and significance of "operational" criteria of application for the terms used in empirical science, and references to further literature on the subject, may be found in C. G. Hempel, *Fundamentals of Concept Formation in Empirical Science* (University of Chicago Press, 1952), sections 5-8; and in the symposium papers on the present state of operationalism by G. Bergmann, P. W. Bridgman, A. Grunbaum, C. G. Hempel, R. B. Lindsay, H. Margenau, and R. J. Seeger, which form chapter II of Philipp G. Frank, ed., *The Validation of Scientific Theories* (Boston: The Beacon Press, 1956).

33. Merton, *op. cit.*, p. 52.

34. Malinowski, *A Scientific Theory of Culture, and other Essays, op. cit.*, p. 159.

35. Malinowski, *ibid.*, p. 90.

very plausibly, a considerable number of different needs, which fall into two major groups: primary biological needs and derivative cultural ones; the latter include "technological, economic, legal, and even magical, religious, or ethical"[36] needs. But if every single one of these needs did actually represent not only a necessary condition of survival but also a sufficient one, then clearly the satisfaction of just one need would suffice to ensure survival, and the other needs could not constitute necessary conditions of survival at all. It seems reasonable to assume, therefore, that what Malinowski intended was to construe the needs of a group as a set of conditions which are individually necessary and jointly sufficient for its survival.[37]

However, this correction of a minor logical flaw does not remedy a more serious defect of Malinowski's definition, which lies in the deceptive appearance of clarity of the phrase "survival of group and organism." In reference to a biological organism, the term 'survival' has a fairly clear meaning, though even here, there is need for further clarification. For when we speak of biological needs or requirements—e.g., the minimum daily requirements, for human adults, of various vitamins and minerals—we construe these, not as conditions of just the barest survival but as conditions of persistence in, or return to, a "normal," or "healthy" state, or to a state in which the system is a "properly functioning whole." For the sake of objective testability of functionalist hypotheses, it is essential, therefore, that definitions of needs or functional prerequisites be supplemented by reasonably clear and objectively applicable criteria of what is to be considered a healthy state or a normal working order of the systems under consideration; and that the vague and sweeping notion of survival then be construed in the relativized sense of survival in a healthy state as specified. Otherwise, there is definite danger that different investigators will use the concept of functional prerequisite—and hence also that of function—in different ways, and with valuational overtones corresponding to their diverse conceptions of what are the most "essential" characteristics of "genuine" survival for a system of the kind under consideration.

Functional analyses in psychology, sociology, and anthropology are even

---

36. Malinowski, *ibid.*, p. 172; see also *ibid.*, pp. 91 ff.

37. In some of his statements Malinowski discards, by implication, even the notion of function as satisfaction of a condition that is at least *necessary* for the survival of group or organism. For example, in the essay containing the two passages just quoted in the text, Malinowski comments as follows on the function of some complex cultural achievements: "Take the airplane, the submarine, or the steam engine. Obviously, man does not need to fly, nor yet to keep company with fishes, and move about within a medium for which he is neither anatomically adjusted nor physiologically prepared. In defining, therefore, the function of any of those contrivances, we can not predicate the true course of their appearance in any terms of metaphysical necessity." (*Ibid.*, pp. 118-19.)

more urgently in need of objective empirical criteria of the kind here referred to; for the characterization of needs as necessary conditions of psychological or emotional survival for an individual, or of survival of a group is so vague as to permit, and indeed invite, quite diverse subjective interpretations.

Some authors characterize the concept of functional prerequisite or the concept of function without making use of the term 'survival' with its misleading appearance of clarity. Merton, for example, states: "*Functions* are those observed consequences which make for the adaptation or adjustment of a given system; and *dysfunctions*, those observed consequences which lessen the adaptation or adjustment of the system."[38] And Radcliffe-Brown characterizes the function of an item as its contribution to the maintenance of a certain kind of unity of a social system, "which we may speak of as a functional unity. We may define it as a condition in which all parts of the social system work together with a sufficient degree of harmony or internal consistency, i.e., without producing persistent conflicts which can neither be resolved nor regulated."[39] But like the definitions in terms of survival, these alternative characterizations, though suggestive, are far from giving clear empirical meanings to the key terms of functional analysis. The concepts of adjustment and adaptation, for example, require specification of some standard; otherwise, they have no definite meaning and are in danger of being used tautologically or else subjectively, with valuational overtones.

Tautological use could be based on construing *any* response of a given system as an adjustment, in which case it becomes a trivial truth that any system will adjust itself to any set of circumstances. Some instances of functional analysis seem to come dangerously close to this procedure, as is illustrated by the following assertion: "Thus we are provided with an explanation of suicide and of numerous other apparently antibiological effects as so many forms of relief from intolerable suffering. Suicide does not have *adaptive* (survival) value but it does have *adjustive* value for the organism. Suicide is *functional* because it abolishes painful tension."[40]

Or consider Merton's formulation of one of the assumptions of functional analysis: ". . . when *the net balance of the aggregate of consequences* of an existing social structure is clearly dysfunctional, there develops a strong and insistent pressure for change."[41] In the absence of clear empirical criteria of adaptation and thus of dysfunction, it is possible to treat this formulation as a covert tautology and thus to render it immune to empirical disconfirmation. Merton

---

38. Merton, *op. cit.*, p. 51. (Author's italics.)
39. Radcliffe-Brown, *op. cit.*, p. 181.
40. Murray and Kluckhohn, *op. cit.*, p. 15 (Author's italics.)
41. Merton, *op. cit.*, p. 40.

is quite aware of such danger: in another context he remarks that the notion of functional requirements of a given system "remains one of the cloudiest and empirically most debatable concepts in functional theory. As utilized by sociologists, the concept of functional requirement tends to be tautological or *ex post facto*."[42] Similar warnings against tautological use and against *ad hoc* generalizations about functional prerequisites have been voiced by other writers, such as Malinowski[43] and Parsons.[44]

In the absence of empirical criteria of adjustment or adaptation, there is also the danger of each investigator's projecting into those concepts (and thus also into the concept of function) his own ethical standards of what would constitute a "proper" or "good" adjustment of a given system—a danger which has been pointed out very clearly by Levy.[45] This procedure would obviously deprive functionalist hypotheses of the status of precise objectively testable scientific assertions. And, as Merton notes, "If theory is to be productive, it must be sufficiently *precise* to be *determinate*. Precision is an integral element of the criterion of *testability*."[46]

It is essential, then, for functional analysis as a scientific procedure that its key concepts be explicitly construed as relative to some standard of survival or adjustment. This standard has to be specified for each functional analysis, and it will usually vary from case to case. In the functional study of a given system s, the standard would be indicated by specifying a certain class or range R of possible states of s, with the understanding that s is to be considered as "surviving in proper working order," or as "adjusting properly under changing conditions" just in case s remains in, or upon disturbance returns to, some state within the range R. A need, or functional requirement, of system s relative to R is then a necessary condition for the system's remaining in, or returning to, a state in R; and the function, relative to R, of an item i in s consists in i's effecting the satisfaction of some such functional requirement.

In the field of biology, Sommerhoff's analysis of adaptation, appropriateness, and related concepts, is an excellent illustration of a formal study in which the relativization of the central functionalist concepts is entirely explicit.[47] The

42. Merton, *op. cit.*, p. 52.

43. See, for example, Malinowski, *A Scientific Theory of Culture, and Other Essays, op. cit.*, pp. 169-70; but also compare this with pp. 118-19 of the same work.

44. See, for example, T. Parsons, *The Social System* (New York: The Free Press, 1951), 29, n. 4.

45. Marion J. Levy, Jr., *The Structure of Society* (Princeton: Princeton University Press, 1952), 76ff.

46. R. K. Merton, "The Bearing of Sociological Theory on Empirical Research" in Merton, *Social Theory and Social Structure, op. cit.*, pp. 85-101; quotation from 98. (Author's italics)

47. See G. Sommerhoff, *Analytical Biology* (New York: Oxford University Press, 1950).

need of such relativization is made clear also by Nagel, who points out that "the claim that a given change is functional or dysfunctional must be understood as being relative to a specified G (or sets of G's)"[48], where the G's are traits whose preservation serves as the defining standard of adjustment or survival. In sociology, Levy's analysis of the structure of society[49] clearly construes the functionalist key concepts as relative in the sense just outlined.

Only if the key concepts of functional analysis are thus relativized can hypotheses involving them have the status of determinate and objectively testable assumptions or assertions; only then can those hypotheses enter significantly into arguments such as those schematized in (4.1), (4.3), and (5.1).

But although such relativization may give definite empirical content to the functionalist hypotheses that serve as premises or conclusions in those arguments, it leaves the explanatory and predictive import of the latter as limited as we found it in sections 4 and 5; for our verdict on the logical force of those arguments depended solely on their formal structure and not on the meaning of their premises and conclusions.

It remains true, therefore, even for a properly relativized version of functional analysis, that its explanatory force is rather limited; in particular, it does not provide an explanation of why a particular item i rather than some functional equivalent of it occurs in system s. And the predictive significance of functional analysis is practically nil—except in those cases where suitable hypotheses of self-regulation can be established. Such a hypothesis would be to the effect that within a specified range C of circumstances, a given system s (or: any system of a certain kind S, of which s is an instance) is self-regulating relative to a specified range R of states; i.e., that after a disturbance which moves s into a state outside R, but which does not shift the internal and external circumstances of s out of the specified range C, the system s will return to a state in R. A system satisfying a hypothesis of this kind might be called *self-regulating with respect to R*.

Biological systems offer many illustrations of such self-regulation. For example, we mentioned earlier the regenerative ability of a hydra. Consider the case, then, where a more or less large segment of the animal is removed and the rest grows into a complete hydra again. The class R here consists of those states in which the hydra is complete; the characterization of range C

48. Nagel, "A Formalization of Functionalism," *op. cit.*, p. 269. See also the concluding paragraph of the same essay (pp. 282-83).

49. Levy speaks of eufunction and dysfunction of a unit (i.e., a system) and characterizes these concepts as relative to "the unit as defined." He points out that relativization is necessary "because it is to the definition of the unit that one must turn to determine whether or not 'adaptation or adjustment' making for the persistence or lack of persistence of the unit is taking place." (Levy, *ibid.*, pp. 77-78).

would have to include (i) a specification of the temperature and the chemical composition of the water in which a hydra will perform its regenerative feat (clearly, this will not be just one unique composition, but a class of different ones: the concentrations of various salts, for example, will each be allowed to take some value within a specified, and perhaps narrow, range; the same will hold of the temperature of the water); and (ii) a statement as to the kind and size of segment that may be removed without preventing regeneration.

It will no doubt be one of the most important tasks of functional analysis in psychology and the social sciences to ascertain to what extent such phenomena of self-regulation can be found, and can be represented by corresponding laws.

7. FUNCTIONAL ANALYSIS AND TELEOLOGY

Whatever specific laws might be discovered by research along these lines, the kind of explanation and prediction made possible by them does not differ in its logical character from that of the physical sciences.

It is true that hypotheses of self-regulation, which would be the results of successful functionalist research, appear to have a teleological character since they assert that within specified conditions systems of some particular kind will tend toward a state within the class R, which thus assumes the appearance of a final cause determining the behavior of the system.

But, first of all, it would be simply untenable to say of a system s which is self-regulating with respect to R that the future event of its return to (a state in) R is a "final cause" which determines its present behavior. For even if s is self-regulating with respect to R and if it has been shifted into a state outside R, the future event of its return to R may never come about: in the process of its return toward R, s may be exposed to further disturbances, which may fall outside the permissible range C and lead to the destruction of s. For example, in a hydra that has just had a tentacle removed, certain regenerative processes will promptly set in; but these cannot be explained teleologically by reference to a final cause consisting in the future event of the hydra being complete again. For that event may never actually come about since in the process of regeneration, and before its completion, the hydra may suffer new, and irreparably severe, damage, and may die. Thus, what accounts for the present changes of a self-regulating system s is not the "future event" of s being in R, but rather the *present disposition* of s to return to R; and it is this disposition that is expressed by the hypothesis of self-regulation governing the system s.

Whatever teleological character may be attributed to a functionalist explanation or prediction invoking (properly relativized) hypotheses of self-regulation lies merely in the circumstance that such hypotheses assert a tendency of certain systems to maintain, or return to, a certain kind of state. But such laws attributing, as it were, a characteristic goal-directed behavior to systems of specified kinds are by no means alien to physics and chemistry. On the contrary, it is these latter fields which provide the most adequately understood instances of self-regulating systems and corresponding laws. For example, a liquid in a vessel will return to a state of equilibrium, with its surface horizontal, after a mechanical disturbance; an elastic band, after being stretched (within certain limits), will return to its original shape when it is released. Various systems controlled by negative feedback devices, such as a steam engine whose speed is regulated by a governor, or a homing torpedo, or a plane guided by an automatic pilot, show, within specifiable limits, self-regulation with respect to some particular class of states.

In all of these cases, the laws of self-regulation exhibited by the systems in question are capable of explanation by subsumption under general laws of a more obviously causal form. But this is not even essential, for the laws of self-regulation themselves are causal in the broad sense of asserting that for systems of a specified kind, any one of a class of different "initial states" (any one of the permissible states of disturbance) will lead to the same kind of final state. Indeed as our earlier formulations show, functionalist hypotheses, including those of self-regulation, can be expressed without the use of any teleological phraseology at all.[50]

There are, then, no systematic grounds for attributing to functional analysis a character *sui generis* not found in the hypotheses and theories of the natural sciences and in the explanations and predictions based on them. Yet, psychologically, the idea of function often remains closely associated with that of purpose, and some functionalist writing has no doubt encouraged this association, by using a phraseology which attributes to the self-regulatory behavior of a given system practically the character of a purposeful action. For example, Freud, speaking of the relation of neurotic symptoms to anxiety, uses strongly teleological language when he says that "the symptoms are created in order to remove or rescue the ego from the situation of danger";[51] the quotations given in section 3 provide further illustrations. Some instructive examples of sociological and anthropological writings which confound the concepts of function

---

50. For illuminating discussions of further issues concerning "teleological explanation," especially with respect to self-regulating systems, see R. B. Braithwaite, *Scientific Explanation* (Cambridge: Cambridge University Press, 1953), chapter X; and E. Nagel, "Teleological Explanation and Teleological Systems" in S. Ratner, ed., *Vision and Action: Essays in Honor of Horace Kallen on His Seventieth Birthday* (New Brunswick, N.J.: Rutgers University Press, 1953); reprinted in H. Feigl and M. Brodbeck, eds., *Readings in the Philosophy of Science* (New York: Appleton-Century-Crofts, Inc., 1953).

51. Freud, *op. cit.*, p. 112.

and purpose are listed by Merton, who is very explicit and emphatic in rejecting this practice.[52]

It seems likely that precisely this psychological association of the concept of function with that of purpose, though systematically unwarranted, accounts to a large extent for the appeal and the apparent plausibility of functional analysis as a mode of explanation; for it seems to enable us to "understand" self-regulatory phenomena of all kinds in terms of purposes or motives, in much the same way in which we "understand" our own purposive behavior and that of others. Now, explanation by reference to motives, objectives, or the like may be perfectly legitimate in the case of purposive behavior and its effects. An explanation of this kind would be causal in character, listing among the causal antecedents of the given action, or of its outcome, certain purposes or motives on the part of the agent, as well as his beliefs as to the best means available to him for attaining his objectives. This kind of information about purposes and beliefs might even serve as a starting point in explaining a self-regulatory feature in a human artifact. For example, in an attempt to account for the presence of the governor in a steam engine, it may be quite reasonable to refer to the purpose its inventor intended it to serve, to his beliefs concerning matters of physics, and to the technological facilities available to him. Such an account, it should be noted, might conceivably give a probabilistic explanation for the presence of the governor, but it would not explain why it functioned as a speed-regulating safety device: to explain this latter fact, we would have to refer to the construction of the machine and to the laws of physics, not to the intensions and beliefs of the designer. (An explanation by reference to motives and beliefs can be given as well for certain items which do not, in fact, function as intended; e.g., some superstitious practices, unsuccessful flying machines, ineffective economic policies, etc.). Furthermore—and is this the crucial point in our context—for most of the self-regulatory phenomena that come within the purview of functional analysis, the attribution of purposes is an illegitimate transfer of the concept of purpose from its domain of significant applicability to a much wider domain, where it is devoid of objective empirical import. In the context of purposive behavior of individuals or groups, there are various methods of testing whether the assumed motives or purposes are indeed present in a given situation; interviewing the agents in question might be one rather direct way, and there are various alternative "operational" procedures of a more indirect character. Hence, explanatory hypotheses in terms of purposes are here capable of reasonably objective test. But such empirical criteria are lacking in other cases of self-regulating systems, and the attribution of purposes to them has therefore no scientific meaning. Yet, it tends to encourage the illusion that a profound understanding is achieved, that we gain insight into the nature of these processes by likening them to a type of behavior with which we are thoroughly familiar from daily experience. Consider, for example, the law of "adaptation to an obvious end" set forth by the sociologist L. Gumplowicz with the claim that it holds both in the natural and the social domains. For the latter, it asserts that "every social growth, every social entity, serves a definite end, however much its worth and morality may be questioned. For the universal law of adaptation signifies simply that no expenditure of effort, no change of condition, is purposeless on any domain of phenomena. Hence, the inherent reasonableness of all social facts and conditions must be conceded."[53] There is a strong suggestion here that the alleged law enables us to understand social dynamics in close analogy to purposive behavior aimed at the achievement of some end. Yet that law is completely devoid of empirical meaning since no empirical interpretation has been given to such key terms as 'end,' 'purposeless', and 'inherent reasonableness' for the contexts to which it is applied. The "law" asserts nothing whatever, therefore, and cannot possibly explain any social (or other) phenomena.

Gumplowicz's book antedates the writings of Malinowski and other leading functionalists by several decades, and certainly these more recent writers have been more cautious and sophisticated in stating their ideas. Yet, there are certain quite central assertions in the newer functionalist literature which are definitely reminiscent of Gumplowicz's formulation in that they suggest an understanding of functional phenomena in the image of deliberate purposive behavior or of systems working in accordance with a preconceived design. The following statements might illustrate this point: "[Culture] is a system of objects, activities, and attitudes in which every part exists as a means to an end,"[54] and "The functional view of culture insists therefore upon the principle that in every type of civilization, every custom, material object, idea and belief fulfills some vital function, has some task to accomplish, represents an indispensable part within a working whole."[55] These statements express what Merton, in a critical discussion, calls the postulate of universal functionalism.[56] Merton qualifies this postulate as premature;[57] the discussion presented in the previous section shows that, in the absence of a clear empirical interpre-

53. L. Gumplowicz, *The Outlines of Sociology;* translated by F. W. Moore (Philadelphia: American Academy of Policical and Social Science, 1899), pp. 79–80.

54. Malinowski, *A Scientific Theory of Culture, and Other Essays, op. cit.,* p. 150.

55. Malinowski, "Anthrolopogy," *op. cit.,* p. 133.

56. Merton, "Manifest and Latent Functions," *op. cit.,* pp. 30ff.

57. *Ibid.,* p. 31.

tation of the functionalist key terms, it is even less than that, namely, empirically vacuous. Yet formulations of this kind may evoke a sense of insight and understanding by likening sociocultural developments to purposive behavior and in this sense reducing them to phenomena with which we feel thoroughly familiar. But scientific explanation and understanding are not simply a reduction to the familiar: otherwise, science would not seek to explain familiar phenomena at all; besides, the most significant advances in our scientific understanding of the world are often achieved by means of new theories which, like quantum theory, assume some quite unfamiliar kinds of objects or processes which cannot be directly observed, and which sometimes are endowed with strange and even seemingly paradoxical characteristics. A class of phenomena has been scientifically understood to the extent that they can be fitted into a testable, and adequately confirmed, theory or a system of laws; and the merits of functional analysis will eventually have to be judged by its ability to lead to this kind of understanding.

8. THE HEURISTIC ROLE OF FUNCTIONAL ANALYSIS

The preceding considerations suggest that what is often called "functionalism" is best viewed, not as a body of doctrine or theory advancing tremendously general principles such as the principle of universal functionalism, but rather as a program for research guided by certain heuristic maxims or "working hypotheses." The idea of universal functionalism, for example, which becomes untenable when formulated as a sweeping empirical law or theoretical principle, might more profitably be construed as expressing a directive for inquiry, namely to search for specific self-regulatory aspects of social and other systems and to examine the ways in which various traits of a system might contribute to its particular mode of self-regulation (A similar construal as heuristic maxims for empirical research might be put upon the "general axioms of functionalism" suggested by Malinowski, and considered by him as demonstrated by all the pertinent empirical evidence.[58])

In biology, for example, the contribution of the functionalist approach does not consist in the sweeping assertion that all traits of any organism satisfy some need and thus serve some function; in this generality, the claim is apt to be either meaningless or covertly tautologous or empirically false (depending on whether the concept of need is given no clear empirical interpretation at all, or is handled in a tautologizing fashion, or is given a specific empirical interpretation). Instead, functional studies in biology have been aimed at showing, for example, how in different species, specific homeostatic and regenerative processes contribute to the maintenance and development of the

58. Malinowski, *A Scientific Theory of Culture, and Other Essays, op. cit.*, p. 150.

living organism; and they have gone on (i) to examine more and more precisely the nature and limits of those processes (this amounts basically to establishing various specific empirical hypotheses or laws of self-regulation), and (ii) to explore the underlying physiological or physicochemical mechanisms, and the laws governing them, in an effort to achieve a more thorough theoretical understanding of the phenomena at hand.[59] Similar trends exist in the study of functional aspects of psychological processes, including, for example, symptom formation in neurosis.[60]

Functional analysis in psychology and in the social sciences no less than in biology may thus be conceived, at least ideally, as a program of inquiry aimed at determining the respects and the degrees in which various systems are self-regulating in the sense here indicated. This conception is clearly reflected in Nagel's essay, "A Formalization of Functionalism,"[61] which develops an analytic scheme inspired by, and similar to, Sommerhoff's formal analysis of self-regulation in biology[62] and uses it to exhibit and clarify the structure of functional analysis, especially in sociology and anthropology.

The functionalist mode of approach has proved illuminating, suggestive, and fruitful in many contexts. If the advantages it has to offer are to be reaped in full, it seems desirable and indeed necessary to pursue the investigation of specific functional relationships to the point where they can be expressed in terms of reasonably precise and objectively testable hypotheses. At least initially, these hypotheses will likely be of quite limited scope. But this would simply parallel the present situation in biology, where the kinds of self-regulation, and the uniformities they exhibit, vary from species to species. Eventually, such "empirical generalizations" of limited scope might provide a basis for a more general theory of self-regulating systems. To what extent these objectives can be reached cannot be decided in *a priori* fashion by logical analysis or philosophical reflection: the answer has to be found by intensive and rigorous scientific research.

59. An account of this kind of approach to homeostatic processes in the human body will be found in Walter B. Cannon, *The Wisdom of the Body* (New York: W. W. Norton & Company, Inc.; revised edition 1939).

60. See, for example, J. Dollard and N. E. Miller, *Personality and Psychotherapy* (New York: McGraw-Hill Book Company, Inc., 1950), chapter XI, "How Symptoms are Learned," and note particularly pp. 165-66.

61. Nagel, "A Formalization of Functionalism," *op. cit.* See also the more general discussion of functional analysis included in Nagel's paper, "Concept and Theory Formation in the Social Sciences," in *Science, Language, and Human Rights;* American Philosophical Association, Eastern Division, Volume 1 (Philadelphia: University of Pennsylvania Press, 1952), pp. 43-64. Reprinted in J. L. Jarrett and S. M. McMurrin, eds., *Contemporary Philosophy* (New York: Henry Holt & Co., Inc., 1954).

62. Sommerhoff, *op. cit.*

MECHANISTIC EXPLANATION AND ORGANISMIC BIOLOGY*

Vitalism of the substantival type sponsored by Driesch and other biologists during the preceding and early part of the present century is now a dead issue in the philosophy of biology—an issue that has become quiescent less, perhaps, because of the methodological and philosophical criticism that has been levelled against the doctrine than because of the infertility of vitalism as a guide in biological research and because of the superior heuristic value of alternative approaches for the investigation of biological phenomena. Nevertheless, the historically influential Cartesian conception of biology as simply a chapter of physics continues to meet resistance; and outstanding biologists who find no merit in vitalism believe there are conclusive reasons for maintaining the irreducibility of biology to physics and for asserting the intrinsic autonomy of biological method. The standpoint from which this thesis is currently advanced commonly carries the label of "organismic biology"; and though the label covers a variety of special biological doctrines that are not all mutually compatible, those who fall under it are united by the common conviction that biological phenomena cannot be understood adequately in terms of theories and explanations which are of the so-called "mechanistic type." It is the aim of the present paper to examine this claim.

It is, however, not always clear what thesis organismic biologists are rejecting when they declare that "mechanistic" explanations are not fully satisfactory in biology. In one familiar sense of "mechanistic," a theory is mechanistic if it employs only such concepts which are distinctive of the science of mechanics. It is doubtful, however, whether any professed mechanist in biology would today explicate his position in this manner. Physicists themselves have long since abandoned the seventeenth-century hope that a universal science of nature would be developed within the framework of the fundamental conceptions of mechanics. And no one today, it is safe to say, subscribes literally to the Cartesian program of reducing all the sciences to the science of mechanics and specifically to the mechanics of contact-action. On the other hand, it is not easy to state precisely what is the identifying mark of a mechanistic explanation if it is not to coincide with an explanation that falls within the science of mechanics. In a preliminary way, and for lack of anything better and clearer, I shall adopt in the present paper the criterion proposed long ago by

---

* A lecture given on May 13, 1950, to the Fullerton Club at Bryn Mawr College in commemoration of the 300th anniversary of Descartes's death.

Jacques Loeb, according to whom a mechanist in biology is one who believes that all living phenomena "can be unequivocally explained in physico-chemical terms,"—that is, in terms of theories that have been originally developed for domains of inquiry in which the distinction between the living and nonliving plays no role, and that by common consent are classified as belonging to physics and chemistry.

As will presently appear, this brief characterization of the mechanistic thesis in biology does not suffice to distinguish in certain important respects mechanists in biology from those who adopt the organismic standpoint; but the above indication will do for the moment. It does suffice to give point to one further preliminary remark which needs to be made before I turn to the central issue between mechanists and organismic biologists. It is an obvious commonplace, but one that must not be ignored if that issue is to be justly appraised, that there are large sectors of biological study in which physico-chemical explanations play no role at present, and that a number of outstanding biological theories have been successfully exploited which are not physico-chemical in character. For example, a vast array of important information has been obtained concerning embryological processes, though no explanation of such regularities in physico-chemical terms is available; and neither the theory of evolution even in its current form, nor the gene theory of heredity is based on any definite physico-chemical assumptions concerning living processes. Accordingly, organismic biologists possess at least some grounds for their skepticism concerning the inevitability of the mechanistic standpoint; and just as a physicist may be warranted in holding that some given branch of physics (e.g., electro-magnetic theory) is not reducible to some other branch (e.g., mechanics), so an organismic biologist may be warranted in holding an analogous view with respect to the relation of biology and physico-chemistry. If there is a genuine issue between mechanists and organismic biologists, it is not *prima facie* a pseudo-question.

However, organismic biologists are not content with making the obviously justified observation that only a relatively small sector of biological phenomena has thus far been explained in physico-chemical terms; they also maintain that *in principle* the mode of analysis associated with mechanistic explanations is inapplicable to some of the major problems of biology, and that therefore mechanistic biology cannot be adopted as the ultimate ideal in biological research. What are the grounds for this contention and how solid is the support which organismic biologists claim for their thesis?

The central theme of organismic biology is that living creatures are not assemblages of tissues and organs functioning independently of one another, but are integrated structures of parts. Accordingly, living organisms must be studied as "wholes," and not as the mere "sums" of parts. Each part, it is maintained, has physico-chemical properties; but the interrelation of the parts involves a distinctive organization, so that the study of the physico-chemical properties of the parts taken in isolation of their membership in the organized whole which is the living body fails to provide an adequate understanding of the facts of biology. In consequence, the continuous adaptation of an organism to its environment and of its parts to one another so as to maintain its charactertistic structure and activities, cannot be described in terms of physical and chemical principles. Biology must employ categories and a vocabulary which are foreign to the sciences of the inorganic, and it must recognize modes and laws of behavior which are inexplicable in physico-chemical terms.

There is time to cite but one brief quotation from the writings of organismic biologists. I offer the following from E. S. Russell as a typical statement of this point of view:

> Any action of the whole organism would appear to be susceptible of analysis to an indefinite degree—and this is in general the aim of the physiologist, to analyze, to decompose into their elementary processes the broad activities and functions of the organism.
>
> But . . . by such a procedure something is lost, for the action of the whole has a certain unifiedness and completeness which is left out of account in the process of analysis . . . . In our conception of the organism we must . . . take into account the unifiedness and wholeness of its activities [especially since] the activities of the organism all have reference to one or other of three great ends [development, maintenance, and reproduction], and both the past and the future enter into their determination . . . .
>
> . . . It follows that the activities of the organism as a whole are to be regarded as of a different order from physico-chemical relations, both in themselves and for the purposes of our understanding . . . .
>
> . . . Bio-chemistry studies essentially the *conditions* of action of cells and organisms, while organismal biology attempts to study the actual modes of action of whole organisms, regarded as conditioned by, but irreducible to, the modes of action of lower unities . . . (*Interpretation of Development and Heredity*, pp. 171-2, 187-8)

Accordingly, while organismic biology rejects every form of substantival vitalism, it also rejects the possibility of physico-chemical explanation of vital phenomena. But does it, in point of fact, present a clear alternative to physico-chemical theories of living processes, and, if so, what types of explanatory theories does it recommend as worth exploring in biology?

(1) At first blush, the sole issue that seems to be raised by organismic biology is that commonly discussed under the heading of "emergence" in other branches of science, including the physical sciences; and, although other questions are involved in the organismic standpoint, I shall begin with this aspect of the question.

The crux of the doctrine of emergence, as I see it, is the determination of the conditions under which one science can be reduced to some other one—i.e., the formulation of the logical and empirical conditions which must be satisfied if the laws and other statements of one discipline can be subsumed under, or explained by, the theories and principles of a second discipline. Omitting details and refinements, the two conditions which seem to be necessary and sufficient for such a reduction are briefly as follows. Let S_1 be some science or group of sciences such as physics and chemistry, hereafter to be called the "primary discipline," to which a second science, S_2, for example biology, is to be reduced. Then (i) every term which occurs in the statements of S_2 (e.g., terms like "cell," "mytosis," "heredity," etc.) must be either explicitly definable with the help of the vocabulary specific to the primary discipline (e.g., with the help of expressions like "length," "electric charge," "osmosis"); or well-established empirical laws must be available with the help of which it is possible to state the sufficient conditions for the application of all expressions in S_2, exclusively in terms of expressions occurring in the explanatory principles of S_1. For example, it must be possible to state the truth-conditions of a statement of the form "x is a cell" by means of sentences constructed exclusively out of the vocabulary belonging to the physico-chemical sciences. Though the label is not entirely appropriate, this first conditio will be referred to as the condition of definability. (ii) Every statement in the secondary discipline, S_2, and especially those statements which formulate the laws established in S_2, must be derivable logically from some appropriate class of statements that can be established in the primary science, S_1—such classes of statements will include the fundamental theoretical assumptions of S_1. This second condition will be referred to as the condition of derivability.

It is evident that the second condtion cannot be fulfilled unless the first one is, although the realization of the first condition does not entail the satisfaction of the second one. It is also quite beyond dispute that in the sense of reduction specified by these conditions biology has thus far not been reduced to physics and chemistry, since not even the first step in the process of reduction has been completed—for example, we are not yet in the position to specify exhaustively in physico-chemical terms the conditions for the occurrence of cellular division.

Accordingly, organismic biologists are on firm ground if what they maintain is that all biological phenomena are not explicable thus far physico-chemically, and that no physico-chemical theory can possibly explain such phenomena until the descriptive and theoretical terms of biology meet the condition of definability. On the other hand, nothing in the facts surveyed up to this point warrants the conclusion that biology is *in principle* irreducible to physico-chemistry. Whether biology is reducible to physico-chemistry is a question that only further experimental and logical research can settle; for the supposition that each of the two conditions for the reduction of biology to physico-chemistry may some day be satisfied involves no patent contradiction.

(2) There are, however, other though related considerations underlying the organismic claim that biology is intrinsically autonomous. A frequent argument used to support this claim is based on the fact that living organisms are hierarchically organized and that, in consequence, modes of behavior characterizing the so-called "higher levels" of organization cannot be explained in terms of the structures and modes of behavior which parts of the organism exhibit on lower levels of the hierarchy.

There can, of course, be no serious dispute over the fact that organisms do exhibit structures of parts that have an obvious hierarchical organization. Living cells are structures of cellular parts (e.g., of the nucleus, cytoplasm, central bodies, etc.), each of which in turn appears to be composed of complex molecules; and, except in the case of uni-cellular organisms, cells are further organized into tissues, which in turn are elements of various organs that make up the individual organism. Nor is there any question but that parts of an organism which occupy a place at one level of its complex hierarchical organization stand in relations and exhibit activities which parts occupying positions at other levels of organization do not manifest: a cat can stalk and catch mice, but though its heart is involved in these activities, that organ cannot perform these feats; again, the heart can pump blood by contracting and expanding its muscular tissues, but no tissue is able to do this; and no tissue is able to divide by fission, though its constituent cells may have this power; and so on down the line. If such facts are taken in an obvious sense, they undoubtedly support the conclusion that behavior on higher levels of organization is not explained by merely citing the various behaviors of parts on lower levels of the hierarchy. Organismic biologists do not, of course, deny that the higher level behaviors occur only when the component parts of an organism are appropriately organized on the various levels of the hierarchy; but they appear to have reason on their side in maintaining that a knowledge of the behavior of these parts when these latter are not component elements in the structured living organism, does not suffice as a premise for deducing anything about the behavior of the whole organism in which the parts do stand in certain specific and complex relations to one another.

But do these admitted facts establish the organismic thesis that mechanistic explanations are not adequate in biology? This does not appear to be the case, and for several reasons. It should be noted, in the first place, that various forms of hierarchical organization are exhibited by the materials of physics and chemistry, and not only by those of biology. On the basis of current theories of matter, we are compelled to regard atoms as structures of electric charges, molecules as organizations of atoms, solids and liquids as complex systems of molecules; and we must also recognize that the elements occupying positions at different levels of the indicated hierarchy generally exhibit traits and modes of activity that their component parts to no possess. Nonetheless, this fact has not stood in the way of establishing comprehensive theories for the more elementary physical particles, in terms of which it has been possible to explain some, if not all, of the physico-chemical properties exhibited by things having a more complex organization. We do not, to be sure, possess at the present time a comprehensive and unified theory which is competent to explain the whole range of physico-chemical phenomena at all levels of complexity. Whether such a theory will ever be achieved is certainly an open question. But even if such an inclusive theory were never achieved, the mere fact that we can now explain some features of relatively highly organized bodies on the basis of theories formulated in terms of relations between relatively more simply structured elements—for example, the specific heats of solids in terms of quantum theory or the changes in phase of compounds in terms of the thermodynamics of mixtures—should give us pause in accepting the conclusion that the mere fact of the hierarchical organization of biological materials precludes the possibility of a mechanistic explanation.

This observation leads to a second point. Organismic biologists do not deny that biological organisms are complex structures of physico-chemical processes, although like everyone else they do not claim to know in minute detail just what these processes are or just how the various physico-chemical elements (assumed as the ultimate parts of living creatures) are related to one another in a living organism. They do maintain, however, (or appear to maintain) that even if our knowledge in this respect were ideally complete, it would still be impossible to account for the characteristic behavior of biological organisms—their ability to maintain themselves, to develop, and to reproduce—in mechanistic terms. Thus, it has been claimed that even if we were able to describe in full detail in physico-chemical terms what is taking place when a fertilized egg segments, we would, nevertheless, be unable to explain mechanistically the fact of segmentation—in the language of E. S. Russell, we would then be able to state the physico-chemical *conditions* for the occurrence of segmentation, but we would, still be unable to "explain the *course* which development takes." Now this claim seems to me to rest on a misunderstanding, if not on a confusion. It is entirely correct to maintain that a knowledge of the physico-chemical composition of a biological organism does not suffice to explain mechanistically its mode of action—anymore than an enumeration of the parts of a clock and a knowledge of their distribution and arrangement suffices to explain and predict the mode of behavior of the time piece. To do the latter one must *also* assume some theory or set of laws (e.g., the theory of mechanics) which formulates the way in which certain elementary objects behave when they occur in certain initial distributions and arrangements, and with the help of which we can calculate and predict the course of subsequent development of the mechanism. Now it may indeed be the case that our information at a given time may suffice to describe physico-chemically the constitution of a biological organism; nevertheless, the established physico-chemical theories may not be adequate, even when combined with a physico-chemical description of the initial state of the organism, for deducing just what the course of the latter's development will be. To put the point in terms of the distinction previously introduced, the condition of definability may be realized without the condition of derivability being fulfilled. But this fact must not be interpreted to mean that it is possible under any circumstances to give explanations without the use of some theoretical assumptions, or that because one body of physico-chemical theory is not competent to explain certain biological phenomena it is *in principle impossible* to construct and establish mechanistic theories which might do so.

(3) I must now examine the consideration which appears to constitute the main reason for the negative attitude of organismic biologists toward mechanistic explanations. Organismic biologists have placed great stress on what they call the "unifiedness," the "unity," the "completeness," or the "wholeness" of organic behavior; and, since they believe that biological organisms are complex systems of mutually determining and interdependent processes to which subordinate organs contribute in various ways, they have maintained that organic behavior cannot be analyzed into a set of independently determinable component behaviors of the parts of an organism, whose "sum" may be equated to the total behavior of the organism. On the other hand, they also maintain that mechanistic philosophies of organic action are "machine theories" of the organism, which assume the "additive point of view" with respect to biological phenomena. What distinguishes mechanistic theories from organismic ones, from this perspective, is that the former do while the latter do not regard an organism as a "machine," whose "parts" are separable and can be studied in isolation from their actual functioning in the whole living organism, so that the latter may then be understood and explained as an aggregate of such independent parts. Accordingly, the fundamental reason for the dissatisfaction which organismic biologists feel toward mechanistic theories is the "additive point of view" that allegedly characterizes the latter. However, whether this argument has any merit can be decided only if the highly ambiguous and metaphorical notion of "sum" receives at least partial clarification; and it is to this phase of the question that I first briefly turn.

(i) As is well known, the word "sum" has a large variety of different uses, a number of which bear to each other certain formal analogies while others are so vague that nothing definite is conveyed by the word. There are well-defined senses of the term in various domains of pure mathematics—e.g., arithmetical sum, algebraic sum, vector sum, and the like; there are also definite uses established for the word in the natural sciences—e.g., sum of weights, sum of forces, sum of velocities, etc. But with notable exceptions, those who have employed it to distinguish wholes which are sums of their parts from wholes which supposedly are not, have not taken the trouble to indicate just what would be the sum of parts of a whole which allegedly is not equal to that whole.

I therefore wish to suggest a sense for the work "sum" which seems to me relevant to the claim of organismic biologists that the total behavior of an organism is not the sum of the behavior of its parts. That is, I wish to indicate more explicitly than organismic biologists have done—though I hasten to add that the proposed indication is only moderately more precise than is customary—what it is they are asserting when they maintain, for example, that the behavior of the kidneys in an animal body is more than the "sum" of the behaviors of the tissues, blood stream, blood vessels, and the rest of the parts of the body involved in the functioning of the kidneys.

Let me first state the suggestion in schematic, abstract form. Let T be a definite body of theory which is capable of explaining a certain indefinitely large class of statements concerning the simultaneous or successive occurrence of some set of properties $P_1, P_2, \ldots P_k$. Suppose further that it is possible with the help of the Theory T to explain the behavior of a set of individuals i with respect to their manifesting these properties P when these individuals form a closed system s_1 under circumstances C_1; and that it is also possible with the help of T to explain the behavior of another set of individuals j with respect to their manifesting these properties P when the individuals j form a closed system s_2 under circumstances C_2. Now assume that the two sets of individuals i and j form an enlarged closed system s_3 under circumstances C_3, in which they exhibit certain modes of behavior which are formulated in a set of laws L. Two cases may now be distinguished: (a) It may be possible to deduce the laws L from T conjoined with the relevant initial conditions which obtain in C_3; in this case, the behavior of the system s_3 may be said to be the sum of the behaviors of its parts s_1 and s_2; or (b) the laws L cannot be so deduced, in which case the behavior of the system s_3 may be said *not* to be the sum of the behaviors of its parts.

Two examples may help to make clearer what is here intended. The laws of mechanics enable us to explain the mechanical behaviors of a set of cog-wheels when they occur in certain arrangements; those laws also enable us to explain the behavior of freely-falling bodies moving against some resisting forces, and also the behavior of compound pendula. But the laws of mechanics also explain the behavior of the system obtained by arranging cogs, weights, and pendulum in certain ways so as to form a clock; and, accordingly, the behavior of a clock can be regarded as the sum of the behavior of its parts. On the other hand, the kinetic theory of matter as developed during the nineteenth century was able to explain certain thermal properties of gases at various temperatures, including the relations between the specific heats of gases; but it was unable to explain the relations between the specific heats of solids—that is, it was unable to account for these relations theoretically when the state of aggregation of molecules is that of a solid rather than a gas. Accordingly, the thermal behavior of solids is not the sum of the behavior of its parts.

Whether the above proposal to interpret the distinction between wholes which are and those which are not the sums of their parts would be acceptable to organismic biologists, I do not know. But, while I am aware that the suggestion requires much elaboration and refinement to be an adequate tool of analysis, in broad outline it represents what seems to me to be the sole intellectual content of what organismic biologists have had to say in this connection. However, if the proposed interpretation of the distinction is accepted as reasonable, then one important consequence needs to be noted. For, on the above proposal, the distinction between wholes which are and those which are not sums of parts is clearly *relative to some assumed body of theory T;* and, accordingly, though a given whole may not be the sum of its parts relative to one theory, it may indeed be such a sum relative to another. Thus, though the thermal behavior of solids is not the sum of the behavior of its parts relative to the classical kinetic theory of matter, it is such a sum relative to modern quantum mechanics. To say, therefore, that the behavior of an organism is not the sum of the behavior of its parts, and that its total behavior cannot be understood adequately in physico-chemical terms even though the behavior of each of its parts is explicable mechanistically, can only mean that no body of general theory is now available from which statements about the total behavior of the organism are derivable. The assertion, even if true, does *not* mean that it is *in principle* impossible to explain such total behavior mechanistically, and it supplies no competent evidence for such a claim.

(ii) There is a second point related to the organismic emphasis on the "wholeness" of organic action upon which I wish to comment briefly. It is frequently overlooked, even by those who really know better, that no theory, whether in the physical sciences or elsewhere, can explain the operations of any concrete system, unless various restrictive or boundary conditions are placed on the generality of the theory and unless, also, specific initial conditions, relevantly formulated, are supplied for the application of the theory. For example, electro-static theory is unable to specify the distribution of electric charges on the surface of a given body unless certain special information, not deducible from the fundamental equation of the theory (Poisson's equation) is supplied. This information must include statements concerning the shape and size of the body, whether it is a conductor or not, the distribution of other charges (if any) in the environment of the body, and the value of the dialectric constant of the medium in which the body is immersed.

But though this point is elementary, organismic biologists seem to me to neglect it quite often. They sometimes argue that though mechanistic explanations can be given for the behaviors of certain parts of organisms when these parts are studied in abstraction or isolation from the rest of the organism, such explanations are not possible if those parts are functioning conjointly and in mutual dependence as actual constituents of a living organism. This argument seems to me to have no force whatever. What it overlooks is that the initial and boundary conditions which must be supplied in explaining physico-chemically the behavior of an organic part acting in isolation are, in general, *not sufficient* for explaining mechanistically the conjoint functioning of such parts. For when these parts are assumed to be acting in mutual dependence, the environment of each part no longer continues to be what it was supposed to be when it was acting in isolation. Accordingly, a necessary requirement for the mechanistic explanation of the unified behavior of organisms is that boundary and initial conditions bearing on the actual relations of parts as parts of living organisms be stated in *physico-chemical* terms. Unless, therefore, appropriate data concerning the physico-chemical constitution and arrangement of the various parts of organisms are specified, it is not surprising that mechanistic explanations of the total behavior of organisms cannot be given. In point of fact, this requirement has not yet been fulfilled even in the case of the simplest forms of living organisms, for our ignorance concerning the detailed physico-chemical constitution of organic parts is profound. Moreover, even if we were to succeed in completing our knowledge in this respect—this would be equivalent to satisfying the condition of definability stated earlier—biological phenomena might still not be all explicable mechanistically: for this further step could be taken only if a comprehensive and independently warranted physico-chemical theory were available from which, together with the necessary boundary and initial conditions, the laws and other statements of biology are derivable. We have certainly failed thus far in finding mechanistic explanations for the total range of biological phenomena, and we may never succeed in doing so. But, though we continue to fail, then if this paper is not completely in error, the reasons for such failure are not the *a priori* arguments advanced by organismic biology.

(4) One final critical comment must be added. It is important to distinguish the question whether mechanistic explanations of biological phenomena are possible, from the quite different though related problem whether living organisms can be effectively synthesized in a laboratory out of nonliving materials. Many biologists apparently deny the first possibility because of their skepticism concerning the second, even when their skepticism does not extend to the possibility of an artificial synthesis of every chemical compound that is normally produced by biological organisms. But the two questions are not related in a manner so intimate; and though it may never be possible to create living organisms by artificial means, it does not follow from this assumption that biological phenomena are incapable of being explained mechanistically. We do not possess the power to manufacture nebulae or solar systems, though we do have available physico-chemical theories in terms of which the behaviors of nebulae and solar systems are tolerably well understood; and, while modern physics and chemistry is beginning to supply explanations for the various properties of metals in terms of the electronic structure of their atoms, there is no compelling reason to suppose that we shall one day be able to manufacture gold by putting together artificially its subatomic constituents. And yet the general tenor, if not the explicit assertions, of some of the literature of organismic biology is that the possibility of mechanistic explanations in biology entails the possibility of taking apart and putting together in overt fashion the various parts of living organisms to reconstitute them as unified creatures. But in point of fact, the condition for achieving mechanistic explanations is quite different from that necessary for the artificial manufacture of living organisms. The former involves the construction of factually warranted *theories* of physico-chemical processes; the latter depends on the availability of certain physico-chemical substances and on the invention of effective techniques of control. It is no doubt unlikely that living organisms will ever be synthesized in the laboratory except with the help of mechanistic theories of organic processes—in the absence of such theories, the artificial creation of living things would at best be only a fortunate accident. But, however this may be, these conditions are logically independent of each other, and either might be realized without the other being satisfied.

(5) The central thesis of this paper is that none of the arguments advanced by organismic biologists establish the inherent impossibility of physico-chemical explanations of vital processes. Nevertheless, the stress which organismic biologists have placed on the facts of the hierarchical organization of living things and on the mutual dependence of their parts is not without value. For though organismic biology has not demonstrated what it proposes to prove, it has succeeded in making the heuristically

valuable point that the explanation of biological processes in physico-chemical terms is not a necessary condition for the fruitful study of such processes. There is, in fact, no more good reason for dissatisfaction with a biological theory (e.g., modern genetics) because it is not explicable mechanistically than there is for dissatisfaction with a physical theory (e.g., electro-magnetism) because it is not reducible to some other branch of that discipline (e.g., to mechanics). And a wise strategy of research may, in fact, require that a given discipline be cultivated as an autonomous branch of science, at least during a certain period of its development, rather than as a mere appendage to some other and more inclusive discipline The protest of organismic biology against the dogmatism frequently associated with mechanistic approaches to biology is salutary.

On the other hand, organismic biologists sometimes write as if any analysis of living processes into the behaviors of distinguishable parts of organisms entails a radical distortion of our understanding of such processes. Thus Wildon Carr, one proponent of the organismic standpoint, proclaimed that "Life is individual; it exists only in living beings, and each living being is indivisible, a whole not constituted of parts." Such pronouncements exhibit a tendency that seems far more dangerous than is the dogmatism of intransigent mechanists. For it is beyond serious question that advances in biology occur only through the use of an abstractive method, which proceeds to study various aspects of organic behavior in relative isolation of other aspects. Organismic biologists proceed in this way, for they have no alternative. For example, in spite of his insistence on the indivisible unity of the organism, J. S. Haldane's work on respiration and the chemistry of the blood did not proceed by considering the body as a whole, but by studying the relations between the behavior of one part of the body (e.g., the quantity of carbon dioxide taken in by the lungs) and the behavior of another part (the chemical action of the red blood cells). Organismic biologists, like everyone else who contributes to the advance of science, must be selective in their procedure and must study the behavior of living organisms under specialized and isolating conditions—on pain of making the free but unenlightening use of expressions like "wholeness" and "unifiedness" substitutes for genuine knowledge.

ERNEST NAGEL.

COLUMBIA UNIVERSITY.

STUDIES IN THE FOUNDATIONS OF GENETICS

J. H. WOODGER
University of London, London, England

In what follows a fragment of an axiom system is offered — a fragment because it is still under construction. One of the ends in view in constructing this system has been the disclosure, as far as possible, of *what is being taken for granted* in current genetical theory, in other words the discovery of the hidden assumptions of this branch of biology. In the following pages no attempt will be made to give a comprehensive account of all the assumptions of this kind which have so far been unearthed; attention will be chiefly concentrated on one point — the precise formulation of what is commonly called Mendel's First Law, and its formal derivation from more general doctrines, no step being admitted only because it is commonly regarded as intuitively obvious. Mendel's First Law is usually disposed of in a few short sentences in text-books of genetics, and yet when one attempts to formulate it quite explicitly and precisely a considerable wealth and complexity of hidden assumptions is revealed. Another and related topic which can be dealt with by the axiomatic method is the following. Modern genetics owes its origin to the genius of Mendel, who first introduced the basic ideas and experimental procedures which have been so successful. But it is time to inquire how far the Mendelian hypotheses may now be having an inhibiting effect by restricting research to those lines which conform to the basic assumptions of Mendel. It may be profitable to inquire into those assumptions in order to consider what may happen if we search for regions in which they do not hold. The view is here taken that the primary aim of natural science is discovery. Theories are important only in so far as they promote discovery by suggesting new lines of research, or in so far as they impose an order upon discoveries already made. But what constitutes a discovery? This is not an easy question to answer. It would be easier if we could identify observation and discovery. But the history of natural science shows abundantly that such an identification is impossible. Christopher Columbus sailed west from Europe and returned with a report that he had found land. What made this a discovery was the fact that subsequent travellers after sailing west from Europe also returned with reports which agreed with that of Columbus. If the entire American continent had quietly sunk beneath the wave as soon as Columbus's back was turned we should not now say that he had discovered America, even although he had observed it. If an astronomer reported observing a new comet during a certain night, but nobody else did, and neither he nor anybody else reported it on subsequent nights, we should not say that he had made a discovery, we should say that he had made a mistake. Observations have also been recorded which have passed muster for a time but have finally been rejected, so that these were not discoveries. Moreover, there have been observations (at least in the biological sciences) which have been ignored for nearly fifty years before they have been recognized as discoveries. Theories play an important part in deciding what is a discovery. Under the influence of the doctrine of preformation, in the early days of embryology, microscopists actually reported seeing little men coiled up inside spermatozoa. Under the influence of von Baer's germ-layer theory the observations of Julia Platt on ecto-mesoderm in the 1890s were not acknowledged as discoveries until well into the twentieth century. Such considerations raise the question: is Mendelism now having a restricting effect on genetical research?

The distinction between records of observations and formulations of discoveries is particularly sharp in genetics; as we see when we attempt to formulate carefully Mendel's observations on the one hand and the discoveries attributed to him on the other. It will perhaps make matters clearer if we first of all distinguish between accessible and inaccessible sets. Accessible sets are those whose members can be handled and counted in the way in which Mendel handled and counted his tall and dwarf garden peas. Inaccessible sets, on the other hand, are those to which reference is usually being made when we use the word 'all'. The set of *all* tall garden peas is inaccessible because some of its members are in the remote past, some are in the (to us) inaccessible future, and some are in inaccessible places. No man can know its cardinal number. But observation records are statements concerning accessible sets and formulations of discoveries are statements concerning inaccessible sets. The latter are therefore hypothetical in a sense and for a reason which does not apply to the former statements, But there are other kinds of statements about inaccessible sets in addition to 'all'-statements. In fact, from the point of view of discoveries, the latter can be regarded as a special case of a more general kind of statement, namely those statements which give expression to hypotheses concerning the *proportion* of the members of one set, say X, which belong to a second set Y. When that proportion reaches unity we have the special case where *all* Xs are Ys. In the system which is given in the following pages the notation 'pY' is used to denote the set of all classes X which have a proportion p of their members belonging to Y, p being a fraction such that $0 \leqslant p \leqslant 1$. This notation can be used in connexion with both accessible and inaccessible sets. In the latter case it is being used to formulate statements which cannot, from the nature of the case, be known to be true. Such a statement may represent a leap in the dark from an observed proportion in an accessible set, or it may be reached deductively on theoretical grounds. In either case the continued use of a particular hypothesis of this kind depends on whether renewed observations continue to conform to it or not. Statistical theory provides us with tests of significance which enable us to decide which of two hypotheses concerning an inaccessible set accords better with a given set of observations made on accessible sub-sets of the said inaccessible set. In the present article we are not concerned with the questions of testing but with those parts of genetical theory which are antecedent to directly testable statements. At the same time it must be admitted that more is assumed in the hypothesis than that a certain inaccessible set contains a proportion of members of another set. As observations take place in particular places, at particular times, must there not be an implicit reference to times and places in the hypotheses concerning inaccessible sets, if such hypotheses are to be amenable to testing against observations? Consider, for example, the hypothesis that half the human children at the time of birth are boys. This would be the case if all children born in one year were boys and all in the next year were girls, and so on with alternate years, provided the same number of children were born in each year. But clearly a more even spread over shorter intervals of time is intended by the hypothesis. Again, there cannot be an unlimited time reference, because according to the doctrine of evolution there will have been a time when no children were born, and if the earth is rendered uninhabitable by radio-activity a time will come when no more children are born. Thus a set which has accessible sub-sets during one epoch may be wholly inaccessible in another.

In what follows no attempt will be made to solve all these difficult problems; we shall follow the usual custom in natural science and ignore them. Attention will be confined to the *one* problem of formulating Mendel's First Law. In the English translation of Mendel's paper of 1866, which is given in W. BATESON's Mendel's *Principles of Heredity*, Cambridge 1909, we read (p. 338):

> Since the various constant forms are produced in *one* plant, or even in *one* flower of a plant, the conclusion appears to be logical that in the ovaries of the hybrids there are as many sorts of egg cells, and in the anthers as many sorts of pollen cells, as there are possible constant combinations of forms, and that these egg and pollen cells agree in their internal composition with those of the separate forms.
>
> In point of fact it is possible to demonstrate theoretically that this hypothesis would fully suffice to account for the development of the hybrids in the separate generations, if we might at the same time assume that the various kinds of egg and pollen cells were formed in the hybrids on the average in equal numbers.

Bateson adds, in a foot-note to the last paragraph: 'This and the preceding paragraph contain the essence of the Mendelian principles of heredity.' It will be shown below that much more must be assumed than is explicitly stated here. L. Hogben, in *Science for the Citizen*, London, 1942, in speaking of Mendel's Second Law mentions the first in the following passage (p. 982):

> It is not, however, a law in the same sense as Mendel's First Law, of *segregation*, which we have deduced above, for it is only applicable in certain cases, and as we shall see later, the exceptions are of more interest than the rule.

But surely, Mendel's First Law is also only applicable in certain cases, and if this is not generally recognized it is because the law is never so formulated as to make clear what those cases are. We cannot simply say that if we interbreed *any* hybrids the offspring will follow the same rules as were reported in Mendel's experiments with garden peas, because it would be possible to quote counter-examples. It is hoped that the following analysis will throw some light on this question and that in this case also the exceptions may prove to be of at least as much theoretical interest as the rule. It will be shown that the condition referred to in the second of the above two paragraphs from Mendel's 1866 paper is neither necessary nor sufficient to enable us to derive the relative frequencies of the kinds of offspring obtainable from the mating of hybrids. It is *not sufficient* because it is also necessary to assume (among other things) that the union of the gametes takes place as random. It is *not necessary* because if the random union of the gametes is assumed the required frequencies can be derived without the assumption of equal proportions of the kinds of gametes. At the same time it will be seen that a number of other assumptions are necessary which are not usually mentioned and thus that a good deal is being taken for granted which may not always be justified.

When we are axiomatizing we are primarily interested in ordering the statements of a theory by means of the relation of *logical consequence*; but where theories of natural science are concerned we are also interested in another relation between statements, a relation which I will call the relation of *epistemic priority*. A theory in natural science is like an iceberg — most of it is out of sight, and the relation of epistemic priority holds between a statement A and a statement B when A speaks about those parts of the iceberg which are out of water and B about those parts which are out of sight; or A speaks about parts which are only a little below the surface and B about parts which are deeper. In other words: A is less theoretical, less hypothetical, assumes less than B. If A is the statement

Macbeth is getting a view of a dagger

and B is the statement

Macbeth is seeing a dagger

then A is epistemically prior to B. Macbeth was in no doubt about A, but he was in serious doubt about B and his doubts were confirmed when he tried to touch the dagger but failed to get a feel of it. Again, if A is the statement

Houses have windows so that people inside can see things

and B is the statement

Houses have windows in order to let the light in

then A is epistemically prior to B.

We not only say that Columbus discovered America, but also that J. J. Thomson discovered electrons. In doing so we are clearly using the word 'discovered' in two distinct senses. What J. J. Thomson discovered in the first sense was what we may expect to observe when an electrical discharge is passed through a rarified gas. He then *introduced* the word 'electron' into the language of physics in order to formulate a hypothesis from which would follow the generalizations of his discoveries concerning rarified gases. It will help to distinguish the two kinds of discoveries if we call statements which are generalizations from accessible sets to inaccessible sets *inductive* hypotheses, and statements which are introduced in order to have such hypotheses among their logical consequences *explanatory* hypotheses. Then we can say that to every explanatory hypothesis S_1 there is at least one inductive hypothesis S_2 such that S_2 is a consequence of S_1 (or of S_1 in conjunction with other hypotheses) and is epistemically prior to it. Were this not so S_1 would not be testable. But, as we shall see later, it is also possible to have an explanatory hypothesis S_3 and an inductive hypothesis S_2, which is *not* a consequence of S_3 although it is epistemically prior to it, *both* of which are consequences of the same explanatory hypothesis S_1. If what you want to say can be expressed just as well by a statement A as by a statement B then, if A is epistemically prior to B, it will (if no other considerations are involved) be better to use A. In what follows I shall try to formulate all the statements concerned in the highest available epistemic priority. Statements concerning parents and offspring only are epistemically prior to statements which also speak about gametes and zygotes; and statements about gametes and zygotes are epistemically prior to statements which speak also about the parts of gametes and zygotes. The further we go from the epistemically prior inductive hypotheses the more we are taking for granted and the greater the possibility of error. The following discussion of Mendel's First Law will be in terms of parents, offspring, gametes, zygotes and environments.

The foregoing remarks may now be illustrated by a brief reference to Mendel's actual experiments. Suppose X and Y are accessible sets of parents. Let us denote the set of all the offspring of these parents which develop in environments belonging to the set E by

$$f_E(X, Y)$$

If all members of X resemble one another in some respect (other than merely all being members of X) and all members of Y resemble one another is some *other* respect (also other than merely all being members of Y), so that the respect in which members of X resemble one another is distinct from that in which members of Y resemble one another, then $f_E(X, Y)$ constitutes an accessible set of *hybrids*. We also need $f_E{}^2(X, Y)$ which is defined as follows:

$$f_E{}^2(X, Y) = f_E(f_E(X, Y), f_E(X, Y))$$

Mendel experimented with seven pairs of mutually exclusive accessible sets and the hybrids obtained by crossing them. It will suffice if we consider one pair. Let 'A' denote the pea plants with which Mendel began his experiments and which were tall in the sense of being about six feet high; and let 'C' denote the peas which he used and which were dwarf in the sense of being only about one foot high. Let us use 'T' to denote the inaccessible set of *all* tall pea plants and 'D' to denote the inaccessible set of *all* dwarf pea plants. Thus we have

$$A \subset T \text{ and } C \subset D$$

let us use 'B' to denote the set of all environments in which Mendel's peas developed. Mendel first tested his As and Cs to discover whether they bred true and found that they did because

$$f_B(A, A) \subset T \text{ and } f_B{}^2(A, A) \subset T$$
$$f_B(C, C) \subset D \text{ and } f_B{}^2(C, C) \subset D$$

He next produced hybrids and reported that

$$f_B(A, C) \subset T$$
$$f_B{}^2(A, C) \in \frac{787}{1064} T \cap \frac{277}{1064} D$$

Finally, he took 100 of the tall members of $f_B{}^2(A, C)$ and self fertilized them. From 28 he obtained only tall plants and from 72 he obtained some tall and some dwarf. This indicated that about one third of the tall plants of $f_B{}^2(A, C)$ were pure breeding talls like $f_B(A, A)$ and two thirds were like the hybrid talls or $f_B(A, C)$.

Closely similar results were obtained in the other six experiments, although the respects in which the plants differed were in those cases not concerned with height but with colour or form of seed or pod or the position of the flowers on the stem. In each case the hybrids all resembled only one of the parental types, which Mendel accordingly called the dominant one. The parental type which was not represented in the first hybrid generation, but which reappeared in the second, he called the recessive one. Mendel took the average of the seven experiments and sums up as follows:

> If now the results of the whole of the experiments be brought together, there is found, as between the number of forms with the dominant and recessive characters, an average ratio of 2.98 to 1, or 3 to 1.

So long as we assert that the average ratio is 2.98 to 1 we are dealing with accessible sets and have no law or explanatory hypothesis. But what does

Mendel's addition 'or 3 to 1' mean? Presumably these few words express the leap from an observed proportion in an accessible set to a hypothetical proportion in an inaccessible set. This represents Mendel's discovery as opposed to his observations. At the same time there is no proposal to extend this *beyond* garden peas. This extension was done by Mendel's successors who, on the basis of many observations, extended his generalization regarding the proportions of kinds of offspring of hybrids over a wide range of inaccessible sets not only of plants but also of animals, In addition to this Mendel also left us his explanatory hypothesis, the hypothesis namely that the hybrids produce gametes of two kinds — one resembling the gametes produced by the pure dominant parents, and the other resembling those produced by the recessive parents. He also assumed that these two kinds of gametes were produced in equal numbers. We have now to consider what is the minimum theoretical basis for deriving this hypothesis as a theorem in an axiom system.

A GENETICAL AXIOM SYSTEM

(In what follows the axiom system is given in the symbolic notation of set-theory, sentential calculus and the necessary biological functors (the last in bold-face type). Accompanying this is a running commentary in words intended to assist the reading of the system; but it must be understood that this commentary forms no part of the system itself.)

The following primitives suffice for the construction of a genetical axiom system expressed on the level of epistemic priority here adopted; for cyto-genetics (and even perhaps for extending the present system) additional primitives are necessary.

(i) 'uFx' for 'u is a gamete which fuses with another gamete to form the zygote (fertilized egg) x'.

(ii) '$dlz\ xyz$' for 'x is a zygote which develops in the environment y into the life z.'

(iii) '$u\ \textbf{gam}\ z$' for 'u is a gamete produced by the life z.'

(iv) δ is the class of all male gametes.

(v) $\female$ is the class of all female gametes.

(vi) '$\textbf{phen}$' is an abbreviation for 'phenotype'

The following postulates are needed for the derivation of the theorems which are to follow:

POSTULATE 1 $(u)(v)(x):uFx.vFx.u \neq v. \supset . \sim(\exists w).wFx.w \neq u.w \neq v$

327

This asserts that not more than two gametes fuse to form each zygote.

POSTULATE 2 $(x)(w):(Eu).uFx.uFw. \supset .x = w$

This asserts that if a gamete unites with another to form a zygote then there is no other zygote for which this is true.

POSTULATE 3 $(u)(v)(x):.uFx.vFx.u \neq v. \supset :u \in \male .v \in \female .v.u \in \female .v \in \male$

This asserts that of the two gametes which unite to form any zygote one is a male gamete and the other a female gamete.

POSTULATE 4 $\male \cap \female = \Lambda$

This asserts that no gamete is both male and female.

POSTULATE 5 $(x)(y)(z)(u)(v):dlz \, xyz.dlz \, uvz. \supset .x = y.u = v$

This asserts that every life develops in one and only one environment from one and only one zygote.

POSTULATE 6 $(x)(y)(z)(x')(y')(z'):dlz \, xyz.dlz \, x'y'z'.$

$$(\exists u).u \, gam \, z.u \, gam \, z'. \supset .x = x'$$

This asserts that if there is a gamete produced by a life z and the same gamete is produced by a life z' then the zygote from which z develops is identical with the zygote from which z' develops. This may seem strange until it is explained that by 'a life' is here meant something with a beginning and an end in time and a fixed time extent. The expression is thus being used in a way somewhat similar to the way in which it is used in connexion with life insurance. Suppose a zygote is formed at midnight on a certain day; suppose it develops for say ten days and on that day death occurs. Then the whole time-extended object of ten days duration 'from fertilization to funeral' is a life which is *complete in time*. But suppose we are only concerned with what happens during the first ten *hours*; then that also is a life, in the sense in which the word is here used, and one which is a proper part of the former one. Now if a gamete is said to be produced by the shorter life it is also produced by the longer one of which that shorter one is a part; we cannot identify the two lives but we can say that they both develop from the same zygote. As here understood the time-length of a life fixes its environment; because the environment of a life is the sphere and its contents which has the zygote from which development begins as its centre and a radius which is equal in light-years to the length of the life in years. But no time-metric is needed for the present system and many complications are therefore avoided.

All the primitive notions of this system are either relations between individuals or are classes of individuals. But the statements of genetics with which we are concerned in what follows do not speak of individual lives, individual environments, individual zygotes or individual gametes but of *classes* of individuals and of relations between such classes. But the classes we require are definable by means of the primitives.

DEFINITION 1 $x \in U(\alpha, \beta) . \equiv :(\exists u)(\exists v) . u \in \alpha . v \in \beta . u \neq v . uFx . vFx$

We thus use '$U(\alpha, \beta)$' to denote the class of all zygotes which are formed by the union of a gamete belonging to the class α with one belonging to the class β.

DEFINITION 2 $z \in L_E(Z) . \equiv :(\exists x)(\exists y) . dlz \, xyz . x \in Z . y \in E$

'$L_E(Z)$' is used to denote the class of all lives which develop from a zygote belonging to the class Z in an environment belonging to the class E.

DEFINITION 3 $u \in G_E(X) . \equiv :(\exists x)(\exists y)(\exists z) . dlz \, xyz . y \in E . z \in X . u \, gam \, z$
'$G_E(X)$' thus denotes the class of all gametes which are produced by lives belonging to the class X when they develop in environments belonging to the class E.

DEFINITION 4 $z \in Fil_{K,M,E}(X, Y) . \equiv :(\exists x)(\exists y)(\exists u)(\exists v) . u \in G_K(X) .$

$$v \in G_M(Y) . uFx . vFx . u \neq v . y \in E . dlz \, xyz$$

The letters '*Fil*' are taken from the word 'filial'. The above definition provides a notation for the class of all offspring which develop in environments belonging to the class E and having one parent belonging to the class X and developing in an environment belonging to the class K and the other parent belonging to the class Y and developing in an environment belonging to the class M. For Mendelian contexts only one environmental class need be considered; provision for this simplification is made below.

The above four definitions suffice for most purposes. But it frequently happens that we need to substitute one of the above expressions for the variables of another and in that way very complicated expressions may arise. In order to avoid this the following abbreviations are introduced by definition:

DEFINITION 5 $D(\alpha, \beta, E) = L_E(U(\alpha, \beta))$

DEFINITION 6 $G(\alpha, \beta, E) = G_E(L_E(U(\alpha, \beta)))$

DEFINITION 7 $F'_{K,M,E}(\alpha, \beta; \gamma, \delta) = Fil_{K,M,E}(D(\alpha, \beta, K), D(\gamma, \delta, M))$

DEFINITION 8 $F_E(\alpha, \beta; \gamma, \delta) = F'_{E,E,E}(\alpha, \beta; \gamma, \delta)$

All the foregoing notions are general and familiar ones. We must now turn to some of a more special and novel kind. If our present inquiry were not confined to the single topic of Mendel's First Law we should at this stage introduce the notion of a genetical system, and we should maintain that genetical systems as then intended constitute the proper objects of genetical investigations. But for the present purpose it suffices if we speak of a specially simple kind of genetical system which we shall call *genetical units*. A genetical unit is a set of three classes: one is a phenotype, another is a class of gametes and the third is a class of environments; — provided certain conditions are satisfied. Suppose $\{P, \alpha, E\}$ is a candidate for the title of genetical unit; then it must be *developmentally closed*, that is to say $D(\alpha, \alpha, E)$ must be a non-empty class and it must be included in the phenotype P; next it must be *genetically closed*, that is to say $G(\alpha, \alpha, E)$ must be non-empty and must be included in α. Thus neither the process of development nor that of gamete-formation takes us out of the system; it thus 'breeds true'. The official definition is:

DEFINITION 9 $S \in \textbf{\textit{genunit}} \equiv :(\exists P)(\exists \alpha)(\exists E) . P \in \textbf{\textit{phen}} . S = \{P, \alpha, E\} .$

$D(\alpha, \alpha, E) \neq \Lambda . D(\alpha, \alpha, E) \subseteq P . G(\alpha, \alpha, E) \neq \Lambda .$

$G(\alpha, \alpha, E) \subseteq \alpha$

The genetical systems with which Mendel worked were genetical units, sums of two genetical units and what may be called set-by-set products of such sums. Thus if $\{P, \alpha, E\}$ and $\{Q, \beta, E\}$ are genetical units with the phenotype P dominant to the phenotype Q we shall have $D(\alpha, \beta, E) \neq \Lambda$ and $D(\alpha, \beta, E) \subseteq P$, so that $\{P, Q, \alpha, \beta, E\}$, the sum of the two units, is developmentally closed; if we also have $G(\alpha, \beta, E) \neq \Lambda$ and $G(\alpha, \beta, E) \subseteq$. $\alpha \cup \beta$, then the sum is also genetically closed. As we shall see shortly these assumptions do not suffice to enable us to infer that the sum will behave according to the Mendelian generalizations. If $\{R, \gamma, E\}$ and $\{S, \delta, E\}$ are two more genetical units so that $\{R, S, \gamma, \delta, E\}$ is their sum, then the set-by-set product of this sum and the former one will be

$$\{P \cap R, Q \cap R, P \cap S, Q \cap S, \alpha \cap \gamma, \beta \cap \gamma, \alpha \cap \delta, \beta \cap \delta, E\}$$

and if it is developmentally and genetically closed this will constitute yet
another type of genetical system which was studied by Mendel and with
which his Second Law was concerned.

Before we can proceed with the biological part of our system we must
now say something about the set-theoretical framework within which it is
being formulated and on the basis of which proofs of theorems are carried
out. We begin with two important definitions, one of which has already
been mentioned. (The definitions and theorems of this part of the system
will have Roman numerals assigned to them in order to distinguish them
from biological definitions and theorems).

DEFINITION I $X \in pY \ . \equiv . \ \dfrac{N(X \cap Y)}{N(X)} - p \ . \ 0 \leqslant p \leqslant 1$

pY is thus the set of all classes which have a proportion p of their members
belonging to Y. $N(X)$ is the cardinal number of the class X.

DEFINITION II $Z \in [X, Y] \equiv : (\exists u)(\exists v) . u \in X . v \in Y . u \neq v . Z = \{u, v\}$

$[X, Y]$ is the pair-set of the classes X and Y, that is to say it is the set of
all pairs (unordered) having one member belonging to the class X and the
other to the class Y.

No attempt is made here to present the set-theoretical background
axiomatically. We simply list, for reference purposes, the following
theorems which can be proved within (finite) set theory and arithmetic.

THEOREM I $N(X) = 0 . \equiv . X = \Lambda$

THEOREM II $X \subseteq Y . \supset . N(X) \leqslant N(Y)$

THEOREM III $N(X \cup Y) = N(X \cap \bar{Y}) + N(X \cap Y) + N(\bar{X} \cap Y)$

THEOREM IV $X \cap Y = \Lambda . \supset . N(X \cup Y) = N(X) + N(Y)$

THEOREM V $N([X, Y]) = N(X \cap \bar{Y}) . N(\bar{X} \cap Y) + N(X \cap Y) .$
$$[N(X \cap \bar{Y}) + N(\bar{X} \cap Y) + N(X \cap Y) - 1]$$

THEOREM VI $X \cap Y = \Lambda . \supset . N([X, Y]) = N(X) . N(Y)$

THEOREM VII $X \neq \Lambda . X \subseteq Y . \equiv . X \in 1Y$

THEOREM VIII　　$X \in pY \cap qY . \supset . p = q$

THEOREM IX　　$X \neq \Lambda . \supset . (\exists p) . \dfrac{N(X \cap Y)}{N(X)} = p . 0 \leqslant p \leqslant 1$

THEOREM X　　$X \neq \Lambda . X \subseteq Y \cup Z . Y \cap Z = \Lambda . \equiv . (\exists p) .$

$$X \in pY \cap (1 - p)Z$$

THEOREM XI　　$X \cap Y = \Lambda . \supset . (pX \cap qY) \subseteq (p + q)(X \cup Y)$

THEOREM XII　　$Y \subseteq P . Z \subseteq Q . P \cap Q = \Lambda . \supset . (pY \cap (1 - p)Z)$

$$\subseteq (pP \cap (1 - p)Q)$$

THEOREM XIII　　$Y \subseteq A . Z \subseteq B . W \subseteq C . A \cap B = B \cap C = C \cap A =$

$$= \Lambda . \supset . (pY \cap qZ \cap (1 - p - q)W) \subseteq (pA \cap qB \cap (1 - p - q)C)$$

THEOREM XIV　　$A \subseteq P . B \subseteq P . C \subseteq Q . A \cap B = B \cap C = C \cap A = P \cap Q = \Lambda . \supset .$

$$(pA \cap qB \cap (1 - p - q)C) \subseteq (p + q)P \cap (1 - p - q)Q$$

THEOREM XV　　$X \cap Y = \alpha \cap \beta = \Lambda . X \in p\alpha \cap (1 - p)\beta . Y \in q\alpha \cap (1 - q)\beta .$

$$\supset . [X, Y] \in pq[\alpha, \alpha] \cap (p(1 - q) + q(1 - p))$$

$$[\alpha, \beta] \cap (1 - p)(1 - q)[\beta, \beta]$$

THEOREM XVI　　$X \cap Y = \alpha \cap \beta = \Lambda . \supset : X \in \tfrac{1}{2}\alpha \cap \tfrac{1}{2}\beta . Y \in \tfrac{1}{2}\alpha \cap \tfrac{1}{2}\beta . \equiv .$

$$N(X \cap \beta, Y \cap \alpha) = N(X \cap \alpha, Y \cap \beta) . X \subseteq \alpha \cup \beta . Y \subseteq \alpha \cup \beta$$

We can now return to the biological part of our system. In genetical statements the notion of randomness frequently occurs. It will be required in two places in the present context. In both of these it means persistence of certain relative frequencies during a process. It means the absence of selection or favouritism.

We shall say that a set S which is the sum of two genetical units is *random with respect to U* or that *the union of the gametes is random in S* if and only if X and Y being any classes of gametes of the form $G(\alpha, \beta, E)$, α and β being any gamete classes and E the environment class of S, whenever we have

$$[X, Y] \in p[\gamma, \delta]$$

we also have　　　　　　　$U(X, Y) \in pU(\gamma, \delta)$

γ and δ also being gamete-classes of S. The following definition covers cases where S has an additional phenotype because there is no dominance.

DEFINITION 10 $S \in rand \ U. \equiv :(\overset{\prime}{\alpha})(\beta)(\gamma)(\delta)(\zeta)(\theta):(E)(p)(\exists S_1)(\exists S_2):$
$S_1, S_2 \in genunit. S = S_1 \cup S_2. v. (\exists R). R \in phen.$
$S = S_1 \cup S_2 \cup \{R\}. \alpha, \beta, \gamma, \delta, \zeta, \theta, E \in S. [G(\alpha, \beta, E),$
$G(\gamma, \delta, E)] \in p[\zeta, \theta]. \supset.$
$U(G(\alpha, \beta, E), G(\gamma, \delta, E)) \in p U(\zeta, \theta)$

Analogously we can say that such a set S is *random with respect to* $D(E)$
or that *development in members of the environment class E of S is random*
if and only if, whenever we have

$$U(G(\alpha, \beta, E), G(\gamma, \delta, E)) \in p U(\zeta, \theta)$$

we also have

$$D(G(\alpha, \beta, E), G(\gamma, \delta, E), E) \in p D(\zeta, \theta, E)$$

the Greek letters all being variables whose values are the gamete classes
of S_1 and 'E' being a variable whose single value (in Mendelian cases) is the
environmental class of S.

DEFINITION 11 $S \in rand \ D(E). \equiv :. (\alpha)(\beta)(\gamma)(\delta)(\zeta)(\theta)(E)(p):(\exists S_1)(\exists S_2):$
$S_1, S_2 \in genunit. S = S_1 \cup S_2. v. (\exists R). R \in phen.$
$S = S_1 \cup S_2 \cup \{R\}. \alpha, \beta, \gamma, \delta, \zeta, \theta, E \in S. U(G(\alpha, \beta, E),$
$G(\gamma, \delta, E)) \in p U(\zeta, \theta) \supset. D(G(\alpha, \beta, E), G(\gamma, \delta, E), E) \in$
$p D(\zeta, \theta, E)$

We now give a list of biological theorems which are provable from the
postulates and definitions and are used in the proofs of the major theorems
to follow. On the right hand side of each theorem are indicated the
postulates (P), definitions (D) or theorems (T) required for its proof.

THEOREM 1 $U(\alpha, \beta) = U(\beta, \alpha)$ [D1.

THEOREM 2 $U(X, X) = U(X \cap \delta, X \cap ♀)$ [D1, P3.

THEOREM 3 $\alpha \cap \beta = \Lambda. \supset. U(\alpha, \alpha) \cap U(\beta, \beta) = \Lambda$ [P1, D1.

THEOREM 4 $\alpha \cap \beta = \Lambda. \supset. U(\alpha, \alpha) \cap U(\alpha, \beta) = \Lambda$ [P1, D1.

THEOREM 5 $E \cap K = \Lambda. v. Z \cap W = \Lambda. \supset. L_E(Z) \cap L_K(W) = \Lambda$ [D2, P2.

THEOREM 6 $U(\alpha, \alpha) \cap U(\beta, \beta) =$

$= \Lambda . \supset . D(\alpha, \alpha, E) \cap D(\beta, \beta, E) = \Lambda$ [D5, D2, P2.

THEOREM 7 $U(\alpha, \alpha) \cap U(\alpha, \beta) =$

$= \Lambda . \supset . D(\alpha, \alpha, E) \cap D(\alpha, \beta, E) = \Lambda$ [D5, T5.

THEOREM 8 $\alpha \cap \beta = \Lambda . \supset . D(\alpha, \alpha, E) \cap D(\beta, \beta, E) = \Lambda$ [T3, T6.

THEOREM 9 $\alpha \cap \beta = \Lambda . \supset . D(\alpha, \alpha, E) \cap D(\alpha, \beta, E) = \Lambda$ [T4, T7

THEOREM 10 $D(X \cap \male, X \cap \female, E) = D(X, X, E)$ [D5, T2.

THEOREM 11 $\alpha \cap \beta = \Lambda . \supset . G(\alpha, \beta, E) \cap G(\beta, \beta, E) = \Lambda$ [D6, D3, D2,
                                                                     P5, P6, T4.

THEOREM 12 $G(\alpha, \alpha, E) \subseteq \alpha . \supset .$

$D(G(\alpha, \alpha, E), G(\alpha, \alpha, E), E) \subseteq D(\alpha, \alpha, E)$ [D1, D2, D5.

THEOREM 13 $F_E(\alpha, \beta; \gamma, \delta) =$

$D(G(\alpha, \beta, E), G(\gamma, \delta, E), E)$ [D8, D7, D4, D5, D6, D1, D2

THEOREM 14 $\{P, \alpha, E\} \in \mathbf{genunit} . \supset . F_E(\alpha, \alpha; \alpha, \alpha) \subseteq P$ [T13, D9, T12.

By a *mating description* is meant a statement of the form $X \subseteq Y$ or $X \in pY$ where 'X' is an expression denoting a set of offspring, e.g. '$F_E(\alpha, \beta; \alpha, \beta)$' and '$Y$' denotes a phenotype. We turn now to the task of discovering what must be assumed in order to derive the characteristic Mendelian mating descriptions, beginning with that which asserts the relative frequencies of dominants and recessives in the offspring of hybrids when these are mated with one another. For reference purposes it will be convenient if we use abbreviations for groups of the various separate hypotheses which enter into the antecedents of the following theorems. Let us therefore put:

H 1.   for: $\{P, \alpha, E\}, \{Q, \beta, E\} \in \mathbf{genunits} . P \cap Q = \alpha \cap \beta = \Lambda$ ($\{P, \alpha, E\}$ and $\{Q, \beta, E\}$ are genetical units and P and Q, and α and β, are mutually exclusive)

H 2.   for: $D(\alpha, \beta, E) \subseteq P$

(the hybrids are included in the phenotype P)

II 3. for: $(\exists R) . R \in \mathbf{phen} . D(\alpha, \beta, E) \subseteq R$

(this covers cases where there is no dominance but the hybrids are included in a third phenotype R)

H 4a. for: $G(\alpha, \beta, E) \cap \male \in \frac{1}{2}\alpha \cap \frac{1}{2}\beta$. $G(\alpha, \beta, E) \cap \female \in \frac{1}{2}\alpha \cap \frac{1}{2}\beta$

(This is one form of Mendel's own hypothesis. He assumed that in the gametes of the hybrids the two kinds occured in equal numbers both in the case of male and in the case of female gametes. Theorem XVI shows that the above form is equivalent to this).

H 4b. for: $G(\alpha, \beta, E) \cap \male \neq \Lambda . G(\alpha, \beta, E) \cap \male \subseteq \alpha \cup \beta$.

$G(\alpha, \beta, E) \cap \female \neq \Lambda . G(\alpha, \beta, E) \cap \female \subseteq \alpha \cup \beta$

(This is a weaker form of H 4a because it only assumes non-emptyness and inclusion).

H 4c. for: $G(\alpha, \beta, E) \neq \Lambda . G(\alpha, \beta, E) \subseteq \alpha \cup \beta$

(This is weaker still because it does not make separate statements regarding the gametes of different sex).

H 5. for: $S = \{P, \alpha, E\} \cup \{Q, \beta, E\} . S \in rand\ D \cap rand\ U(E)$

(This is the hypothesis that the system in question is the sum of two genetical units (H 1) and is random both with respect to the union of the gametes and also with respect to the development of the resulting zygotes in the environments belonging to E.

H 5a. for: $S = \{P, \alpha, E\} \cup \{Q, \beta, E\} \cup \{R\}$ and $S \in rand\ U \cap rand\ D(E)$

(This is to cover the cases when there is no dominance).

The following theorems are asserted for all values of the variables P, Q, R, α, β, E.

THEOREM 15 states that if we have H 1, H 2, H 4a and H 5 we also have three quarters of the offspring of the hybrids belonging to the dominant and the remaining quarter to the recessive phenotype.

THEOREM 15 $H 1 . H 2 . H 4a . H 5 . \supset . F_1(\alpha, \beta; \alpha, \beta) \in \frac{3}{4}P \cap \frac{1}{4}Q$

In order to make all the steps explicit we give the following derivation of this theorem:

(1) Using 'X' as an abbreviation of '$G(\alpha, \beta, E)$' we have, by H 1 and P 4

$$(X \cap \male) \cap (X \cap \female) = \alpha \cap \beta = \Lambda$$

(2) By (1), H 4a and T XV we can write:

$$[X \cap \male, X \cap \female] \in \frac{1}{2} . \frac{1}{2}[\alpha, \alpha] \cap 2(\frac{1}{2} . \frac{1}{2})[\alpha, \beta] \cap \frac{1}{2} . \frac{1}{2}[\beta, \beta]$$

(3)　From (2), H 5 and D 10 we are now able to obtain:

$$U(X \cap \male, X \cap \female) \in \tfrac{1}{4}U(\alpha, \alpha) \cap \tfrac{1}{2}U(\alpha, \beta) \cap \tfrac{1}{4}U(\beta, \beta)$$

(4)　We next obtain from (3), H 5 and D 11:

$$D(X \cap \male, X \cap \female, E) \in \tfrac{1}{4}D(\alpha, \alpha, E) \cap \tfrac{1}{2}D(\alpha, \beta, E) \cap \tfrac{1}{4}D(\beta, \beta, E)$$

(5)　From H 1, D 9 and H 2 we have:

$$D(\alpha, \alpha, E) \subseteq P \text{ and } D(\alpha, \beta, E) \subseteq P \text{ and } D(\beta, \beta, E) \subseteq Q$$

(6)　From H 1 we have $\alpha \cap \beta = \Lambda$ and so with the help of T 8 and T 9 we get:

$$D(\alpha, \alpha, E) \cap D(\alpha, \beta, E) = D(\alpha, \beta, E) \cap D(\beta, \beta, E) =$$
$$= D(\beta, \beta, E) \cap D(\alpha, \alpha, E) = \Lambda$$

(7)　From (5) and (6) with the help of T XIV we now get:

$$\tfrac{1}{4}D(\alpha, \alpha, E) \cap \tfrac{1}{4}(D(\alpha, \beta, E) \cap \tfrac{1}{4}D(\beta, \beta, E) \subseteq \tfrac{3}{4}P \cap \tfrac{1}{4}Q$$

(8)　By T 10 we have:

$$D(X \cap \male, X \cap \female, E) = D(X, X, E)$$

(9)　From (4), (7) and (8) we obtain:

$$D(X, X, E) \in \tfrac{3}{4}P \cap \tfrac{1}{4}Q$$

(10)　Putting '$G(\alpha, \beta, E)$' for 'X' in (9) in accordance with (1):

$$D(G(\alpha, \beta, E), G(\alpha, \beta, E), \in \tfrac{3}{4}P \cap \tfrac{1}{4}Q$$

(11)　By substitution of 'α' for 'γ' and 'β' for 'δ' in T 13 we get:

$$F_E(\alpha, \beta; \alpha, \beta) = D(G(\alpha, \beta, E), G(\alpha, \beta, E), E)$$

(12)　Finally from (10) and (11) we obtain the required result:

$$F_E(\alpha, \beta; \alpha, \beta) \in \tfrac{3}{4}P \cap \tfrac{1}{4}Q$$

Before commenting on this we shall give the remaining theorems.

THEOREM 16 is concerned with the offspring of hybrids when mated with the recessive parents; a mating type commonly called a back-cross. It is stated here in a somewhat unusual from and with the weakest possible antecedent. It states that if the hypotheses H 1, H 2, H 4c and H 5 are adopted then we should expect the proportions of the two phenotypes in the offspring to be identical with the proportions of the two kinds of gametes in the gametes produced by the hybrids. If, therefore, we assume, on the basis of samples, that $F_E(\beta, \beta; \alpha, \beta) \in \frac{1}{2}P \cap \frac{1}{2}Q$ we must also assume that $G(\alpha, \beta, E) \in \frac{1}{2}\alpha \cap \frac{1}{2}\beta$. The first of these hypotheses is epistemically prior to the second and yet they both occur together in the consequent of this theorem.

THEOREM 16 $H\,1.H\,2.H\,4c.H\,5.\supset.(\exists p).F_{E,}(\beta, \beta; \alpha, \beta) \in pP\cap(1-p)Q.$

$$G(\alpha, \beta, E) \in p\alpha \cap (1 - p)\beta$$

The derivation of Theorem 16 requires: T 13, T 11, T X, T XV, T VII, D 9 D 10, D 11 and T XII.

In the next theorem we have the same antecedent as in Theorem 15 except that nothing is assumed about the relative proportions of the two kinds of gamete in the gametes produced by the hybrids.

THEOREM 17 $H\,1.H\,2.H\,4b.H\,5.\supset.(\exists p)(\exists q).F_E(\alpha, \beta; \alpha, \beta) \in$

$$(p - pq + q)P \cap (1 - p)(1 - q)Q.$$

$$G(\alpha, \beta, E) \cap \male \in p\alpha \cap (1 - p)\beta.\,G(\alpha, \beta, E)\cap\female \in q\alpha\cap(1-q)\beta$$

In this case, if we assume, as a result of sampling, that $(p-pq+q) = \frac{3}{4}$ and $(1 - p)(1 - q) = \frac{1}{4}$ we cannot determine the value of p and of q. But if p has first been ascertained with the help of THEOREM 16 and sampling then (at least when p = q) the result can be applied to THEOREM 17. For the derivation of this theorem we require T X, P 4, T XV, D 10, D 11, T 2, D 5, D 9, T 9, T 8, T 13, T XIV. The next next theorem is the theorem corresponding to THEOREM 17 in systems where there is no dominance.

THEOREM 18 $H\,1.H\,4b.H\,5a.\supset.(\exists p)(\exists q).F_E(\alpha, \beta; \alpha, \beta) \in pqP \cap$

$$(p(1 - q) + q(1 - p))R \cap (1 - p)(1 - q)Q.\,G(\alpha, \beta, E)\cap$$

$$\cap \male \in p\alpha \cap (1 - p)\,\beta.\,G(\alpha, \beta, E) \cap \female \in q\alpha \cap (1 - q)\beta$$

In this case, if on the basis of sampling we assign a value to pq and to $(p(1 - q) + q(1 - p))$, then we can determine the values of p and q. The theorem requires: T X, P 4, T XV, D 10, D 11, T 2, D 5, T 9, T 8, T XIII, T 13.

Finally a theorem will be given which might have been known to Mendel. It is an example of a system which includes only *one* genetical unit. Suppose F and M are the females and males respectively of some species, suppose further that g and h are two mutually exclusive classes of gametes and H a class of environments all satisfying the following conditions: (i) $\{F, g, H\}$ is a genetical unit; (ii) $D(g, h, H) \neq \Lambda$. $D(g, h, H) \subseteq M.G(g, h, H) \neq \Lambda.G(g, h, H) \subseteq g \cup h$. (iii) $D(h, h, H) = \Lambda$ (therefore $\{M, h, H\}$ is *not* a genetical unit); (iv) $S = \{F, M, g, h, H\}$ and S is *rand* $U \cap$ *rand* $D(H)$. If these conditions are satisfied we shall have:

$$F_H(g, g; g, h) \in \tfrac{1}{4}F \cap \tfrac{1}{4}M \text{ if and only if } G(g, h, H) \in \tfrac{1}{4}g \cap \tfrac{1}{4}h$$

THEOREM 19 $F \cap M = g \cap h = \Lambda.\{F, g, H\{ \in genunit.D(g, h, H) \neq \Lambda.$

$D(g, h, H) \subseteq M.G(g, h, H) \neq \Lambda.G(g, h, H) \subseteq g \cup h.S =$

$= \{F, M, g, h, H\}.S \in rand\ U \cap rand\ D(H). \supset.$

$F_H(g, g; g, h) \in \tfrac{1}{4}F \cap \tfrac{1}{4}M. \equiv .G(g, h, H) \in \tfrac{1}{4}g \cap \tfrac{1}{4}h$

This theorem requires for its derivation T 11, T XV, T VII, D 10, D 11, T XII, T 13.

We can now see clearly what was Mendel's discovery in the Christopher Columbus sense and what was his discovery in the J. J. Thomson sense distinguished above. His discovery in the first sense (inductive hypothesis) was the $\tfrac{1}{4}P \cap \tfrac{1}{4}Q$ frequencies in the offspring when hybrids are mated, if this is understood as being asserted (as above) for inaccessible sets. This is expressed in THEOREM 15. Mendel's discovery in the second sense (explanatory hypothesis) is the hypothesis that is expressed in H 4a. But we have seen that in this form it is unnecessary. The much weaker form of H 4c suffices, especially if we begin with T 16 and then, using its results with the value of p determined by sampling (coupled with the additional hypothesis: $p = q$), we pass to T 17. Where there is no dominance (in Mendel's experiments one phenotype is in each case dominant to the other) p and q can be determined independently of T 16 with the help of T 18. Thus the convenient minimum assumption is H 4b. It could be argued that the assumption of two kinds among the gametes of hybrids is not so much a discovery of the second kind as a special application of a general *causal principle* to embryology and genetics. But this does not mean that it cannot be discussed.

It is often said that Mendel discovered what is called particulate inheritance. But, except in the sense in which gametes are particles, Mendel did not specifically speak of particles. Strictly speaking a hypothesis involving cell parts only becomes important when we consider the breakdown of Mendel's Second Law. The whole of Mendel's work can be expressed with the help of $D(\alpha, \beta, E)$, $G(\alpha, \beta, E)$ and $F_B(\alpha, \beta; \gamma, \delta)$ and thus in terms of gamete and environment classes, the classes of zygotes which can be formed with them and the classes of lives which develop from the zygotes in the environments.

The above analysis has shown the central role which is played by the hypotheses of random union of the gametes and of random development in obtaining the Mendelian ratios (see especially steps (2), (3) and (4) in the proof of Theorem 15). These do not receive the attention they deserve in genetical books. Sometimes they are not even mentioned. This is particularly true of the hypothesis of random development. That Mendel was aware of it is clear from the following passage in the translation from which we have already quoted (p. 340):

> A perfect agreement in the numerical relations was, however, not to be expected, since in each fertilization, even in normal cases, some egg cells remain undeveloped and subsequently die, and many even of the well-formed seeds fail to germinate when sown.

In addition to the special hypotheses H 1 to H 5 there are also the six postulates to be taken into consideration. Any departure from these could affect the result. This provides plenty of scope for reflexion. But perhaps the most striking feature of the Mendelian systems is the fact that only one class of environments is involved and is usually not even mentioned. Some interesting discoveries may await the investigation of multi-environmental systems. Provision for this is made in Definitions 4, 7 and 11 and a variable having classes of environments as its values accompanies all the above biological functors. At the same time attention should be drawn to the fact that no provision is made, either here or in current practice, for mentioning the environments of the gametes. And yet it is not difficult to imagine situations in which the necessity for this might arise.

It will be noticed that no use has here been made of the words 'probability', 'chance', or 'independent', although these words are frequently used in genetical books with very inadequate explanation. Here the term 'random' has been used but its two uses have been explained in detail In passing it may be mentioned that 'S is random with respect to F_B' is also definable along analogous lines and then the Pearson-Hardy law is derivable.

In conclusion I should like to draw attention to the way in which the foregoing analysis throws into relief the genius of Mendel, which enabled him to see his way so clearly through such a complicated situation. I also wish to express my thanks to Professor John Gregg of Duke University and to my son Mr Michael Woodger of the National Physical Laboratory for their help in the preparation of this article.

The British Journal for the Philosophy of Science

VOLUME XI   AUGUST, 1960   NO. 42

BIOLOGY AND PHYSICS *

J. H. WOODGER

I HAVE been asked to take part in a discussion in which biology is to be compared with physics.   But as I am too ignorant of modern physics to do this I must confine myself chiefly to stating what appear to me to be the most general peculiarities of biology.   At the same time I shall presuppose no knowledge of biology on the part of my audience beyond that which is inevitably picked up in the course of everyday life.   I shall try to show that the features which I have selected for mention are all traceable to a single general principle which can be understood by any one.   At the same time, as I am not writing an article for the *Encyclopedia Brittanica*, I shall make no claim to cover the whole subject.   For example I am not at all sure that the remarks that follow apply to bacteria or to viruses.

What I am going to call the basic principle of biology is the statement which asserts that living things have *parts* which stand in the relation of *existential dependence* to one another.   What is meant by this will, I hope, be clear from the following examples.   When a man's head is cut off the head and the trunk both perish.   Head and trunk are mutually existentially dependent.   When a man's leg is cut off the leg will perish and so will the trunk if bleeding does not stop in time or the wound becomes fatally infected.   The leg is existentially dependent upon the trunk but the trunk is not so dependent upon the leg.   An example on the social level is provided by the existential dependence of a new born infant on its mother.   It must not be supposed that existential dependence ceases to hold when a substitute for the normal second term of the relation can be found.   An infant does not cease to be existentially dependent because the duties of its mother are taken over by a foster-parent.   Similarly, parts which can be kept alive in

* Given at the Fourth Annual Conference in Philosophy of Science, Cambridge, 25–27 September, 1959.

G   89

some artificial medium do not cease thereby to be existentially dependent; in this case the biologist who provides the artificial medium, and other care, is the analogue of the foster-parent; so is the surgeon who, in the case of the amputated leg, stops bleeding and takes antiseptic precautions. Dependence of roots on leaves and leaves off roots provides an example from the plant world. The study of the mutual inter-dependence of parts and of parts on environmental factors constitutes the task of one of the primary subdivisions of biology, namely physiology.

I turn now to a class of relations which I call hierarchy-generating relations which also appear to be highly characteristic of biology, and the exemplifications of which are connected with the occurrence of existential dependence between parts. But before I can explain what I mean by a hierarchy-generating relation I must first explain what I mean by a hierarchy. And by a hierarchy I mean any relation which is one-many and such that its converse domain is identical with the whole set of terms to which the beginner of the relation stands in some power of the relation. This is a purely abstract definition because the notion of hierarchy as used here is not one belonging to any particular empirical science. It is a purely set-theoretical notion. At the same time it does not occur in those sections of Whitehead and Russell's *Principia Mathematica* which are devoted to the theory of relations. Nevertheless a class of relations is defined there which is a *sub-class* of the class of all hierarchies, namely the class of *progressions*. A progression is a relation which is one-one (i.e. many-one as well as one-many) and such that its *domain* consists of its beginner and all the terms to which that beginner stands in some power of the relation. A progression is thus a hierarchy which is many-one and has no last term. A relation which differs from a progression only in having a last term may be called, in relation theory, a *line*. It is most neatly definable as a hierarchy whose converse is also a hierarchy.[1]

To illustrate these general remarks and especially to emphasise what is excluded by the notion of hierarchy I give some arrow-figures. And first I give (Fig. 1) a diagram of domain, converse domain, and field of a relation to illustrate some words I shall use which are not used in *Principia Mathematica*. Members of the field of a relation which do not belong to its converse domain are called the *beginners* of the relation in *Principia Mathematica*. In the case of a

[1] I have used this notion in *Biology and Language*, London, 1952, p. 222. I owe the above improved definition to Dr A. Lindenmayer.

90

hierarchy this class has only one member. Members which belong to the converse domain but not to the domain are there called beginners of the converse, but I shall call them the *terminals* of the relation.

FIELD

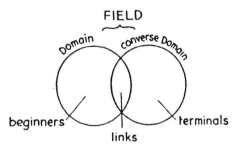

FIG. 1. Diagram of the three mutually exclusive subsets of the field of a two-termed relation: *beginners*, things which stand in the relation to something but have nothing standing in the relation to them; *terminals*, things which have something standing in the relation to them, but do not stand in it to anything themselves; *links*, things which both stand in the relation to something and have something standing in it to them.

Members of the field which belong both to the domain and to the converse domain I shall call *links* of the relation. In an arrow-figure beginners will have an arrow running from them but not to them, terminals will have an arrow running to them but not from them, and links will have at least one arrow running to them and at least one running from them. A diagram of a regular and one of an irregular hierarchy are given (Fig. 2); the former is such that every member of

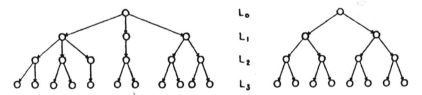

FIG. 2. Arrow-figures of hierarchies; irregular on the left, regular on the right. Numbering of levels indicated by L_0, L_1, etc.

the domain stands in the relation to exactly the same number of members of the converse domain as every other member. Arrow-figures of relations which are *excluded* by the definition of hierarchy are also given (Fig. 3): thus A is not a hierarchy because it has no beginner although it has a finite field. B has no beginner because its domain is included in its converse domain; the dots are intended to indicate an

91

infinite regress. C is not a hierarchy because it is not one-many. D is not a hierarchy because, although it has one and only one beginner, this beginner does not stand in some power of the relation to every member of the converse domain. E is not a hierarchy because it has two beginners. F represents a progression and G a line. A useful notion in connection with hierarchies is that of a *level*. A level of a hierarchy is a set of members of its field to which the beginner stands in the same power of the relation and which includes all the members of the field to which the beginner stands in this power of the relation.

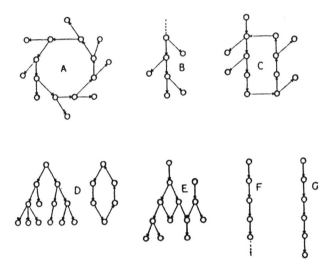

FIG. 3. A to E arrow-figures of relations which are *not* hierarchies: A, because it has no beginner, although it has a finite field; B, because its domain is included in its converse domain; C, because it is not one-many, having a member of its field to which *two* members stand in the relation; D, because although it has but one beginner this does not stand in some power of the relation to *every* member of the converse domain; E, because it has more than one beginner *and* is not one-many. F is an arrow-figure of a progression and G an arrow-figure of a line in the sense explained in the text.

Thus the beginner itself is the sole member of zero level; the terms to which it stands in the relation constitute level one; the terms to which they stand in the relation form level two, and so on.[1]

We can now proceed to hierarchy-generating relations. A relation is hierarchy-generating if and only if by taking any member

[1] For further information about hierarchies and for explanations of technical terms belonging to relation theory which have been used above see the author's *Axiomatic Method in Biology*, 1937. For a good explanation of the notion of being related by some power of a relation see Quine's *Mathematical Logic*, 1951, p. 215.

92

of its domain and limiting its field to that member and all the terms to which it stands in some power of the relation, the relation so limited satisfies the definition of hierarchy. Thus if the story of Adam in the book of Genesis is accepted as quite literally true then the relation of fatherhood is a hierarchy because there will be one and only one father who is not a child. But even if fatherhood is not a hierarchy it is a hierarchy-generating relation. For if we take any man who has children and limit the field of fatherhood to this man and all his descendants, the resulting relation will be a hierarchy. The female members of its field will all be terminals, together with all the childless males.

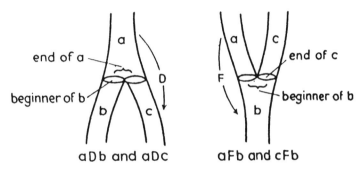

FIG. 4. Space-time diagram of the modes of origin of cells, depicting cells related by **D** and **F**.

Examples of hierarchy-generating relations in biology (apart from the one just given) are provided by certain relations between *cells*. All that it is necessary to know about cells in order to understand what follows is given in the following statements:

(1) Every cell has a beginning and an end in time;
(2) In no cell do beginning and end coincide;
(3) The beginning of every cell *either* (i) is a proper part of the end of a previously existing cell, *or* (ii) is the result of the *fusion* of the ends of *two* cells.
(4) Every cell possesses existentially dependent parts.
(5) Every cell has an environment.

Let us denote by ' **D** ' the relation in which a cell *a* stands to a cell *b* (see Fig. 4) when the beginning of *b* is a proper part of the end of *a*. And let us denote by ' **F** ' the relation in which a cell *a* stands to a cell *b* when the end of *a* is a proper part of the beginning of *b*. Then

93

we can say that every cell belongs to the domain or the converse domain of at least one of these relations (see Fig. 5) and that **D** and the converse of **F** (which can be denoted by ' F^{-1} ') are hierarchy-generating relations. These two relations differ enormously in the size of their fields. The number of links of **D** and of terminals of **D** is colossal and so are the powers of **D** which exist. But links of **F** are comparatively rare; they occur in the embryo sacs of flowering plants but apparently nowhere else; and powers of **F** above F^2 are empty. The importance of **F** seems to lie in the way in which it provides a means whereby terminals belonging to two **D**-hierarchies may unite to form the beginner of another.

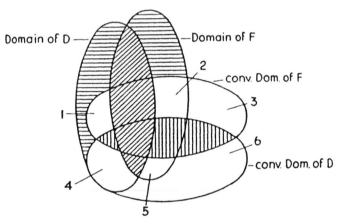

FIG. 5. Diagram of the constituents of the union of the fields of D and F. Sets having no members are represented by shaded areas. Non-empty sets: (1) beginners of D which arise by fusion (called zygotes in animals); (2) links of F (fused polar cells in the embryo-sacs of flowering plants); (3) terminals of F (in animals, zygotes which fail to divide); (4) links of D; (5) beginners of F (gametes in animals); (6) terminals of D other than gametes.

It is not difficult to see a connection between existential dependence and the occurrence of the **D**-relation between cells. If you have a bicycle and you want another but cannot get parts to put together to form one you have no recourse but to saw your bicycle into two halves —splitting each wheel, bar, pedal, etc., longitudinally. If cells have existentially dependent parts new cells will not begin by the coming together of parts but by the splitting of parts of existing cells. Cells are tender things and would have disappeared long ago had they not possessed the power of division. But division alone will not suffice— it would merely yield smaller and smaller cells. We should expect division to be preceded by a process of *duplication* of parts within cells from raw materials *ingested* by the cell from its environment.[1]

Now we must turn to another hierarchy-generating relation among biological objects. Consider first a square. If we bisect each side and join opposite bisecting points by straight lines we have four smaller squares which together compose the square with which we began. The same process can now be repeated with each of the four smaller squares producing sixteen still smaller squares, and so on. Let us denote by ' Sq ' the two-termed relation between any square and each of the four smaller squares into which it is divisible in the above way. Then we can say that, although Sq is not a hierarchy it is a hierarchy-generating relation. For, if we take any square—call it *s*—and limit the field of Sq to *s* and all the squares to which *s* stands in some power of Sq the resulting relation is a hierarchy of which *s* is the beginner. Every square is the beginner of an Sq-hierarchy. Moreover, these will be regular hierarchies; because for any natural number *n*, the number of members in level *n* of such a hierarchy will be 4^n. Now, something analogous to this, but much less simple and regular, is to be found in the biological world. The adult human body is divisible into certain parts (called *organ-systems*), for example, the alimentary, the respiratory, the excretory, and other systems, which collectively and in their specific and very complicated anatomical relations compose the body. Each organ in turn is analysable into certain parts (organs), the alimentary system, for example, is analysable into such parts as mouth, pharynx, gullet, stomach, intestine, liver, pancreas, which together in their anatomical relations compose it. Each organ again is resolvable into certain regions or layers which exemplify what are called *tissues*. Thus the intestine has lining layer, a muscular layer, a nervous layer and a covering layer. But of these layers only the lining is specifically concerned with alimentation, the muscular layer belongs to the muscular system which here, so to speak, combines with the alimentary system to ensure movement of the food along the canal. Each tissue is composed of cells of a characteristic kind, together with (in some cases) certain cell-products like fibres or tissue fluids. Each cell is next analysable into two major

---

[1] The process of duplication prior to division is beautifully illustrated by the protozoan *Euglypha alveolata* which is covered by small shell-plates except for a small hole for the protrusion of pseudopodia. Prior to division a duplicate set of shell-plates is formed round the nucleus. See E. A. Minchin: *An Introduction to the Study of the Protozoa*, 1912, p. 112, Fig. 59.

95

parts: a central nucleus (some cells have more than one) surrounded by a part called the cytoplasm. Each nucleus is composed chiefly of chromosomes, although there are also non-chromosomal parts; and each cytoplasm has such parts as mitochondria, Golgi bodies, secretion granules, etc. And so on. Certain of the spatial parts of the human body thus fall into *levels* in a spatial hierarchy. Let us call the relation in which the whole body stands to each of its organ-systems or the relation in which an organ system stands to each of its organs or any part to each of its next-level parts (like the relation between square and each quarter-square) the relation **S**. Then although **S** is not a hierarchy each whole body will be the beginner of an **S**-hierarchy.

But instead of talking about bodies we shall speak of *lives* to remind ourselves that we are speaking about time-extended things. Some cells are lives and some are not lives, but the cells which are not lives are proper parts of lives which are not cells. Lives are objects to which *taxonomic names* are applicable, but to the spatial parts of lives such names are not applicable. Thus we say that a man is a member of the species *Homo sapiens*, but we do not say that his big toe or his liver is a member of this or any other taxonomic group. Consequently cells which are lives are also given taxonomic names, like *Amoeba proteus*, the microscopic life which is found in ponds. But cells which are spatial parts of lives which are not cells are not given such names. We must distinguish between unfinished lives and finished lives. A life *a* which is a proper part of another life *b* is an unfinished life if its end is before that of *b* in time. Thus a child of one year old which continues living until it is two years old is an unfinished life. A finished life is one whose end is the beginning of a *corpse*.

In addition to these remarks about lives it is also necessary, in order to understand what follows, to know what is given in the following statements:

(1) Every life has a beginning and an end in time.

(2) In no life is the beginning coincident with the end.

(3) The beginning of every life which is not a cell is *either* (i) the beginning of a cell which is a life, *or*
(ii) a cluster of beginnings of cells which are not lives.

(4) If the beginning of a life is the beginning of a cell that cell belongs to the converse domain either of **D** or of **F** (see Fig. 5).

(5) If the beginning of a life *b* is the beginning of a cell belonging to the converse domain of **D**, then there is a life *a* which is not a cell

96

from which that cell has been derived by division; the life *a* is called the asexual parent of *b* (Fig. 6 (ii)).

(6) If the beginning of a life *c* is the beginning of a cell *c* belonging to the converse domain of **F**, then there is at least one life *a* which has produced a cell which has fused with another cell to form *c*, and *a* is called a *sexual* parent of the life *c* (and of any life of which *c* is a proper part). (Fig. 6 (iii)).

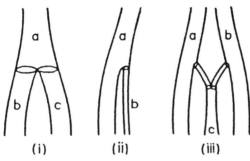

FIG. 6. The three modes of origin of lives which are not cells: (i) division; (ii) budding (if beginning with many cells), or parthenogenesis (if beginning with one cell); (iii) sexual reproduction. In (i) *a* is asexual parent of both *b* and *c*; in (ii) *a* is asexual parent of *b*; in (iii) *a* and *b* are both sexual parents of *c*; *a* and *b* may be one, when both gametes are produced by the same life.

(7) If the beginning of a life is a cluster of beginnings of cells which are not lives there are again two possibilities: (i) a life which is not a cell *divides*, i.e. it ends in *two* cell-clusters *each* of which is the beginning of a life which is not a cell. In this case the first life is asexual parent of *both* of the resulting lives (Fig. 6 (i)); (ii) a life which is not a cell separates off a cluster of beginnings of cells which forms the beginning of the new life but does not itself end until later. In this case also the first life is asexual parent of the second (Fig. 6 (ii)). In case (i) it is customary to speak of one life *dividing* to form two (just as with division of cells); in case (ii) we speak of the parental life *budding* to produce offspring (if the life *b* begins from a cluster of cells).

(8) The cells which are parts of a life which is not a cell fall into at least two mutually exclusive classes such that the members of one class are existentially dependent upon members of the other classes.

(9) Every life which is not a cell has an environment.

In connection with the different ways in which lives which are not cells may begin it may be of interest to mention that in what is

97

called identical twinning a combination of *two* processes is believed to occur. First a life begins as a result of a *sexual* process and then this life *divides* to form two lives—the identical twins—which are thus the immediate result of an *asexual* process. It is a consequence of this that (contrary to our usual way of speaking) the woman who gives birth to a pair of identical twins is *not* (biologically speaking) their mother but their *grandparent*; for she stands to them in a relation which is the relative product of *two* parental relations. First there is the sexual and then this is followed by the asexual parental relation, and this is a form of the grandparental relation, although neither grandmother nor grandfather.

Returning to our principal theme it will be seen that there are many ways of overcoming the difficulties presented by the occurrence of existentially dependent parts in lives which are not cells. If the spatial hierarchy is not too complicated, so that division does not separate too many cells of one kind from others of another kind upon which they are existentially dependent then reproduction by division or by budding is possible. But among animals it is comparatively rare for lives that are not cells to reproduce in those ways.

Failing division or budding we have the alternative of beginning with a single cell. When a life which is not a cell begins as a cell we say that a process of *development* takes place. This is a characteristic biological process again necessitated—if new lives which are not cells are to appear—by the occurrence of existentially dependent parts in such lives. It consists essentially of the following processes: first *division* repeated until many cells have been formed. This is followed by *deployment* of these cells into masses arranged in anticipation of the establishment of the ground plan characteristic of the type of life with which we are dealing. This primary deployment may be followed by subsidiary processes of similar kind establishing parts of parts, and parts of parts of parts, in the manner of a spatial hierarchy. Finally a process of *differentiation* takes place among the cells, resulting in the production (in the appropriate relations) of secretory cells, muscle cells, nerve cells, covering cells, blood cells, etc. When this is complete division among many D-lines will have ceased and the resulting terminal cells will form the adult body, except in places where there is loss of cells, such as in the skin or among blood-cells. There a stock of link cells must remain from which losses can be made good. Under pathological conditions cells which would normally be terminals may become link cells and result in the formation of cancers and new growths. The

98

study of the conditions which normally lead to the inhibition of division is thus of great importance.

The occurrence of differentiation among the cells of a **D**-hierarchy means that a cell-process distinct from duplication and the distribution of duplicates among new cells takes place. There is a process which we can call *elaboration* yielding cell-parts which can be called *elaborates*—parts which are *not* duplicates of parts occurring in cells standing in some power of **D** to the cell in which elaboration is taking place, but appear now for the first time in the life concerned. The production of different kinds of elaborates in separate cells—muscle-cell elaborates in some, nerve-cell elaborates in others, and so on, leads to differentiation and the establishment of mutually existentially dependent groups of cells. An aggregation of cells all of the same kind does not lead to a spatial hierarchy of a higher level than the cell because there is no existential dependence among the cells. This is well illustrated by the filamentous algae which are simply threads of similar cells. Existential dependence thus results in a kind of cycle of processes:

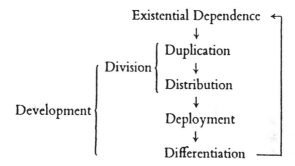

Before concluding a brief reference must be made to two kinds of parts of lives which are not cells which have not yet been mentioned. These may be called *secretions* and *accretions*. Secretions are parts which begin as parts of cells and end outside cells. These are exemplified by tears, drops of digestive secretions and the hard parts of bones. In most cases the shells of animals are formed in this way. But cases are known where shells are formed from materials picked up from the environment. Such parts would exemplify what is here meant by accretions. The clothing, the houses and the tools of human beings are also accretions; they are parts which have not begun inside the cells of the animal which uses them. Machines are also accretions; they are also existentially dependent upon the persons who make them.

99

Perhaps the persons who make them are also becoming existentially dependent upon them.

It was pointed out above how the branch of biology called physiology was concerned with existential dependence. Two other characteristic branches of biology are morphology and genetics. Morphology is concerned with the comparative study of spatial hierarchies, from the structural point of view. Genetics owes its existence to the existence of the parental relation.

In conclusion we may say that biology is concerned with objects whose parts (or some of them) exhibit hierarchical order in space and are themselves ordered by hierarchies whose fields are extended in time; this type of order being connected with the occurrence in these biological objects of parts which are existentially dependent upon one another.

Middlesex Hospital Medical School
London W. 1.

ACKNOWLEDGMENTS

Hempel, Carl G. "Geometry and Empirical Science." *American Mathematical Monthly* 52 (1945): 7–17. Reprinted with the permission of the Mathematical Association of America.

Reichenbach, Hans. "The Philosophical Significance of the Theory of Relativity." In P.A. Schlipp, ed., *Albert Einstein: Philosopher-Scientist* (New York: Harper, 1949): 289–311. Reprinted with the permission of editor, Paul A. Schlipp.

Frank, Philipp. "Introduction to the Philosophy of Physical Science, on the Basis of Logical Empiricism." *Synthese* 8 (1950): 28–45. Reprinted with the permission of Kluwer Academic Publishers.

Reichenbach, Hans. "The Logical Foundations of Quantum Mechanics." In Maria Reichenbach and Robert S. Cohen, eds., *Selected Writings, 1909–1953, Vol.II* (Dordrecht, Reidel, 1978): 237–78. Reprinted with the permission of D. Reidel Publishing Company.

Grünbaum, Adolf. "Logical and Philosophical Foundations of the Special Theory of Relativity." *American Journal of Physics* 23 (1955): 450–64. Reprinted with the permission of the American Association of Physics Teachers.

Carnap, Rudolf. "Psychology in Physical Language." *Erkenntnis* (1932): 165–98. Reprinted with the permission of Kluwer Academic Publishers.

Hempel, Carl G. "The Logical Analysis of Psychology." In N. Black, ed., *Readings in the Philosophy of Psychology* (Cambridge: Harvard University Press, 1980): 14–23. Reprinted with the permission of the author.

Schlick, Moritz. "On the Relation between Psychological and Physical Concepts." In Henk L. Mulder, ed., *Philosophical Papers, Vol.II* (Dordrecht: Reidel, 1979): 420–36. Reprinted with the permission of Oxford University Press.

Feigl, Herbert. "The Mind-Body Problem in the Development of Logical Empiricism." *Revue Internationale de Philosophie* 4 (1950): 64–83. Reprinted with the permission of Editions Universa.

Feigl, Herbert. "Functionalism, Psychological Theory, and the Uniting Sciences: Some Discussion Remarks." *Psychological Review* 62 (1955): 232–35. Reprinted with the permission of the American Psychological Association.

Neurath, Otto. "Sociology and Physicalism." *Logical Positivism* (Glencoe: Free Press, 1959): 282–317. Reprinted with the permission of Kluwer Academic Publishers.

Hempel, Carl G. "The Function of General Laws in History." *Journal of Philosophy* 39 (1942): 35–48. Reprinted with the permission of the Journal of Philosophy, Inc., Columbia University, and the author.

Grünbaum, Adolf. "Historical Determinism, Social Activism, and Predictions in the Social Sciences." *British Journal for the Philosophy of Science* 7 (1956): 236–40. Reprinted with the permission of the *British Journal for the Philosophy of Science*.

Watkins, J.W.N. "Historical Explanation in the Social Sciences." *British Journal for the Philosophy of Science* 8 (1957): 104–17. Reprinted with the permission of the *British Journal for the Philosophy of Science*.

Hempel, Carl G. "The Logic of Functional Analysis." In Carl G. Hempel, *Aspects of Scientific Explanation* (New York: The Free Press, 1965): 297–330. Reprinted with the permission of the author.

Nagel, Ernest. "Mechanistic Explanation and Organismic Biology." *Philosophy and Phenomenological Research* 11 (1951): 327–38. Reprinted with the permission of Brown University.

Woodger, J.H. "Studies in the Foundations of Genetics." In L. Henkin, P. Suppes, and A. Tarski, eds., *The Axiomatic Method, with Special Reference to Geometry and Physics* (Amsterdam: North-Holland, 1959): 408–28. Reprinted with the permission of North Holland Publishers.

Woodger, J.H. "Biology and Physics." *British Journal for the Philosophy of Science* 11 (1960): 89–100. Reprinted with the permission of the *British Journal for the Philosophy of Science*.

Printed and bound by CPI Group (UK) Ltd, Croydon, CR0 4YY

17/10/2024

01775685-0009